Hariolf Grupp (Ed.)

Dynamics of Science-Based Innovation

With 50 Figures

Springer-Verlag Berlin Heidelberg GmbH

Dr. Hariolf Grupp
Fraunhofer-Institut für Systemtechnik
und Innovationsforschung
Leiter der Abteilung Technischer Wandel
Breslauer Str. 48
D-7500 Karlsruhe 1
Germany

ISBN 978-3-642-86469-8 ISBN 978-3-642-86467-4 (eBook)
DOI 10.1007/978-3-642-86467-4

Originally published by Springer-Verlag Berlin · Heidelberg in 1992
Softcover reprint of the hardcover 1st edition 1992

2142/7130-543210 - Printed on acid-free paper

PREFACE

This volume intends to give an insight into progress in the field of studies on modern science and technology. Researchers from Sweden, Japan and Germany began a "three-country comparative study" in 1984. One of the primary aims of this study group was to better take account of the increasing importance of Japan in both analytical work and technology policy. To this end, *researchers* from the Research Policy Institute (RPI) at the University of Lund, the Graduate School of Policy Science at Saitama University in Urawa, and the Fraunhofer Institute for Systems and Innovation Research in Karlsruhe met almost every year with *policy makers* from the three countries, in order to see how well the scientific debate is reflected in the interests of practitioneers in the related policies.

The cooperation with the Swedish Board for Technical Development (STU)[1], the Japanese Ministry of Education, Science and Culture (Monbusho), and the German Federal Ministry for Research and Technology (BMFT) brought about numerous "grey" papers, publications and two volumes of seminar proceedings. The first book[2] deals with the problems of measuring technological change and summarizes tentative research plans from our first meetings. I concluded then, in November 1986, that "quantitative results are to be checked in a qualitative discursive process with the involved people. (...) The interaction of various indicators raises the pressure of argument and credibility. Case studies in dynamic fields of technology ideally supplement quantitative approaches."[3]

---

1 From July 1st, 1991, STU is renamed The National Board for Industrial and Technical Development (NUTEK).

2 Grupp, H. (ed.), *Problems of Measuring Technological Change*, Verlag TÜV Rheinland, Köln.

3 op. cit., p. 34.

Where do we stand now, about five years further down the track? The first results of our joint research were edited last year by Jon Sigurdson, covering everything we did on the measurement of the dynamics of *technological change*.[4] In his preface, Jon also covered the beginnings and short history of our study group in review. The significant changes in national research and development (R&D) strategies are the focal point of the volume mentioned which suggests new ways of applying quantitative indicators to study current shifts. Also covered are the difficulties in applying new methods to insufficient data bases and finally the importance of the environmental and social contexts of change.

That publication did not terminate our cross-national engagement to understand more of science, innovation and technology policy. At our sixth conference hold in Karlsruhe, Germany, in August 1990 we tried to expand the focal point of our cooperation a little bit. During our joint work in the first five years we realised several times that it is quite difficult to compare science and technology strategies in the bigger countries like Japan and Germany to a smaller country like Sweden. The scales "of everything" differ a great deal. As a result, we renamed the "three-country comparative study" and since 1989 call ourselves the "Network of research groups in the field of R&D dynamics". In the August 1990 meeting, colleagues from the Centre for Science and Technology Studies from Leiden University in the Netherlands participated for the first time. This centre introduces a great deal of expertise and know-how in the studies of science and technology into the network. The Dutch Ministry of Education and Science became involved. We also invited guest speakers from France, East Germany, the United Kingdom and the United States of America, as well as observers from German science supporting agencies (DFG, Wissenschaftsrat, MPG, Volkswagen-Stiftung). The outcome of the 1990 meeting is contained in this book.

Is this volume a conference proceedings book? The answer is possibly "No, not exactly"; I am really not certain. In distinction to customary conference proceedings our meeting brought only invited speakers together. The frames of their contributions were discussed beforehand and selected in such a way that they fitted well into the overall theme of measuring the dynamics of science-based innovation. We mutually commented on each others' oral contributions and altered the subsequent contributions to this book accordingly after the conference. The chapters were organized systematically and interlinked by the editor (who would alone be responsible for any cross references misinterpreting the original intentions of the authors). Thus, this

---

4 Sigurdson, J. (ed.), (1990), *Measuring the Dynamics of Technological Change*, Pinter Publishers, London.

volume is half-way a collectively authored monography on the dynamics of science-based innovation, the complexity of its internal evolution, and its growing importance to science *and* industrial R&D management. May the reader of this volume judge by his or her own which of the labels are more appropriate.

Last, but not least, we want to thank the sponsoring organizations which have made joint research possible since 1984. Among the ministerial observers, Dr Frank Schlie-Roosen (BMFT) was not just a most interested discussant, but he also made his minister pay for part of our bills during the joint research work last year. We hope that this volume not only stimulates new research, but also provides new insights into the exciting and multi-facetted field of science-based innovation to all policy makers.

Particularly, I want to thank all those who helped me with the conference and the manuscripts of this volume. I just mention Sybille Hinze, Monika Silbereis, Renate Klein and Ilse Gottschalg from my own institute and Michael Gärtner, a student trainee from Australia, who worked on good English in some chapters, knowing that many more people from the other institutes involved also supported and helped with this volume.

Hariolf Grupp
Karlsruhe
May 1991

CONTENTS

PART I

CONCEPTS AND NEW METHODOLOGY

1 BY WAY OF INTRODUCTION: ALLIANCES BETWEEN SCIENCE RESEARCH AND INNOVATION RESEARCH

Hariolf Grupp, Edo Albrecht and Knut Koschatzky

1.1 HISTORIC BACKGROUND AND RESEARCH QUESTIONS NOWADAYS

In the 1970s the development of the world economy was increasingly characterised by symptoms of crisis and phenomena of stagnation. This affected countries of different economic and social orders. Under the impression of these difficulties, a broad international discussion about possible causes and response strategies was triggered off.

It is generally known that the first response was the postulation of absolute limits to economic growth (Meadows et al. 1972). The so-called world patterns from which these conclusions were derived did, however, constrain the variables for technologies and techniques, as we know today (Freeman 1984, pp. 489-507). In the meantime this postulate has been largely abandoned. Those limits which manifested themselves in the 1970s were identified as limits of a certain technological-economic paradigm.

Freeman is known to have criticised the so-called world models. He said: "They are based essentially on the extrapolation of the characteristics of one phase of growth and do not allow for the possibility of paradigm changes in the mode of growth. (...) The mistake (...) was to confuse the 'limits' of one particular development paradigm with the 'limits to growth' of the system in general." (op.cit., p. 492). The consequences of the new approach of thinking were self-evident: Economic growth continues to be ensured if new technologies, so-called high and key technologies, are introduced, or, a fortiori, if there is a shift-over to a new technological-economic paradigm.

Freeman's criticism of the world models is essentially based on the concept of long waves, the so-called Kondratieff cycles. This concept is marked by a changeable history which cannot be set forth here in greater detail. Meanwhile it has been shown consistantly by filter theory that any waves vary strongly with periods between 20 and 60 years or so, being non-uniform in mode and wavelength (Gerster 1988). I.e., for prediction of future cycle periods the Kondratieff waves are *not* useful. It is, however, worth emphasizing that the phenomenon was holisticly interpreted by Freeman in terms of innovation theory (Freeman 1987). This also involved a resort to the ideas of Schumpeter. He is known to have realised the central significance of innovations as a driving force of economic growth and set this thesis against the traditional economy of balance with the former saying that economic growth takes place not only continuously but also in discrete steps.

Such a thesis has won recognition also in science theory, which is interesting from the point of view of the history of theory. Predominantly, the cumulative description of science developments has been criticised primarily by Kuhn (1967) and Lakatos (1974). Thus, the notion of paradigm has gained central importance within the framework of discrete patterns both of the development of science and economy. Elsewhere the theoretical and conceptual aspects of a discrete pattern of economic growth was already dealt with in greater detail (Albrecht 1988). It may be sufficient to refer to some selected problems relevant to this volume.

Within the framework of a discrete pattern of economic growth technologies are regarded as variables. In other words: the abandonment of old technologies and the introduction of new technologies manifest themselves under certain conditions as necessary prerequisites of economic growth. This process is designated as technological revolution. "A vital characteristic of this (...) type of technical change", Freeman writes, "is that is must have *pervasive* effects throughout the economy, i. e., it must not only lead to the emergence of a new ranges of products and services in its own right, it must also affect every other branch of the economy by changing the input cost structure and conditions of production and distribution throughout the system." (Freeman 1984, p. 498). The following effects are particularly emphasized:

- drastic cost reduction in many products and services,
- considerable improvement of the technological characteristics of many products and processes,
- social and political acceptance,
- reduction in the load to the environment,
- pervasive effects on the economic system as a whole.

The term of new technology has been generally defined as comprising basic technologies, that is those capable of ensuring economic growth, now and in the future. They are largely part of information technology, biotechnology and material technology (see also Stankiewicz 1990).

Whereas at the beginning main attention was drawn to the introduction of new technologies, now one goes one step further. The change of economic mechanisms and of the production conditions is also considered a constituent element of the change of the (scientific) paradigm. That is why in the meantime we speak of a new technological-economic paradigm: The revolution is not only of technological but also of economic nature.

Modern economy means an economy of learning curves. "High-tech products tend to have two things in common: They fall in price rapidly, as production builds up (they possess steep learning curves), and they get replaced fairly frequently (they have short life cycles). The trend in high-tech is towards things becoming steeper and shorter." (The Economist 1986, p. 48n). The economic effect depends on new criteria, including growing variety of products, high product quality, shorter throughput and resetting times, increasing innovation rate etc.

This concept outlined here cannot explain all for sure, but explains at least essential features of the growth strategy which was successfully implemented by the industrially developed countries in the 1980s. What was essentially confirmed is the assumption that new technologies can become a decisive source of economic growth. Those new technologies designed to ensure economic growth are generally designated as "science-based" technologies or high technologies. In other words: they are based - as a rule - on the utilization of the laws of the internal structure of matter. The development, introduction and broad application of new technologies must consequently be prepared by a relevant strategy of basic research. This assertion is based on the assumption that the present course in basic research is insufficient for the development of new technologies.

In this introduction to the book one must abstain from discussing these problems in greater detail. The following ten chapters in this volume offer more insights into the dynamics of science-based innovation. But attention should be drawn to some principal conceptual problems. It seems *imperative that innovation research allies itself with science research*. Such an alliance is implemented in the network of people who came together regularly and who wrote this volume (see preface). One also should take great efforts to bring about such an alliance between science research and innovation research in the unified German landscape of technology and science. Such an alliance is, however, rare. The principle difficulty seems to be that innovation research is categorized in patterns of economics whereas science research traditionally is very close to philosophy or sociology. Both realms so far have had few points of contact.

Nevertheless, the strong impact which science exerted on economic growth after the Second World War made some economists work on the development of science. As an example, Machlup wrote in 1962: "The production of knowledge is an economic activity, an industry, if you like, (...) but economists have neglected to analyze the production of knowledge." (Machlup, 1962, p. 9). In particular, economists oriented towards innovation theory and supporters of the concept of long waves brought about important contributions towards a deeper understanding of science as a determinant of economic growth by analysing interactions between invention and innovation. Among them Freeman et al. literally mentions the importance of basic science as a source of radical technological break-throughs (Freeman et al. 1982, p. 47).

This new way of thinking turned out to be of considerable heuristic significance. It allows to answer old questions in a new mode and to formulate problems not being on the agenda before. One of these items is the relation between scientific and techno-economic paradigms. Freeman & Perez (1988, pp. 40-66) - to give another example - assume that each techno-economic paradigm is coupled to a specific national innovation system which includes both the scientific infrastructure and national R & D.

The new lines of thinking also led to an altered interpretation of the growth of science. In the 1950s there was a comprehensive debate on the growth of science and its limitations. These limits were postulated, however, theoretically, as the considerations were based on cumulative models of science and economic growth (de Solla Price 1965). A completely different picture may be depicted if one models scientific and economic growth in discrete steps. The authors of this introductory chapter would like to venture the thesis that every period of industrial growth corresponds to a certain stationary condition of the scientific potential. The borrowing from physical patterns of thinking is not to be denied in this context.

In quantum theory one proceeds from the assumption that the energy radiation of an atomic system does not take place continuously but in separate emissions. Only when the electrons turn into a new stationary condition are they bound to absorb or emit energy. A hypothesis of that kind, of course, is highly abstract and perhaps cannot be verified empirically. It is however likely to provide a mental model of thinking by which the relationship between science and economic growth might be placed into a concrete context of history. This might be helpful in redefining phenomena associated with the Kontratieff cycles which are known now to be classified in a holistic way (Freeman 1987).

Each cycle is to be marked by specific conditions of economic growth. From that arise *new demands on science* as well as on the structure of the scientific potential. In other words: The structure of the scientific potential must be ever changing in order to be in a position to ensure economic growth in the new period of economic development.

Already a first glance at the periodisation shows that some significant events of the development of science can be classified remarkably well. The transition before World War II, for instance, was linked in Germany (among other countries) with the emergence of an industrial research potential as well as with the creation of facilities for basic and applied research outside the universities. A significant feature of the next Kondratieff cycle was, for example, the creation of large-scale research centres which took place initially in the United States and the USSR and later on in other industrially developed countries. Financial spending on research and development then grew for a limited period in an overproportional manner in relation to the gross national product and, finally, levelled off at a relatively stable value. The contribution made by R & D expenditures relative to the gross national product, for reasons of simplicity, may be considered a good indicator for the assumption of cycles (in addition see also chapter 9 in this volume). Of course this hypothesis cannot be statistically proved for all other cycles.

True, no striking changes seem to take place in the setup of the research potential during transition to the cycles to come, but the following trends are discernible:

- upgrading of basic research;
- revaluation of research programmes;
- stronger development of integrated research and of international research cooperation;
- devaluation of project research of traditional character and a more comprehensive promotion of so-called generic technologies.

These tendencies are significant indeed. In the 1980s, more or less comprehensive discussions about the reorganisation of the landscape of research took place in many developed industrial countries. In conceiving research programmes, political priorities evidently played an important role in the 1950s. In the meantime the spectrum of societal objectives has become broader and more differentiated. So, for instance, in the 1950s the understanding for ecological problems and for dangers arising from the new technologies was hardly developed. It is also worth emphasizing that generic technologies are marked by an increasingly complex character. This is the background for the following chapters in this volume.

1.2 A COGNITIVE MODEL

The complexity discussed above is difficult to represent in a simple model. For many years so-called linear models of science, technology and innovation have been used. In essence they reflect sequential processes mostly starting with a science push and ideally ending in an innovative product. If this simplistic thinking is accepted, then the recipe for technology policy reads: More science leads to more innovation. All the complex selection mechanisms are thus overlooked. The extreme uncertainties under which innovation-related decisions have to be taken are disregarded. In special cases linearity might be true, in particular, if the innovative products are not produced and not traded under the conditions of free markets. If the related research is public responsibility and if government procurement agencies are at the same time the only customer (space products, nuclear products, armaments, etc.) then the linear model might have its historic justification. But also in these cases it is still based on inadmissible simplifications (consider, e.g. nuclear fusion).

In the large field of industrial technology and more so in high-level consumer technology, scientific, technological and economic progress are certainly not linked in a sequential manner (Rosenberg 1976, pp. 68-75). Here, any linear models are not at all applicable. To cope with this more general situation in science and innovation the authors of this chapter developed cyclic or coupled models of science and innovation phases (Kline (1991) calls them chain-linked models). This was done independently of each other and not knowing each other's work mutually because of the isolation of the two former Germanies. The latest versions have been published by Albrecht (1990, p. 823) and Grupp (1990, p. 59). After "unifying" their approaches an improved model is suggested. Figure 1-1 presents a schematic view of the cyclic model. It is of the type of the cognitive models discussed in chapter 2, section 2-2.

Six important types of innovation-related functions are differentiated, not assuming that they are clearly distinct and without overlap. Nevertheless there is a type of research which is dedicated towards a deeper theoretical understanding of phenomena. Theories, formal mathematical models, mental models and physical models (e.g. for crystal structure) are worked out in this phase. Holistic approaches for understanding nature and mankind are included in this type of scientific work. No clear-cut ideas on future exploitability in a technical or economic sense do play a role. Therefore, the question may be asked whether theory and model development is an invention- or innovation-oriented function. If one considers certain classes of innovation processes, namely the genesis of generic technologies then, in the authors' opinion, scientific achievements may not be neglected (see also chapter 4). At least in this century this type of scientific activity is an important source for innovation - amongst others. In nuclear energy, micro-electronics and in the laser field, e.g., decisive technical break-throughs were made by scientists (sometimes working in industrial laboratories) who did not intend to work for a well-defined practical aim.

In terms of technical realisation two functions of innovation-oriented activity are differentiated further: In industrial laboratories new products are developed and new processes introduced. They all aim at commercial success and are funded mostly from business sources. Another type of technical realisation is far from commercial applicability (e.g. fusion research, breeder technology, Zeppelin air-ships, Wankel rotary engines). However, these "innovations" have been realized technically at least as prototypes. Many new military products belong into this category. Their technical utility is tested successfully and they are sometimes used, sometimes stored. But there is no phase of innovation-oriented industrial development with the notion to introduce these products cost-effectively on commercial markets. The early lasers in the 1960s were technically realised without a view towards their broad economic diffusion which took place later on, indeed, after intensive industrial development in the late 1970s and 1980s (see chapter 9 of this volume), the same might be true for genetic engineering with a phase of technical realization in the early 1980s.

Another function is innovation itself, that is the first commercial realisation of new technology on a world-wide scale. Even if the industrial followers of the innovator claim that they have also mastered the same innovation, it should be called imitation. Imitations may be as painful and difficult as pioneer innovations. Adaptations, incremental innovations and diffusion are regarded merely as routine managing business by many economists. In reality improving and diffusing products and processes is seldom if ever a simple task or a replication by unimaginative imitators. From utilisation of diffused innovations, that is their consumption and disposal, incentives for new research and more innovation may be triggered off at least in terms of the cyclic model suggested here. The push on science and technology from environmental problems may serve as a good example.

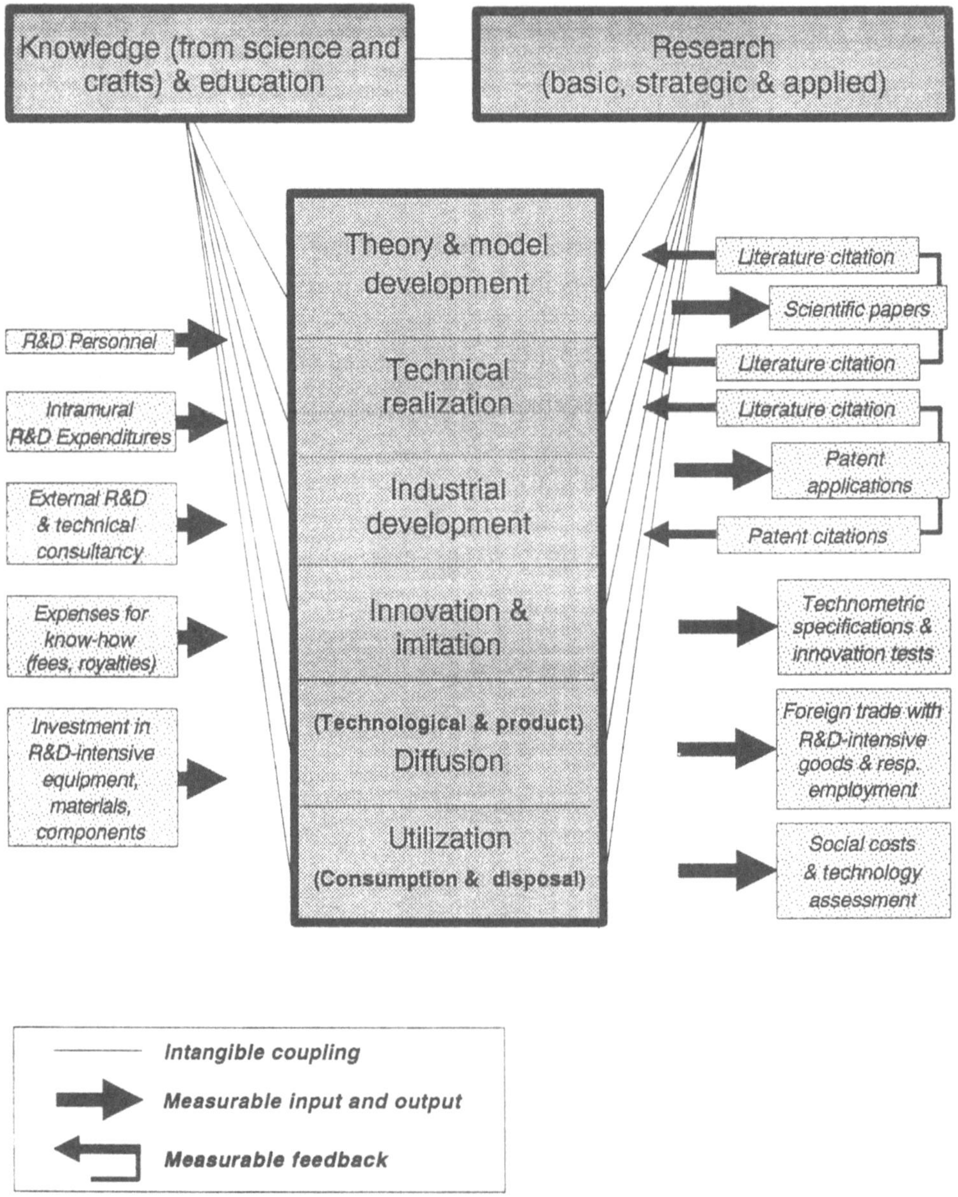

Figure 1-1: A schematic model of the coupled functions of science, technology and innovation

Despite of the distinctions in the innovation-oriented types of activity made in figure 1-1 they have much in common. All six functions are performed by educated and talented people and are thus coupled to the body of knowledge from science and crafts and education. At least because of this coupling to national education and culture, country-specific factors, the national systems or styles of innovation do still play a role. On the other side, *all* types of research are needed to produce innovations be it called fundamental, basic, strategic or applied research. In the coupled model research is seen as a form of problem solving, which may be called upon at any point in the innovation-oriented process. A firm is understood to have an existing knowledge base within which it seeks solutions to the problems which inevitably occur in producing innovations. The science system, on a different level, takes up problems which cannot be resolved within the existing knowledge base, and thereby extends the knowledge base of the firm. "Because research can relate to any phase of the innovation process, it is a complex and internally differentiated activity which potentially forms a wide variety of functions. It is an adjunct to the innovation process, not a precondition for it; indeed many of its activities will be shaped by the innovation process, and many of its problem areas will derive from innovative ideas generated elsewhere." (Smith 1990, p. 23).

In order to elaborate the long-term innovation-oriented type of science discussed above, which is not coupling closely to industrial activities, Krupp (1984) introduced the term "long-term application-oriented basic research". It is understood easily in terms of the cyclic model, when the non-exhaustive selection of research inputs to the innovation function "technical realisation" is considered.

One of the major reasons for introducing the coupled model represented by figure 1-1 is the visualisation of measurement opportunities. When measuring the dynamics of science-based innovation one faces the problem that clear-cut measurement procedures are difficult to define and to validate. The well-known approach by indicators, that is by quantitative information on certain inputs, through-puts and outputs, grasps only parts of the complex and cyclic (feedback) innovation-oriented processes as may be seen in figure 1-1 (non-exhaustive selection of indicators). Bibliometric indicators discussed in this volume in sections 7 and 8 offer a measurement chance of the R&D functions dedicated to theories, modelling and technical realisation. From industrial development more frequent patents than papers come out. When inventors publish papers (section 3), they should have an intermediate function in this scene. Thus the proposed model is also an aid to understanding and classifying the contributions to this volume. It represents a certain way of approaching the complex matter of science-based innovation and offers an opportunity to speak a common language in science and innovation research. Technometric and foreign trade indices are not utilized in this volume but mentioned in figure 1-1. However, in the preceeding volume (Sigurdson 1990) the interdependence of science and technical specifications was investigated (Grupp 1990) as well as the time regime between papers, patents, production and foreign trade (Grupp et al. 1990). In this sense the chapters to follow do not stand completely alone but will try to continue the type of analysis abstracted by Sigurdson (1990, pp. 1-10) from the contributions in the preceeding book.

1.3 FROM MODELS TO SCIENCE POLICY AND INNOVATION MANAGEMENT

This volume attempts to analyse some of the changes in the relations between science and innovation and suggests ways of using various quantitative indicators to study the underlying dynamics.

Among the notable exceptions from this general line of the volume is the paper by Rikard Stankiewicz (chapter 2). In his mainly theoretical arguments he differentiates between specific technology and a complex of activities centered around the production, assembly and organization of knowledge enhancing man's instrumental control of his physical, biological and social environment. Figure 1-1 suggests that Stankiewicz's socio-cognitive subsystem of technology may be characterized by some degrees of autonomy but is not completely distinct from the system of concrete technology. As the various links between knowledge, education and innovative functions of the society are intangible (figure 1-1), Stankiewicz's theoretical approach is the only one presently possible to throw some light on the socio-cognitive aspects of technology.

Part I of this volume deals with concepts like the one by Stankiewicz or the ideas presented in the first sections of this introductory chapter. As far as measurement procedures are concerned, they are going to be established in chapters 3 and 4 which depict promising starting points. Vololona Rabeharisoa explores the meaning of scientific articles published by inventors. She has focused on the narrow case of French fuel cell R&D in order to be able to assess the contents of both papers and patents. Her case analysis displays the intermediate dynamic functions in the science and innovation complex by pointing to the crucial role of, e.g., technical papers. Rabeharisoa's contribution very successfully argues that it will be an unsuccessful task to differentiate scientific from technological researchers or experts, at least in fuel cell R&D. Rabeharisoa's approach will be followed on in other cases. Some of the contributors to this volume will soon present a comparable analysis of laser medicine (Noyons et al. 1992).

The statistical approach described by Grupp & Schmoch (chapter 4) is complementary to chapter 3. Their methodology is not being applied to in-depth case studies (which would nevertheless be possible as well), but tries to differentiate between larger technological fields with respect to the role science plays in different technology sectors. As innovation literature introduced the notion of science-based *industries*, the open question was whether and how science-based *technologies* may be defined. Given the complexities of the cyclic model of innovation and, in particular, the pervasive effects of the science-based technological revolution throughout the economy (Freeman 1984, p. 491) it is indeed questionable to demarcate science-based innovation by traditional industry branches. From this line of arguments the technology-based methodology of Grupp & Schmoch may have a greater potential for future analysis than industry-based economic research. Aside to a presentation of methodology, chapter 4 outlines some first applications as well. They consist of an international comparison between countries and of an institutional sketch of the unified Germany.

Studies of science-based innovation may be broken down by technological cases (see part III of this volume) or by nations. Part II consists in a triptych composed of pictures from two very different countries: East Germany and Japan. Edo Albrecht (in chapter 5) describes the links between science and innovation in East Germany before the unification with West Germany. As Albrecht points out there were surprisingly close linkages between universities and industry. The cooperation experience between the two institutional sectors was good, much better than in West Germany. If such a good cooperation is regarded as an important prerequisite for science-based innovation, as is argued in the beginning of this chapter, then the economic failure of the former German Democratic Republic (GDR) looks like a puzzle. But naturally there are predominant factors apart from science and technology (communication isolation from Western countries, procurement problems of R&D-intensive components, appliances and parts, and so on) which must not be overlooked in science and innovation research.

In contrast to the well developed linkages between university and industry in the former GDR, Shinichi Yamamoto's contribution (chapter 6) on Japan reveals declining research shares of the university sector. More important than formal external research agreements seem to be looser forms of cooperations like donations. Despite the lesser developed linkages in Japan she does very well in science-based innovation. This puzzle

- compared to the GDR - can again only be explained by the political and social system. Obviously the national aspects continue to be very strong and are not overridden by the globalization of modern technology. This is why national case studies are an important tool for science and innovation analysis.

The third contribution to part II of this volume is at the crossroads between country and technological case studies. Lennart Stenberg gives a selective view of Japanese research organization in the field of molecular beam epitaxy (chapter 7). His analysis is the most detailed case study in this volume. It is supported by semi-automatic use of bibliometric data which are manually grouped by institutions. Thus the production of scientific papers is utilized as a proxy of institutional or sectoral activity. Both Yamamoto and Stenberg are studying science-based innovation in Japan. Contrary to the overall picture in chapter 6 Stenberg substantiates an increasing and more vital part of Japanese universities in the particular field of modern epitaxy research. This is why technology case studies are ever so important: The cyclic or coupled model of innovation allows for so many degrees of freedom (in terms of black-box model theory), or, it represents largely social systems and beneath them systems of concrete technology that abstraction should be perceived as being at its infancy. However, contemporary science as well as economic research is overwhelmingly starting from abstract postulations.

Three more technology case studies are presented in part III of this volume. Two of them (chapter 8 and 9) which are interrelated selected laser medicine and polyimide chemistry. These two technological specialities from physics and chemistry have in common that their historic development exposes strong cycles. Thus their analysis gives evidence for the coupled model of innovation. The technology case studies are not specific to selected countries, but provide an internationally comparable analysis.

Ben van Vianen and Anthony van Raan explore the knowledge expansion in applied science and again utilize bibliometric methods (chapter 8). They explain the cyclicity in the development of polyimides by alternating periods of technology leading and of science leading. The principle connections summarized at the beginning of this section are clearly reproduced in this case. Laser medicine, on the other hand, is so highly diffusive presently that it is difficult to locate its actual position in terms of theoretical or model perceptions.

The latter case is put into the perspective of overall laser R&D by Ulrich Schmoch and Hariolf Grupp (chapter 9). Their patent assessment of the expansion of knowledge in applied science suggests that the total laser field should be compared to all polyimide

R&D. On this aggregation level, about the same cyclic trends become visible for both fields of applied science (compare figure 9-1 in chapter 9). In the mid-seventies the "limits to growth" both for laser and for polyimide technology were seemingly reached. But then a sharp recovery of patent applications and, thus, inventions can be observed. The second take-off beyond the old limits to growth is induced by a growing number of marketable inventions which require special technical modifications of the coherent light beam sources or the polyimide materials, respectively. Among the new applications in the laser field, laser medicine appears and stimulates laser core technology to expand in a certain new direction.

Anders Granberg in chapter 10 maps another evolving technology cluster: Factory automation. In contrast to the two previous case studies he defined the clusters in factory automation by cognitive interpretation of selected literature and by asking experts. This more qualitative approach - as Granberg shows convincingly - can produce similarly valid maps of specific technical fields as bibliometric or other statistical methods. These two approaches are not so much different in the end as the large differences in methodology might suggest. The abstraction in bibliographical classification by codes and keywords determined by authors, journal editors or database experts, which the bibliometricians exploit, seems not to be so much different from the direct excerpts from published literature.

The composition and structure of factory automation as Granberg depicts it, is rich insofar as various perspectives of the evolving technology cluster are confronted with each other: the view of users, suppliers, subsuppliers and the overall view. Thus in a different way, than by indicator measurement, the complex model of science-based innovation-oriented processes is further supported.

In this volume some related contributions are missing for various reasons. As a remedy, part IV summarizes not only mutual comments of the contributors to this book, but includes summaries of relevant research work by Francis Narin, Ben Martin and Eric Engelsman and its discussion. In this way, hopefully, the 1990-91 state of the art in science research and in innovation research with respect to measurements of the dynamics of science-based innovation is documented.

This volume attempts to add evidence for a better understanding of the interfaces between science, technology and innovation. Present and future roles of basic research and technology genesis in an ever more competitive world are not treated explicitly. But by understanding a bit more of the basic dynamics of science-based innovation, it is hoped to prevent *R&D policy makers* from creating specific technology policy measures contrary to expectations and principle considerations. In many countries new arguments for an assessment of the impact of increased government spending on basic and fundamental science as opposed to industrial technology development are under scrutiny. But obviously this dichotomy is not so meaningful as the political debaters assume. Now and in the future science-based innovation continues to need both science and innovation.

To *industrial research managers* the search for particularly science-dependent sub-areas among all entrepreneurial business areas is much more important than in the arena of R&D policy. For the company, it is becoming increasingly true that links to science need to be mastered in areas outside their traditional businesses. Probably the central industrial R&D departments must concern themselves with the opportunities for science-linked innovation more so than business-area development laboratories.

REFERENCES

Albrecht, E. (1988), 'Dialektik des ökonomischen Wachstums. Schlüssel- und Hochtechnologien unter den Bedingungen der intensiv erweiterten Reproduktion', *Wissenschaftliche Zeitschrift der Humboldt-Universität*, Reihe Mathematik/Naturwissenschaften, 37, pp. 5-33.

Albrecht, E. and Dohnert, O. (1990), 'Zu einigen theoretischen und methodischen Grundfragen einer vergleichenden Analyse von Forschungs- und Innovationsstrategien', *Wissenschaftliche Zeitschrift der Humboldt-Universität*, Reihe Gesellschaftswissenschaften, 39, pp. 819-834.

Dosi, G., Freeman, Ch., Nelson, R., Silverberg, G. and Soete, L. (eds.) (1988), *Technical Change and Economic Theory*, Pinter Publishers, London.

The Economist (1986), Survey on High Technology, 23rd August.

Freeman, Ch. (1984), 'Prometheus unbound', *Futures*, 16, pp. 494-507.

Freeman, Ch. (1987), *Technology Policy and Economic Performance*, Pinter Publishers, London.

Freeman, Ch., Clark, J. and Soete, L. (1982), *Unemployment and Technical Innovation: a Study of Long Waves and Economic Development*. Pinter Publishers, London.

Freeman, Ch. and Perez, C. (1988), 'Structural crises of adjustment, business cycles and investment behaviour', in G. Dosi et al.(eds.), pp. 40-66.

Gerster, H.J. (1988), *Lange Wellen wirtschaftlicher Entwicklung*, Verlag Peter Lang, Frankfurt.

Grupp, H. (1990), 'Technometrics as a missing link in science and technology indicators', in J. Sigurdson (ed.), pp. 57-76.

Grupp, H., Schwitalla, B., Schmoch, U., and Granberg, A. (1990), 'Developing industrial robot technology in Sweden, West Germany, Japan and the USA', in J. Sigurdson (ed.), pp. 106-129.

Inose, H., Kawasaki, M. and Kodama, F. (eds.) (1991), *Science and technology policy research*, Mita press, Tokyo.

Kline, S. J. (1991), 'Models of Innovation and their Policy Consequences', in H. Inose et al. (eds.), pp. 125-140.

Krupp, H., (1984), *Basic Research in German Research Institutions, Excluding Universities and the Max-Planck Society*, Proceedings of the Japan-Germany Science Seminar, Japan Society for the Promotion of Science, Tokyo, pp. 73-109.

Kuhn, Th., (1967), *Die Struktur wissenschaftlicher Revolutionen*, Suhrkamp Verlag, Frankfurt am Main.

Lakatos, I., (1974), 'Falsifikation und die Methodologie wissenschaftlicher Forschungsprogramme', in *Kritik und Erkenntnisfortschritt*, Braunschweig, pp. 89-189.

Machlup, F. (1962), *The Production and Distribution of Knowledge in the United States*, Princeton, N.J.

Meadows, D., et al., (1972), *The Limits to Growth*, Universe Books, New York.

Noyons, E.C.M., van Raan, A.F.J., Grupp, H. and Schmoch, U., (1992), 'The Science and Technology Interface: Inventor-Author Relations in Laser Medicine Research', to be published.

Rosenberg, N. (1976), *Perspectives on Technology*, Cambridge University Press, Cambridge.

Sigurdson, J. (ed.), (1990), *Measuring the Dynamics of Technological Change*, Pinter Publishers, London.

Sigurdson, J., (1990), 'Challenge and new analytical methods in science and technology policy - an introduction', in J. Sigurdson (ed.), pp. 1-10.

de Solla Price, D.J. (1965), *Little Science, Big Science*, Columbia University Press, New York.

Smith, K. (1990), 'New Innovation Indicators: Conceptal Basis and Practical Problems', Paper presented to the OECD coference on "Consequences of the technology economy programme for the development of indicators", Paris, 2nd-5th July 1990.

Stankiewicz, R., (1990), 'Basic technologies and the innovation process', in J. Sigurdson (ed.), pp. 13-38.

2 TECHNOLOGY AS AN AUTONOMOUS SOCIO-COGNITIVE SYSTEM

Rikard Stankiewicz

2.1 INTRODUCTION

The increasing complexity and dynamism of contemporary technology places great demands on the effectiveness and adaptability of R&D systems. Among the problems which receive intensive attention in the current policy discussions are the appropriate *organizational forms for generic R&D* and the optimum methods for the *internal and external interfacing of various aspects of technological and scientific activity*. At the practical level, attempts are made to solve these problems by experimenting with novel ways of organizing and supporting R&D (research parks, joint industry-university research centres, technology development programmes, etc.). At the analytical level there is a renewed interest in the relationships between science, technology and economy (see, e. g., chapter 1). As the record of scholary debates indicates we are still far from a concensus on what constitutes the appropriate general model for the analysis of those relationships. Various diametrically opposed approaches are being tried out.

At one extreme there has been a tendency, especially in the earlier studies of technological innovation, to view science, technology and economy as a set of causally linked *variables*[1] - rather like the volume, pressure and temperature of a gas in a container. This approach while logically admissible seems heuristically disfunctional. After all science, economy and technology constitute complex social systems. The interactions between them have a logic which is quite different from that applying to simple physical quantities. To comprehend that logic we must understand the inner structure of the systems involved. While it is possible to claim that we posses today a fairly advanced understanding of science and economy, the same can hardly be said

[1] The variables actually used have been, for most part, aggregates such as the spending on science, investment in industrial R&D, manpower levels, economic growth rates, etc. (see figure 1-1 in chapter 1, for instance).

about technology. The reason for it is, largely, the protean character of technology as a social phenomenon. The term is used to denote material artefacts as well as social organizations, abstract knowledge as well as practical skills, major sociotechnical complexes as well as minor everyday implements. Clearly we are not dealing here with a homogeneous phenomenon but rather with one which is highly differentiated and constantly undergoing further mutations. An effective approach to technology requires that these characteristics are recognized and made explicit both at the cognitive and the socio-organizational level.

As a reaction to the earlier simplistic models of technological change new approaches have recently been developed which stress the complexity of technology itself and of its societal ramifications (see Bijker, Hughes, & Pinch 1987). There is, however, a risk that the pendulum is swinging too far in the direction of ideographic approaches which obscure the structural features of the phenomena studied. I have in mind the tendency to view the relationships between technology, science and other social activities as a "seamless web" (Hughes 1986). In my view the risk of imbalance associated with that approach can be diminished if we recognize explicitely that science, technology, and economy constitute *interpenetrating* (Münch 1983) rather than fused systems.

Interpenetration occurs when two systems overlap in some manner (share certain elements or have closely coupled elements) without however loosing their autonomy and identity. For instance: it is generally accepted that what we customarily call "science" contains important elements of "technology", and vice versa (as is shown once again in subsequent chapter 3). Indeed without this sort of overlap the modern science and technology would hardly be possible. Yet we are not prepared to deny each of the systems a separate identity.

The phenomenon of interpenetration is extremely difficult to study[2]. The interactions between interpenetrating systems are different from those which occur between supersystems and their subsystems or between independent systems with sharply defined boundaries. They are much more subtle and difficult to measure. Some of the links are deeply hidden from an external observer, others are weak but have important cumulative effects over longer periods of time, and so on. The study of the modes of interpenetration constitutes therefore a major challenge to the students of the relations between science, technology and economy.

---

2 The concept has been introduced by T. Parsons and often occurs in the writings of neo-functionalists such as Münch and Luhmann. However, it has not been sufficiently elaborated on the theoretical, and, especially methodological level.

It is my opinion that in order to tackle those problems (the complexity and differentiation of technology and the role of interpenetration) we should study the evolution of technology within the theoretical framework of the social systems analysis[3]. In this paper I will attempt to sketch out how such analysis can be developed by conceptualizing the general dynamics of technology as a process of structuring of a complex system through cycles of functional differentiation and integration.

Functional differentiation[4] is a process through which social functions (or combinations of functions) are separated from the original multifunctional structures and transferred to new more specialized ones. The degree of differentiation can vary from a very rudimentary articulation of latent components of existing social roles to the emergence of complex social subsystems characterised by a high level of autonomy. The causes of differentiation are manifold. The most fundamental one is the need of social systems to cope with the growing complexity of their external and internal environments. Generally speaking differentiation means that the systems can increase their *overall* complexity while maintaining the task complexity of their *elements or subsystems* at an acceptable level. Closely related factors contributing to the differentiation are the need to reduce or eliminate disfunctional couplings between different activities, to accommodate divergent interests of the social actors and thus to reduce the social tensions and conflicts.

Ideally the result of differentiation is the enhanced efficiency and adaptiveness of the system as a whole. However, that desirable state of affairs requires that the necessary level of functional and social integration is simultaneously achieved. The degree and character of differentiation of sub-systems must harmonize with the requisite pattern of the over-all systemic integration (its structure and governance). The successful differentiation demands among other things that the emerging structures are sustainable, i.e. internally coherent, capable of maintaining effective transaction with their environment and perceived as legitimate in a wider social context. Many cases of differentiation do not meet these criteria and lead to inefficiency and waste.

Simplifying things greatly, one might say that the main mechanisms of integration among functionally differentiated subsystems fall into four major categories: normatively specified exchanges, organizational hierarchies (power and authority), markets, and various modes of interpenetration. Generally integration is achieved by

---

3 For a general review of the systems analytical approaches see Willke (1987).

4 The processes of social differentiation have been discussed by such writers as Parsons (1966,1971) and Luhmann (1984, 1986). In the field of economics releted phenomena have been analysed, among others, by Chandler (1977) and Williamson (1983).

relaying simultaneously on two or more of these relationships. However the relative importance of the modes of integration varies considerably from one sphere of social activity to another. Particularly important in this context is the process of autonomization of the emerging subsystems. Autonomy of a system is understood here as relative independence from the neigbouring systems due to *self-organisation* and effective *boundary-maintenance*. It does not imply a total isolation from the environment but rather the ability to respond to the environmental stimuli in a coherent and *selective* manner preserving the system's structure and identity.

The advanced forms of social differentiation which are characterized by autonomization of sub-systems appear to weaken (though not eliminate) the role of the hierarchical structures and the normatively regulated exchange relations while enhancing the importance of markets and various forms of interpenetration.

The failure to achieve the balance between differentiation and integration results in decreased adaptiveness of the system as a whole. This can take a number of different forms including: i) arrested differentiation and consequently limited development of the individual functions; ii) unbalanced development in which certain functions become excessively subordinated to others; iii) malfunctioning exchanges between subsystems resulting in restricted flow of resources and information; iv) conflicts between subsystems over resources; and v) loss of legitimacy and resulting drive towards dedifferentiation.

*

Chapter 2 is divided into two main parts. In the first, section 2.2, I discuss the process of differentiation of technology - both external and internal. In particular I pose the question whether the socio-cognitive subsystem of technology is becoming autonomous. I will argue that such indeed is the case and that - somewhat paradoxically - one of the main reasons for it is the increasingly intimate relationship between technology and science.[5] In the second part, section 2.3, we will be chiefly concerned with the means by which the various subsystems of technology become integrated among themselves and with other relevant social subsystems. The chapter ends with some reflections on the desirable directions for future research and on the policy implications of the conclusions drawn so far (section 2.4).

---

5 An analysis of the self-organization of science has been recently presented by Krohn & Küppers (1989, 1989a).

2.2 DIFFERENTIATION OF TECHNOLOGY AS A SOCIO-COGNITIVE SYSTEM

Technology is an instrument of adaptation and as such belongs to a larger cluster of adaptive activities which include the creation and use of artefacts, skills and knowledge as well as of different forms of social organization. The various aspects of that cluster have gradually become differentiated from the surrounding social matrix and from one another. It is of course impossible to deal here with these processes in any detail. Let me therefore mention just two important socio-cultural bifurcations. Parsons (1966) points out to one important division involving the differentiation of the *economic sub-system* from the more general adaptive complex[6]. It has led to a rapid institutional evolution leading to the creation of the sophisticated modern economies. (One interesting side effect of this development has been the tendency of the economists to view technology as an exogenous factor in the economic theory.) Another, from our point of view, fundamental division has been the differentiation of the *socio-cognitive sub-system of technology* from the more diffuse *concrete technological systems.* In this paper I will concentrate primarily on the second of these bifurcations.

The concrete technological systems (TS) are defined here as the material and behavioural complexes (in economic terms: embodied capital) through which the *adaptive activities* (production of goods and services broadly understood) is carried out. Thus Hughes (1987) gives the following description of "technological systems":

> "Among the components in technological systems are physical artifacts, such as turbogenerators, transformers, and transmission lines in electric light and power systems. Technological systems also include organizations, such as manufacturing firms, utility companies, and investment banks, and they incorporate components usually labeled scientific, such as books, articles, and university teaching and research programmes. Legislative artifacts, such as regulatory laws, can also be parts of technological systems. Because they are socially constructed and adapted in order to function in systems, natural resources, such as coal mines, also qualify as system artefacts".

The socio-cognitive system of technology (SCST), on the other hand, is defined here as a complex of activities centred on the production, assembly and organization of *knowledge* enhancing man's *instrumental control* of his physical, biological and social environment. They consist of scientists and engineers, educational institutions, R&D

---

6 Parsons sees technology as a concrete system: "Technological organization .... should be regerded as a boundary structure between the society as a system and the organic-physical environment" (Parsons 1966). Also: "Technology ... is the primarily physical reference of the complex which includes the *economy* as its primary social system reference. The economy is the aspect of of the societal system which functions not just to order technological procedures socially, but more importantly to fit them into the social system and control them in the interests of social units, whether induividual or collective". (ibid.)

departments and research organizations, technical journals, patent offices and professional and scientific associations. Above all they contain a vast body of symbolically coded, stored and communicated knowledge at various levels of generality and organization.

Historically the concrete and cognitive aspects of technology tended to be strongly interlinked if not wholly fused. Thus the production of technical knowledge occured almost exclusively as the outcome of the activities within the concrete technological systems and was consequently guided largely by the logic of these systems. Even today there is a very close linkage between techno-economic praxis and the cognitive evolution of technology. The knowledge needs generated within concrete technological systems tend to be highly specific and context-dependent and the knowledge acquired and used to satisfy them is to a high degree tacit and person-bound. This notwithstanding it seems clear that the process of differentiation of the SCST from the concrete technological systems has been under way for some time and is growing in importance.

The existence of a socio-cognitive system presupposes the presence of a communication network structured in a manner which enhances the production of knowledge in some coherent area. The network *processes* the knowledge produced and supplies the necessary cognitive *inputs* for further knowledge production. The networks can be long-lived or ephemeral, formal or informal. Their stability depends largely on the degree to which they can be differentiated and institutionalized within a larger social system.

The process of institutional differentiation of SCST, which accelerated strongly since the mid-19th century[7], has proceeded at two levels: (i) within the primary adaptive units such as firms, bureaucracies, etc., and (ii) at the societal level. The first of these processes involved the gradual emergence of various technical expert roles within the adaptive units. These roles multiplied and underwent numerous functional subdivisions and recombinations leading eventually to the emergence of modern industrial laboratories and engineering departments. On the societal level, the differentiation of expert roles has been followed by their generalization (detachment from the specific organizational loci) resulting in the emergence of professions. This was followed by the systematization and formalization of training leading up to the academization of technology. These differentiation processes resulted in the emergence of two distinct

---

7 On the conceptual level, the emergence of technology as a distinct *intellectual* activity is of course much earlier. It can be traced at least to the Renaissance. However, its *institutional* embodiment took place first with the emergence of the engineering profession in the 19th century.

types of socio-cognitive systems in technology: (i) the *actor-specific sub-systems* which are strongly integrated within their host organization (firm, bureaucratic agency, etc.) and through them closely linked to the underlying concrete technological systems; and (ii) the *general purpose societal sub-systems* with a somewhat looser relation to the concrete systems. The relative importance in the growth of technology of these two institutional and organizational dimensions is a matter of debate and depends on the general model of technical change one prefers.

Have the above described processes of differentiation led to increased *autonomy* of the SCST? It will be argued here that the autonomous core of the SCST exists and is in fact becoming steadily more important. The *empirical* evidence for this can be seen in the growing significance of "basic" or "generic" technologies, and in the rapid development of "engineering sciences". However, there are also strong *theoretical* reasons for expecting a steady trend towards autonomization of the cognitive system of technology. This kind of development, I would argue, is inherent in the dynamics of any major socio-cognitive system. There are five general reasons for it:

(i) *All sustained learning processes are inherently recursive.* This means that the outputs of the learning processes constitute their inputs. The system is, to a large extent, a cumulative function of its own history. The essence of this phenomenon is well captured by such phrases as "science grows out of science", "technology grows out of technology".

(ii) *Systems consisting of self-organizing sub-systems tend themselves to be self-organising.* This is very much the case with socio-cognitive networks which consist of largely autonomous components such as individual investigators and research groups.[8] Autonomous components are, by definition, resistant to external control. Consequently, a stable ordering of these components will be largely a result of a self-organizing process. An important precondition here is a high density or frequency of recurrent interactions among the components.

(iii) *The density of internal interactions within a cognitive system tends to increase over time.* As the amount of accumulated knowledge increases the interdependence of various specialized learning processes intensifies. The initially individual and particularistic cognitive activities become interconnected and large dense communication networks emerge.

8 For a discussion of self-organization in science see Krohn and Küppers (1989).

(iv) *The increase in the internal complexity of a cognitive system makes self-organization imperative.* The increasing complexity of the accumulated technical knowledge can be reduced in three basic ways: (a) through specialization, i.e. subdivision and segmentation; (b) through functional recombination; and (c) generalization. All these processes depend on intensive knowledge processing within the system and presuppose increasing self-organization. The resulting communication networks become more structured and exclusive.

(v) *The external governability of the cognitive systems tends to decrease due to the growing information or competence asymmetries between such systems and the neighbouring social systems.* An important threshold is crossed when a cognitive system acquires superior *diagnostic competence* as regards the cognitive needs of other social systems[9].

Of course the tendencies towards self-organisation and autonomization of the SCST encounter resistance from various countervailing forces. Among the latter are: (i) the cultural hostility and (ii) the particularism and contextuality of technological praxis .

The notion of technology as autonomous seems to clash head on with the idea that technology is about "means" towards "ends"[10]. Any talk about autonomy of technology is interpreted as leading to the displacement of goals by means, a technocratic pathology, if not a Frankensteinian monstrosity[11]. This sort of fears are largely a result of the failure to distinguish between the concrete technology and its cognitive system. A far going autonomy in the former would be an obvious cause of concern; an autonomy of technology at the cognitive level is a different matter.

The two other factors counteracting the autonomization of SCST - its particularism, and contextuality - are far more important. They appear to bind technologies to the social loci of their use thus reducing the density of recurrent interactions among the knowledge producers. Indeed they seem to constitute the main reason for the resistance within the technological communities against the institutionalization of autonomous SCST.

The judgement regarding the present and future balance between the factors enhancing or counteracting the autonomization of SCST depends on the general model

---

9 This in fact is one of the main sources of power and influence exercised by professions.

10 The widespread adherance to the rigid goals-means paradigm in the analysis of human action is increasingly dubious in the light of recent theoretical developments, especially in the field of systems theory.

11 For example: Ellul (1980).

of technological change one chooses as the basis of analysis. I should like to distinguish between three main categories of such models: the *standard evolutionary*, the *systemic* and the *cognitive* one. They could be viewed as eleborations of certain crucial aspects of SCST which emerge at the consecutive stages of technological development. Thus, they are complementary rather than mutually exclusive. In fact all the three are evolutionary in a broad sense. Nevertheless, the students of technological change differ rather strongly in the relative emphasis they place on these models.

THE STANDARD EVOLUTIONARY MODEL

Many features of technological change can be fairly easily analysed in terms of a conventional evolutionary model. In one sense technology is always a matter of specific adaptations to unique environments. Such adaptation are achieved through a process of variation and selective retention occuring more or less continously at a very large number of points in the social system. Much of technical knowledge produced in this manner is tacit and context-specific. However, some successful variants, in so far as they have more than local significance, will diffuse through trade and imitation. Some of them will become "generalized" in the sense of spreading to new qualitatively different niches. This is usually accompanied by their further "mutations". The diffusion is central because it is the precondition of the process of "recombination" - the creation of new technological solutions based on fusion of existing elements. This process of recombination is regarded as one of the main mechanisms of technological advance (and creativity generally). The process of diffusion is affected by several factors including the structure of the economic system and the existence of appropriate generalized communication systems (particularily the educational system).

In this model the SCST is essentially a "knowledge pool" analogous to "gene pool". Its societal components have an important by essentially derivative function of disseminating information. They appear to have no significant creative role in the sense of making major original contributions to the advance of technology. The educational systems, technical universities and various professional bodies codify the technology developed in the adaptive systems (chiefly firms) and enrich it by information derived from science. The latter contributes elements (methods, concepts, facts) which when introduced to the general reservoir of technological knowledge can enter into new combinations with the prexisting ones.

The type of technology development model has a strong appeal to economists. This is natural: the importance of "learning by doing", the function of users in technological innovation (von Hippel 1988), the creative role of markets in the shaping of technology (Clark 1985), the proprietary character of much of newly produced technological knowledge, its contextual specificity and tacit character (Eliasson 1989 and Pelikan 1990) suggest that technology can largely be viewed as an endogenous economic phenomenon. In fact considerable progress has recently been made along these lines in incorporating technological change within the framework of economic theory.

THE SYSTEMIC MODEL

The evolutionary model outlined above goes a long way towards explaining the general dynamics of technological change. However it is not fully satisfactory as an account for certain characteristic features of contemporary technology. Its main shortcoming is that it pictures technological change as a disjointed and atomistic process more or less wholly determined by the dynamics of the economic environment. This concept has been recently challenged by several new approaches emphasizing the systemic character of technology[12].

In most of its versions the notion of the "technological system" comes very close to what we earlier defined as the "concrete technological system"[13] and includes both material artefacts and social elements. Freeman and Perez (1988), to take one example, have emphasized the institutional components of the technological systems. The generally heterogenous character of these system - the mixing of diverse social, material and symbolic components - has been stressed by Latour (1987), Callon (1987) and Law (1987).

The technological systems appear to have a certain coherence and developmental momentum of their own. The interpretations of the nature and significance of that momentum vary among different authors. Some like Hughes admit only reluctantly that the systems do have a certain internal logic. Others, among them the economists such as Dosi, Sahal, Nelson or Winter, seem more prepared to ascribe to technology a degree of autonomy[14]. The essence of the argument seems to be that all complex systems tend to evolve fairly well defined structures (basic configurations) which constrain the development of their sub-systems or components. Certain kinds of modifications in the sub-systems, desirable as they may seem in themselves, would demand major changes in the basic system architecture, changes which may

---

12 Interestingly this trend coincides with the growing criticisms of the classical neo-Darwinian models of evolution in biology and the ferment in the theory of science following Kuhn's theories. There are also some rather striking similarities to the development in the systems theory.

13 Compare the definition proposed by Hughes (1987) and above.

14 For a "constructivist" critique of these approaches see van den Belt & Rip (1987).

be neither feasible nor desirable. Other modifications, on the other hand, seem to enhance the overall effectiveness of the system and are therefore favoured[15]. Thus the system tends to develop along a trajectory which defines the possible compromises between the internal constraints of the technologies involved and the demands of the socio-economic environment. These gradual and essentially cumulative developmental processes are periodically interrupted by radical transformations as a result of which the basic system configuration is changed or replaced by an emergent alternative. The radical transformations, which are relatively rare, can be influenced - speeded up or retarded - by a variety of intellectual, economic and institutional factors.

The above general concept of technological systems dynamics has its correlates at the level of the SCST. Although most of the "constructivists" tend to view the cognitive components of technology as integral components of the concrete technological systems[16], other researchers like Dosi, Constant or Laudan emphasize the existence of distinct knowledge complexes which encode and govern the technological systems. Dosi (1982) has proposed that these cognitive complexes could be viewed as analogous to scientific paradigms as defined by Kuhn (1970). The paradigms guide the prevailing technological praxis as well as the production of new technological knowledge[17]. Dosi does not work out the resulting dialectical relationship between the concrete technical systems and the paradigms in much detail. However, these questions have been addressed by several historians and sociologists of technology[18]. Thus Constant (1980, 1984) and Laudan (1984) have based their analysis of the cognitive dynamics of technologies on the study of the "communities of practitioners" corresponding to major technological systems. The members of the communities interact through a variety of channels and jointly develop a specific technological paradigm. The dense communication web among the practitioners supported by the educational system and other institutional arrangements gives the SCST a degree of coherence. The communities of practitioners are relatively small and structured hierarchically into specialities corresponding to the various elements of a given concrete technological system.

---

15 Comp. Hughes' (1983) concept of "reverse salients", Rosenbergs(1976) "bottlenecks" or Sahal's (1985)"technological guideposts".

16 See the definition by Hughes cited above

17 Dosi (1982) defines technological paradigm as follows "We shell define 'technological paradigm' broadely in accordance with the epistemological definition as an 'outlook', a set of procedures, a definition of the 'relevant' problems and of the specific knowledge related to their solution. We shell argue also that each 'technological paradigm' defines its own concept of 'progress' based on its specific technological and economic trade-offs. Then we will call a 'technological trajectory' the direction of advance within a technological paradigm."(p. 148)

18 See especially Laudan (1984)

The Kuhnian approach does have a great heuristic value in the studies of technology dynamics. Its conceptual machinery is very useful in the analysis of hierarchically organized cognitive systems and technologies tend to be hierarchically organised at both the concrete and conceptual level. However, just as there are limitations on the applicability of the paradigm theory to science so there are, perhaps to even a greater extent, difficulties with using it in the context of technology.[19] In my view the main weakness of the approaches discussed above is the failure of its proponents to elucidate the precise relationship between the dynamics of the concrete systems and the corresponding technological paradigms. For most part the technological paradigms are defined as a sort of "superstructure" closely linked to the dominant technological practice and necessary for its reproduction and conservation. They are said to guide R&D programmes and other types of inventive activity enhancing the speed and efficiency with which the developmental tendencies inherent in the concrete systems enfold. The institutions such as universities and technical High Schools, professional associations and educational systems generally are essentially the bearers of the established traditions. Thus despite the emphasis on the knowledge dimension of technology, quite strong in the writings of such writers as Dosi, most of the "systems theorists" view the dynamic of the *concrete* technological systems a dominant determinant of the development[20]. In fact, some authors such as Freeman and Perez (1988) prefer to talk about "techno-economic paradigms" rather than technological ones. They see the techno-economic paradigm as a "meta-paradigm" common to several concrete technological systems, which "once established as the dominant influence on engineers, designers and managers, becomes a 'technological regime' for decades" (ibid.).

The SCST is seen as playing a more direct and active role in the technological development only when the concrete systems underlying the paradigm become destabilized or go through a crisis. The crisis triggers divergent search for radically new technical solutions (some of which may in fact have been available for some time but were suppressed). However, here again the causes of the crisis are typically sought in the concrete systems themselves. Laudan (1984) lists various technological objectives which guide the activities of a "community of practitioners"[21]. Only one of them seems

19 For a sympathetic but critical review see Gutting (1984).

20 The manner in which the "practitioners" are integrated into the system is described by Hughes (1987) as follows: "A crucial function of people in technological systems, besides the obvious role of inventing, designing, and developing systems, is to complete the feedback loop between system performance and system goal and in so doing to correct errors in system performance".

21 They include: 1) the unsolved problems "given directly by the environment", 2) the functional failures of current technology, 3) extrapolations from past technological successes; 4) imbalances between related technologies; and 5) perceived potential problems.

to arise in an unequivocal manner from within the cognitive system itself. On the basis of available scientific knowledge, the community of practitioners may be able to predict crises in the technical system before they manifest themselves in practice[22]. Such "presumptive anomalies", to use Constant's terminology, may give rise to technical R&D programmes which are a step or two removed from the daily adaptive problem-solving and no longer reducible to technology refinements along the well established trajectories. This is important since it shows that the developments within the cognitive system of a technology can destabilize the concrete system rather than the other way around. The degree of independence of the SCST thus achieved is made possible by the inputs from another autonomous socio-cognitive system - science. The differentiation and autonomization of the SCST itself carries that development still further.

THE COGNITIVE MODEL

The idea of technology as a rational, theory-based enterprise has been current in the West at least since the Renaissance. Some authors (Münch 1983) attribute the extraordinary success of the occidental technology to this rationalistic orientation. In a strong version this orientation involves the belief that technological problems can be solved through application of general principles. In a softer variant, the claim is made that these principles are a useful guide in the empirical search for technological solutions. These claims have been hotly disputed. The controversy concerns both the very possibility or desirability of the rationalist approach and, if this is admitted, the precise status and origins of the generalized technological knowledge. The adherents (explicit or implicit) of the simple evolutionary model of technological change tend to declare "technology-as-theory" a myth (Fores 1988) or at best a mere device for abstract codification of the empirically accumulated technical knowledge. Those who admit the existence and usefulness of theory in technology fall into two groups: (i) the reductionists (Bunge, Popper) who believe that the (natural) sciences provide all the necessary theoretical underpinings of technology and (ii) the anti-reductionists (Koryé; Layton, Channell) who claim for technology a status of a cognitive system similar to but by no means identical with that of science.

We shall not pursue these debates here. They involve more than purely intellectual issues. They have been (at least since the mid 19th century) a tool of infighting concerning the academic status of science resp. technology. They have also been strongly influenced by the controversies surrounding the highly charged notions of

22 Also Dosi (1982) regards scientific advances as one of the main factors leading to the emergence of new technological paradigm.

"theory" and "science"[23]. It is I believe more fruitful to adopt a more pragmatic position and focus on the empirical fact that important parts of the cognitive system of technology are becoming organized into "fields", i.e. communication systems linking sets of knowledge producers (not necessarily "practitioners") working in coherent technological domains which are defined analytically rather than by direct reference to concrete technical systems. These fields are quite similar in form to the fields found in science[24]. The fundamental difference between scientific fields and technological ones are neither methods nor concepts (in these they become increasingly similar), but rather the basic organizing values: explanation in the case of science and instrumental control in the case of technology. The two types of fields structure knowledge differently and they generate different problem hierarchies.

A major force behind that development is the rapid growth of technological knowledge resulting in a veritable explosion of complexity. In order to be retrievable, transmittable and operationally accessible the unwieldy corpus of technical knowledge has to be structured and whenever possible reduced to generic formulas. This creates an internal meta-technological research agenda based on the reflection on the "state-of-the-art" rather than determined by specific practical needs of the moment. It also leads to the separation of roles of the *engineer* (oriented towards the more or less creative application of the existing technical know-how to specific problems arising within concrete technical systems) and the *technologist* (seeking to extend the limits of technological capability by developing solutions to classes of problems). At the structural level this leads to a differentiation of the departmental structure of technical universities into traditional engineering disciplines, technological fields and applied science fields (Granberg & Stankiewicz 1981a).

The meta-technological research agenda focuses on: (i) the transformation of the application- or object-oriented knowledge into generic principles and models, that is *generalization*; and (ii) systematic elucidation of the relationship between structure and function, i.e. development and articulation of "*functional insight*". This increasingly self-referential character of technological activity leads to the articulation of fundamental concepts and approaches which are then increasingly used in diagnosing and defining the "practical problems". The latter are thus gradually transformed from the primary focus of technological activity to a form of raw material.

---

23 The concept of engineering *science* is probably a cause of many of these difficulties.

24 As examples of emerging technological fields one could mention "cybernetics", "control engineering", "computer sciences", "genetic engineering", etc. (see also chapters 8 and 9).

The above described tendencies lead to a partial decoupling of the technological fields from the concrete technical systems. Overlapping with the "communities of practitioners" there emerge the communities of technologists developing and sharing knowledge relevant to many different communities of practitioners. This is enhanced by the development of new basic technologies (Granberg & Stankiewicz 1981; Stankiewicz 1986). By basic technologies we mean the technologies which focus on specific natural processes or structures or, better, certain classes of such processes and structures (for instance laser technology, protein engineering, or nuclear technology). These technologies tend to be highly generic, i.e. applicable to a variety of functions in a variety of concrete systems. Thus, a laser specialist (to take one example) may partake of several different communities such as the community of *engineers* in some field of application of lasers, the community of laser *technologists* in general and, possibly, the community of laser *scientists*. Each of these communities will evolve different criteria of importance or relevance of various aspects of R&D activity. The technological community is likely to gradually isolate a set of fundamental features of laser phenomena the knowledge and control of which is strategic to the general advance of laser technology. Although in their choice of R&D problems they will be influenced by the needs of the practitioners, the technologists are not wholly governed by these needs. They will tend to *select* from the diversity of such practical problems some which are strategic, i.e. the solution of which will lead to major advance in the generic technical knowledge.

The technological fields are linked together through the communities of practitioners centred around various concrete systems. However, it seems that there also emerge more direct connections between these fields. Due to their scientification, the languages of basic technologies are becoming increasingly similar. Two tendencies are particularily important here: (i) the growing reliance of basic technologies on the molecular and sub-molecular phenomena (which leads to the homogenization of their scientific base); and (ii) the development of formal technical disciplines such as information and control theory. As a result the developments in the hitherto separate technological fields become closely correlated[25] and the SCST acquires growing cores of transdisciplinary generic knowledge.

The developments described above transform, organize and, to an extent, homogenize the chaotic "knowledge pool" of which we spoke while discussing the standard evolutionary model of technological change. The cognitive convergence between various fields of technology greatly increases the fusability of its diverse elements thus

---

[25] As examples of this phenomenon one can mention the interaction between optics and electronics, electronics and biotechnology, biotechnology and chemical technology, etc. (see also chapter 9).

making the "cognitive paradigms" much more fluid and interpenetrating (Stankiewicz 1990). These developments reinforce the autonomy of the cognitive sub-system of technology. The growing body of "functional insight", i.e. generalized yet well articulated knowledge of the crucial functional requirements in various areas of technology, makes technologists capable of identifying important problems without constant feedback from the concrete systems.

*

Let us recapitulate. The three models of technological change described above can be said to represent three stages in the process of differentiation and autonomization of the socio-cognitive system of technology. The first stage, which is well represented by the standard evolutionary model defines the SCST as a "knowledge pool" from which elements are drawn by different actors in their search for the solution of the highly specific problems they are confronted with. In the process of the problem solving additional knowledge is generated which, to varying degrees, is fed back to the pool. Much of the technological knowledge is in fact accumulated locally and does not enter the pool. The degree of differentiation of the SCST is so low that the question of its autonomy hardly arises.

The systemic model captures the complexity and interconnectedness of the contemporary technologies. The system character of many of these technologies at the concrete level endows them with a certain internal momentum. The SCST is then defined in relation to such concrete semi-autonomous systems. However, the autonomy SCST itself is strongly limited. The engineering traditions or, as they are also called "paradigms", constitute differentiated but basically subordinate components of the dominant concrete systems.

The cognitive model emphasizes the emergence of coherent fields of generic technological knowledge which are increasingly autonomous in relation to the concrete systems. This autonomy depends in part on the internally accumulated technological knowledge (functional insight) and partly on the close symbiosis with science. The fields develop internal R&D programmes which spearhead technological development especially in the strategic areas of basic technologies. The autonomous SCST does not constitute an apex of some hierarchical control system nor is it the extreme end of some linear sequence of causes. It rather plays the role of one actor among many who determine the evolution of concrete technologies. But it is an increasingly important actor performing unique functions in the overall process of technological change.

The differentiation and autonomization of the SCST reflects the complex and, in an important sense, contradictory character of technological activity. It is a result of an inevitable tension between two sides of technology: one striving to produce the *specific and context-sensitive solutions* to the adaptive problem of everyday social and economic life and, on the other, aiming at the achievement of maximum *generalized control* over the environment. The former aims at the reduction of undesirable variety and uncertainty. The latter increases the contingency social action. The two dimensions are mutually supportive in a long run, but in a shorter perspective they require a degree of cognitive and organizational differentiation.

2.3 FUNCTIONAL INTEGRATION OF THE SOCIO-COGNITIVE SYSTEM OF TECHNOLOGY

As pointed out earlier the process of functional differentiation of society means that the functions originally performed by multifunctional structures are taken over by new, more specialized ones. When a differentiated sub-system itself achieves a certain degree of complexity the process of autonomization tends to set in. The sub-system becomes increasingly self-organizing, develops a boundary and its transactions with the environment become highly selective. In order to survive the system must fulfil certain conditions: (i) it must be internally integrated and able to undergo an internal segmentation and differentiation without the loss of cohesion; (ii) it must correspond to a vital cluster of societal functions which are hard to perform effectively using other structures; and (iii) it must be organized in a way which, while allowing for effective transactions and coordination with other systems, adequately protects the core activities of the sub-system in question. The long-lived autonomous systems have usually the ability to influence and shape their environment (Luhmann 1986, Krohn & Küppers 1989). The most successful of them become anchored in the culture of a society and cannot be changed in a radical manner without provoking a major societal legitimacy crisis.

The problem of transactions and coordination among autonomous systems is of special importance. A failure to achieve the necessary degree of functional integration among them can lead to various "pathologies" such as the contraction or even disappearance of the system, the loss of self-organizing capacity leading to the system's internal fragmentation, ineffectiveness and an eventual demiss as an autonomous unit, the dominance or predatory behaviour leading to conflicts with other social systems and the weakening of the system's legitimacy within a larger socio-cultural context.

Differentiation with little autonomization poses few serious problems of functional integration. The latter can be achieved by relying on the traditional normative patterns of division of labour or by the recourse to hierarchical organization as in the case of modern bureaucracies. The integration of the highly autonomous systems requires other mechanisms, namely: generalized exchange media or interpenetration. Integration based on exchange relations (especially those mediated through markets) presupposes that transactions between the systems can be organized in an effective manner (comp. Williamson 1983). Among others it is required that the parties can communicate to each other all the relevant information regarding supply and demand in a simple and straightforward manner - preferably through prices. These conditions are hard to achieve in the case of transactions which involve knowledge. Consequently, in the situations where the interaction between autonomous systems requires large decentralized or multichannel information flows interpenetration tends to be a superior mechamism of integration.

As noted earlier interpenetration means that two social systems overlap or share certain elements. These elements constitute the "interpenetration zones". The most important form of interpenetration involves individuals participating in some manner in two autonomous sub-systems. Another involves sharing of certain cultural components (concepts, beliefs, values). Interpenetration is generally enhanced by mobility of persons between systems, multiple socialization of individuals through the educational process, emergence of hybrid roles and ecological proximity. The degree of interpenetration can be enhanced by various contrived means such as the creation of special mediating organizations. The crucial point is however that the contact or interpenetration surfaces must be broad. The opposite of interpenetration is compartmentalization, i.e. a situation where the participation in the systems involved is mutually exclusive and where the interactions between them, if at all permitted, occurs through narrow and strongly controlled channels[26].

The essential feature of interpenetration is that two systems should be able to exercise influence over one another without their autonomy becoming compromised. The flow of information occuring between the systems at a large number of points amplifies certain aspects of each system's behaviour thus creating a kind of "resonance" between them. The dynamics of interaction between science and technology illustrate well the nature of this phenomenon[27]. Throughout their history the two systems have exercised

---

26 Münch (1983) opposes interpenetration to differentiation. Clearly he interprets differentiation as compartmentalisation. This is not the position taken here: Differentiation does not imply compartmentalization and is therefore compatible with interpenetration. Indeed it is the need to promote both that constitutes the main challenge to modern institutional development.

27 See, among others, Böhme 1983, Krohn et al. 1974, Krohn & Küppers 1989, Stankiewicz 1986).

profound influence on each other without either loosing its identity. Technology responds to those scientific developments, old or new, which coincide with its needs and interests. The same is true of the response of science to technology. Attempts to impose the agenda of one of the system's on the other tend to be unproductive and is generally resisted.

Yet it must be clearly understood that successful interpenetration is not easy to achieve and that once achieved it carries with it certain risks. Especially it can lead to a gradual dedifferentiation of the sub-systems which thus loose their self-organizing ability and effectiveness. The risk is that, in the cases where the basic governing values of the two systems are not fully compatible one set of values will eventually become dominant at the expense of the other. An opposite danger is that interpenetration, by threatening the systems' identities actually stimulates destructive competition between them. Here again the story of science and technology provides many illustrations of conflicts in which either of the systems considered its integrity threatened. Indeed, keeping the right balance between differentiation, autonomy and interpenetration constitutes one of the hardest tasks of institutional design. In order to prevent the zones of interpenetration from degenerating into institutional battlefields or gray areas of confusion, the interacting systems must evolve a conscious attitude to one another, certain shared "definitions of the situation" and suitable "rules of the game". Different societies and institutions seem to have different ability to achieve this sort of adjustments. Also the time needed to arrive at a satisfactory pattern may vary considerable and the stability of the solutions found is always relative. It seems that as far as the SCST is concerned we are still far from a stable solution. The reasons are that the process of differentiation of the SCST has not progressed far enough and that its progress has been very uneven across different fields of technology. At the same time the requirements of integration between the SCST and other relevant systems - particularily the economy - are very strong .

The institutional infrastructure of technology today consists of clusters of arrangements centred around four poles: (a) the economy (engineering departments and R&D labs of companies, collective R&D establishments, branch associations, etc); (b) the polity (the state technocracies such as mission agencies, national laboratories, technology policy bodies, etc), (c) the professions (professional associations); and (d) the academic system (educational institutions, research laboratories, associated allocation systems)[28]. Each of these sets of institutions has been undergoing continuous evolution and their

[28] P. Weingart (see Laudan 1984) carries out a similar analysis within a more orthodoxal Parsonian scheme involving thee main sub-systems: market/economy, political system, and culture.

relative roles have been changing over time. As things stand today none of them can be regarded as the undisputable institutional base for the autonomous SCST. More specifically, there are four main contenders to that status are: (i) the universities; (ii) the independent institutes and centres; (iii) the network structures linking industrial labs; and (iv) the super-networks interconnecting all the three components in the shape of various "programmes". The appropriateness of these institutional alternatives must be judged in terms of: 1) the degree to which they can sustain and protect the differentiated core of the autonomous SCST; and 2) their ability to achieve a high degree of interpenetration with other social systems - chiefly economy and science. A proper analysis along these lines would require much more space than is available here. I will limit myself to a few brief comments.

Scientist and engineers in the industrial laboratories constitute a major and essential part of the SCST. However, it is not likely that networks of such labs can function well as the institutional core of the SCST. The autonomy of the industrial R&D units is too limited and the communication in which they can engage too restricted to permit the formation of sufficiently dense and self-organizing networks. Neither do such units allow a sufficient degree of mutual interpenetration. On the contrary they have inherent tendency to interact in an exchange mode (through markets) or through ad hoc hierarchies. This mode can be very efficient for the day-to-day adaptive problems solving but it does not work well in the more diffuse and uncertain areas of knowledge production. The attempts to differentiate long-term R&D within large companies (central labs and research centres) does not solve these problems nor do the collective research programmes and networking. Such measures can contribute to the better functioning of the SCST only if they are used as a complement to a core institutional structure based on different principles.

Can such a structure be provided by independent institutes, national laboratories, etc.? It is clear that these organizations have made important contributions to the development of many technologies. However their limitations are severe. Because of their mission character, units of this type tend to age rapidly and become isolated from other parts of the R&D system. Neither are they particularily suitable from the point of view of interpenetration. Effective technology transfer tends to be their Achilles heel. They can overcome these shortcomings by becoming closely associated with universities. However, this only emphasizes their weakness as the institutional base of technological R&D in their own right.

Generally it is not easy to see how the networking between various heterogeneous units can solve the problem of institutionalization of the core of the SCST. It certainly can contribute to increased interpenetration between different sub-systems but is too ephemeral to create the stable and dense core system of interactions. Such a core can be established only if a network emerges which focuses on a coherent technological domain, develops a "language" of its own and a reward structure which encourages communication within it.

There can be little doubt that the academic system is potentially best suited to provide the institutional base for the core of the SCST[29]. It does so for the same reasons as in the case of science: (i) its provides a general cultural and institutional foundation for a highly differentiated and autonomous intellectual activity; (ii) it institutionalizes a communication system which greatly enhances the density of internal interaction; and (iii) it has a great ability to penetrate virtually all other social systems through its educational function. Finally (iv) the academic system is uniquely suited to promote the interpenetration of technology and science - an essential condition of the emergence of autonomous SCST.

However, even though the academic system may be best place to become the core of the autonomous SCST its success is by no means automatically assured. Just as not all academic systems have been capable of rapidly accommodating science, not all of them will be successful in absorbing technology. The reasons are manifold:

1) The general conservatism of the academic system as regards its internal structure is one problem. This conservatism has been greater in most European countries than in the USA. It may be more serious with respect to technology than to science since the former is in some ways more dynamic than the latter and requires a larger measure of institutional flexibility and innovativeness.

2) There are some serious problems regarding institutional integration of technology with science and other disciplines. The cultural differences between the two traditions are often a serious impediment. The established science fields sometimes exhibit "imperialistic" tendencies in relation to technology and technological R&D is often forced into organizational moulds for which it is not suited.

3) The very mode in which the technological fields have been institutionalized in the universities is often problematic. For perfectly legitimate reasons they tend to be

[29] Among others Nelson (1990) emphasizes strongly the role of universities as the creators and distributors of generic technical knowledge.

strongly determined by the needs of the dominant concrete systems. This is reflected, among other things, in the recruitment of "practitioners" to academic positions. This mechanism increases the interpenetration between the academic technology and applied environments, especially industry, but makes the development of coherent own tradition more difficult. Generally I would claim that there is a degree of tension in the academic system between the role of "technologist" and that of "engineers".

4) The organization of academic technology tends to be fragmented reflecting the need to specialize the teaching functions. The result is multiplication of often small research units which are inferior to their counterparts in industry[30].

These and similar difficulties have to be overcome before the academic system can become the institutional base for the SCST (Stankiewicz 1986). At the same time it must be emphasized that the role of universities and other academic institutions is not to replace the other environments - such as the free-standing institutes or industrial laboratories - or to function as an alternative to the "communities of practitioners". The participation of the scientists and engineers from those environments or communities in the SCST is crucial, especially as regards the development of basic technologies (Stankiewicz 1990). However, the universities are needed to give the SCST the core structure which will allow it to become a dynamic self-organizing producer of knowledge.

The general trend towards the differentiation and autonomization of the SCST has also broader policy implications which cannot be discussed here in full detail. It calls for certain changes in the general style or attitude in dealing with science and technology. Autonomization, whatever its ultimate institutional basis, makes the hierarchical form of the societal control of the SCST awkward. The growing recognition by the policy-makers of this fact has brought about a gradual change to softer and more indirect forms of external governance the SCST. The emphasis is more and more on mechanisms which promote "resonance" between largely autonomous sub-systems. Direct "broad-band" information flows between science, technology and industy are faciliated by various cooperative R&D programmes involving academic R&D as well as industrial, by research parks and centers in association with universities, etc. And there remaines an ample room for further institutional and organizational experimentation along these lines.

30 The tendency for the academic technology to fall behind industrial R&D has been a recurring problem ever since the academization of technology started in the 19th century.

All those developments imply changes in the roles performed by policy infrastructures. Their content-specific functions of priority-makers in substantive areas tend to decrease while the generalist role of managing the interactions in large heterogenous systems is increasing in importance.

2.4 CONCLUDING COMMENTS

As opposed to the relatively sophisticated level of understanding of the socio-cognitive system of science, our knowledge of the cognitive dimension of technology continues to be very incomplete. These deficiences are to a large extent a consequence of the greater complexity and heterogeneity of technology. In this paper I have argued that some of these deficiencies can be overcome by the application to technology of the approaches deriving from the social systems analysis. In particular, I have stressed the importance of the concept of *interpenetration* as a mode of integration among highly differentiated autonomous systems. Systematic examination of this mechanism in the relations between technology, science and other social systems should be fruitful. It ought be pursued using both quantitative and qualitative approaches. In particular, the bibliometric studies (see chapters 3, 4, 7, 8 and 9 in this book) can be very helpful in identifying the emergent zones of interpenetration of science and technology.

At the same time it is important that the studies of interactions between different sub-systems are balanced by sustained analysis of the sub-systems themselves. In particular it is crucial that the dynamics of technological *fields* is better understood. This calls for studies of engineering disciplines and other "applied" fields (chapters 8 and 9).

Finally it is of great importance that the relationship between the institutional structures and the processes of differentiation and integration of the SCST are given adequate attention. It seems quite clear that the ways in which technological R&D is institutionalized differs considerably from country to country (Stankiewicz 1990 and Granberg 1990). It is not inconceivable that the relative success of the countries in the field of technological competition is linked to these differences.

REFERENCES

Bijker, W. E., Hughes, T. H. and Pinch, T. (eds.) (1987), *The Social Construction of Technological Systems. New Directions in the Sociology and History of Technology*, MIT Press, Cambridge.

Bunge, M. (1966), 'Technology as Applied Science', *Technology and Culture*, 7.

Böhme, G, et al., (1983), *Finalization in Science*, D. Reidel Publ. Co., Dordrecht.

Callon, M. (1987), 'Society in the Making: The Study of Technology as a Tool for Sociological Analysis', in Bijker, Hughes, Pinch (eds.) (1987).

Carlsson, B. (ed.) (1989), *Industrial Dynamics, Technological, Organizational and Structural Changes in Industries and Firms*, Kulwer Academic Publishers.

Clark, K. (1985), 'The integration of design hierarchies and market concepts in technological evolution', *Research Policy*, 14.

Constant, E. W. (1980), *The Origins of the Turbojet Revolution*, The John Hopkins University Press, Baltimore.

Constant, E. W. (1984), 'Communities and Hierarchies: Structure in the Practice of Science and Technology', in Laudan (ed.).

Chandler, A. D. Jr. (1977), *The Visible Hand: The Managerial Revolution in American Business*, Harvard University Press, Cambridge.

Channell, D. F. (1982), 'The Harmony of Theory and Practice: The Engineering Science of W. J. M. Rankine', *Technology and Culture*, pp. 39-52.

Channell, D. F. (1988), 'Engineering Science as Theory and Practice', *Technology and Culture*, pp. 98-103.

Dosi, G. (1982), 'Technological Paradigms and Technological Trajectories', *Research Policy*, 11, pp. 147-162.

Eliasson, G. (1989), 'The Dynamics of Supply and Economic Growth', in Carlsson (ed.).

Eliasson, G. (1990), 'The firm as a competent team', *Journal of Economic Behaviour and Organisation*, 13, pp. 275-298.

Ellul, J. (1980), *The Technological System*, Continuum, New York.

Fores, M. (1988), 'Transformations and the Myth of "Engineering Science": Magic in a White Coat', *Technology and Culture*, pp. 62-81.

Freeman, Ch. and Perez, C. (1988), 'Structural criseses of adjustment, business cycles and investment behavior', in Dosi et al. (eds.), *Technical Change and Economic Theory*, Pinter Publishers, London and New York.

Granberg, A. (1990), 'Laser Research in Sweden, West Germany and Japan', in Sigurdson (ed.), pp. 77-105.

Granberg, A. and Stankiewicz, R. (1981), 'The development of generic technologies - the cognitive aspects', in Granstrand and Sigurdson (eds.), *Technological and Industrial Policy in China and Europe*, RPI, Lund.

Granberg, A. and Stankiewicz, R. (1981a), 'Specialiseringstendenser i den högskolebaserade tekniska forskningen (The specialization trends in the university based technical R&D)', RPI, Lund.

Gutting, G. (1984), 'Paradigms, Revolutions, and Technology', in Laudan (ed.) (1984).

Hippel, von, E. (1988), *The Sources of Invention*, Oxford University Press, Oxford.

Hughes, T. P. (1983), *Networks of power. electrification in Western Society, 1880-1930*, The Johns Hopkins University Press, Baltimore and London.

Hughes, T. P. (1986), 'The Seamless Web: Technology, Science, Etcetera, Etcetera', *Social Studies of Science*, 16, pp. 281-92.

Hughes, T. P.(1987), 'The Evolution of Large Technological Systems', in Bijker, Hughes, Pinch, (eds.).

Koyré, A. (1948), 'Les philosophes et la machine. II. Les origines du mechanisme', *Critique* 4, pp. 628.

Krohn, W. et al. (eds.) (1978), *The Dynamics of Science and Technology,* Reidel Publishing Co., Dodrecht.

Krohn, W. and Küppers, G. (1989) *Die Selbstorganisation der Wissenschaft*, Suhrkamp, Frankfurt am Main.

Krohn, W. and Küppers, G. (1989), 'Self-organization: A New Approach to Evolutionary Epistemology', in Hahlweg and Hooker (eds.), *Issues in Evolutionary Epistemology*, State University of New York Press, New York.

Kuhn, T. (1970), *The Structure of Scientific Revolutions*, Chicago University Press, Chicago.

Latour, B. (1987), *Science in Action*, Open University Press.

Laudan, R. (ed.) (1984), *The Nature of Technological Knowledge. Are Models of Scientific Change Relevant?* Reidel Publishing Co., Dodrecht.

Laudan, R. (1984), 'Cognitive Change in Technology and Science', in Laudan, (ed.).

Law, J. (1987), 'Technology and Heterogenous Engineering: The Case of Portuguese Expansion', in Bijker, Hughes, Pinch (eds.).

Layton, E. T. (1974), 'Technology as Knowledge', *Technology and Culture*, 15, pp. 31-41.

Layton, E. T. (1988), 'Science as a Form of Action: The Role of the Engineering Sciences', *Technology and Culture*, pp. 82-97.

Luhmann, N., (1984), *The Differentiation of Society*, Columbia University Press, New York.

Luhmann, N., (1986), *Ökologische Kommunikation*, Opdalen.

Lundvall, B. (1988), 'Innovation as an Interactive Process', in Dosi, et al. (eds.) *Technical Change and Economic Theory*, Pinter Publishers, London.

Münch, R. (1983), 'Modern Science and Technology: Differentiation or interpenetration', *Int. J. of Comparative Sociology*, XXIV, 3-4, pp. 157-175.

Nelson, R. (1990), 'Capitalism as an engine of progress', *Research Policy*, 19, pp. 193-214.

Parsons, T. (1966), *Societies: Evolutionary and Comparative Perspectives*, Prentice-Hall, Englewood Cliffs.

Parsons, T. (1971), *The System of Modern Societies*, Prentice-Hall, Englewood Cliffs.

Pelikan, P. (1988), 'Evolution, Economic Competence, and the Market for Corporate Control', *Journal of Economic Behaviour and Organisation*, 12, pp. 279-303.

Pinch, T. J. and Bijker, W. E. (1984), 'The Social Construction of Facts and Artefacts: or How the Sociology of Science and the Sociology of Technology might Benefit Each Other', *Social Studies of Science*, 14, pp. 399-441.

Rosenberg, N. (1976), *Perspectives on Technology*, Cambridge University Press, Cambridge.

Sahal, D. (1985), 'Technological guideposts and innovative avenues', *Research Policy*, 14.

Sigurdson, J. (ed.) (1990), *Measuring the Dynamics of Technological Change*, Pinter Publishing Ltd., London.

Stankiewicz, R. (1986), *Academics and Entrepreneurs. Developing University-Industry Relations,* Pinter Publishing Ltd., London.

Stankiewicz, R. (1990), 'Basic Technologies and the Innovation Process', in Sigurdson, (ed.) (1990), pp. 13-38.

van den Belt, H. and Rip, A. (1987), 'The Nelson-Winter-Dosi Model and Synthetic Dye Chemistry', in Bijker, Hughes, Pinch (eds.).

Williamson, O. E., (1983), *Markets and Hierarchies*, The Free Press, New York.

Willke, H. (1987), *Systemtheorie. Eine Einführung in die Grundprobleme*, Gustav Fischer Verlag, Stuttgart.

3 A SPECIAL MEDIATION BETWEEN SCIENCE AND TECHNOLOGY: WHEN INVENTORS PUBLISH SCIENTIFIC ARTICLES IN FUEL CELLS RESEARCH

Vololona Rabeharisoa

3.1 INTRODUCTION

Elucidating the relationships between science and technology holds an important place in the socio-economics of innovation. One of the dimensions of this problem can be formulated as follows: What are the scientific contents of innovative technologies?

By describing the links between patents and scientific articles, scientometrics offers interesting answers to this question. Thus, studying the citations of scientific articles made by a body of patents allows us to designate certain articulations between science and technology (Narin et al. 1986; van Vianen et al. 1989; see also chapter 4 in this volume). These citations, however, are not the only indicators of the "science-based" nature of technology. Other links between science and technology get woven once inventor-patent-filers also author scientific articles.

This chapter is precisely a contribution to the study of links between science and technology established by such "inventor-researchers". Practically, we shall focus on fuel cell patents filed in priority in France between 1960 and 1987 and on the articles published by their inventors. By studying titles and summaries of these patents and articles, we compare the scientific and technical themes pertaining to the products of these inventors. Finally, thanks to a brief presentation of the articles cited on the front page of these patents, we can qualify this *cited research* with respect to the *research produced* by the inventors.

This will be our concluding point: we specify how the articles cited by the patents, on the one hand, and the articles published by the inventors, on the other hand, define precise links between science and technology. In other words, we put forth the hypothesis that these links specify themselves as a function of the mediations at stake.

These last few years, the elucidation of the relationships between science, technology and the market has been one of the key questions in innovation studies (see, e. g., chapters 1 and 2). Case studies conducted by sociologists (Pinch et al. 1984; Bijker 1987; Callon 1986; Law et al. 1987) and economists (Rosenberg 1982; von Hippel 1988) have profoundly shaken those theoretical speculations which wanted to place science, technology and the market in clearly separate spheres (Kuhn 1983; Dosi 1982); each sphere contained its own dynamic and they only corresponded with each other via "frontier exchanges". From a more pragmatic point of view, so-called "technological" programmes no longer consist in merely constructing prestigious machines, in the process mobilizing a few top-level scientists. Increasingly, these programmes involve diverse concerns, from scientific excellence to the search for the components needed to manufacture a chip capable of holding its own in the world market (Callon et al. 1990). The managers of R&D thus begin to play the "go-betweens" in order to forge connections between laboratories, companies, financial circles, etc. (see also chapter 2 on the "interpenetration" of science and technology).

We will focus here on the links between science and technology. One of the dimensions of this problem can be formulated as follows: what are the scientific contents of innovative technologies? From our point of view, answering this question consists in identifiying, in the same movement, the modes of interaction between scientists and technologists, and the knowledge and know-how that flow between them.

Articulations between science and technology get created in different ways; there are also several different "intermediaries" involved. For example, in the course of a cooperative contract between a firm and a university laboratory, the skills incorporated by engineers and researchers get mobilized. In the course of a colloquium involving both scientists and technologists, "papers"[1] circulate among them. Within the industrial laboratory of a company, engineers can work on designing new technical devices, patent their inventions, and publish articles about their results[2]. The gradient going from science to technology is therefore neither unique nor fixed once and for all. It has a different shape depending on the "frontiers" that actors establish between their activities. Further elaborations on this issue may be found in chapters 8 and 9 of this

---

1 These "papers", as well as research and study reports, constitute what we now call "grey literature".

2 Obviously, these different modes are not always mutually exclusive.

volume. Thus, showing the links between *science* and *technology* amounts to describing the "splits" and the "bridges" created by actors between these two entities.

This paper addresses a particular "split-bridge" created by researchers and engineers through two of their products: patents and scientific articles. We shall describe a specific articulation joining these two types of documents. This articulation, which we found while studying the domain of fuel cells in France, is due to the dual activity of certain inventors who publish scientific articles in addition to filing patents. First, though, we will examine another possible link between science and technology: that designated by the articles cited by the patents. We will see that the *research cited* and the *research produced* by the inventors each define specific links between science and technology. This allows us to show why it is interesting to obtain a precise description of mediations between science and technology in order to better understand their relationship.

3.2 WHAT FUEL CELL PATENTS FILED IN FRANCE COUNT AND TELL

FROM DEMAND...

The Service de la Programmemation de la Recherche de l'Agence Française pour la Maîtrise de l'Energie (Research Planning Service of the French Agency for Mastering Energy, the AFME), after learning the conclusions of the experts who said that the field of fuel cells in France is "morose", has been wondering what it should do to help matters. This is not the place to detail the mission or means of action of the agency. Let us just say that it is in charge of implementing French policy on mastering energy, and that in order to do so, it has at its disposal the financial and technical means to support products and procedures likely to lead to energy savings in various sectors of economic life throughout the R&D process. Fuel cells are among these products. In order to decide whether or not to act in favour of fuel cells, the AFME has asked us to report on the evolution of the field in the last few years. More specifically, the agency is interested in the scientific research that underlies this field on the one hand, and the *commercial applications* of fuel cells on the other hand. Here we shall concentrate on the question of the *scientific bases* for fuel cells.

... TO THE "SCIENTOMETRIC TEST"

We begin with fuel cell patents[3] filed in France and entered in the WPI(L)[4] database. Indeed, the AFME's request specifically addressed these fuel cells. Thus the patent, a document which designates and deciphers the technical device, seems like an interesting starting point.

What does our exercise consist of? Starting with these patents, we will identify and describe scientific research pertaining to fuel cells. How? By following the firms, inventors, and documents designated by these patents. Indeed, the front page of a patent provides us with D, a body of descriptors of the entities united by the document:

D = { - dates the priority patents were filed
- country in which the priority patents were filed
- list of countries where the patent is filed
- names of inventors
- institutional affiliations of the inventors
- title words of the document
- summary words of the document
- list of patents cited by the document
- number of scientific articles cited by the document}[5].

Analyzing the values taken by these descriptors over the body of patents that we are concerned with should allow us to describe the morphology of the field of fuel cells as outlined by the patents. Two of these descriptors in particular give access to the *science* designated by the patents:

- the names of the inventors, which allow us to gather the articles they publish, if they publish any;
- the number of scientific articles cited by the patents, if any.

At this stage, we would like to make a point that is both *methodological* and *theoretical*. We will limit our analysis to databases of articles and patents. This means that we will not physically go to visit firms or laboratories. The economic dimension is

---

3 The term "patent document" refers either to patents which have been granted and those still in the application process. In the rest of this paper, we use the word "patent" to designate either of these two.

4 We will discuss the choice of this database later.

5 This is in fact the front page of a family of patents entered into WPI(L).

not the only reason we restrict ourselves. Another reason stems from the idea that articles and patents define a certain organization of science and technology (Bastide et al. 1989; Latour 1989), and it is precisely this particular techno-scientific configuration that we wish to describe. Of course, it is possible that our inventors neither cite nor publish articles. If such is the case, this would be less a sign of an *impossibility of going back up to science* through patents than a local indicator qualifying the links between science and technology. In other words, it would mean that for the field of French fuel cells, patents do not establish links with scientific research. It is in this sense that we speak of a "scientometric test": the question of interest to us is knowing whether fuel cell patents filed in France call upon science, and if so, upon which science[6].

CONSTITUTING A SOURCE-FILE B OF PATENT DOCUMENTS

We choose to work with the WPI(L) database. Our main arguments in favour of this database result from:

- First of all, its multidisciplinary character. In as much as we seek to identify patents that refer to fuel cells in any way at all, we do not wish to limit ourselves, a priori, to thematic databases.
- Secondly, its indexing principle. Because WPI(L) turns the title of a document into a body of keywords that specify the document, an investigation based on words is particularly pertinent using this database.

Our list of patents comprises all the documents containing the word "fuel cell" or "fuel cells" on the front page and filed in priority in France. We shall call this list B. We wish to make two comments about this construction of B:

- The merits and limitations of interrogating a database by words or by classification codes have been widely addressed in scientometric literature (for example, see Hyams 1985). We have chosen to interrogate by words because:

6 We would like to rest for a moment on this last point. In accordance with the AFME's request, we are supposed to, as much as possible, describe *science* with respect to fuel cells. We will thus try to report on the subjects addressed by the articles linked to the patents. Thus the analyses which we shall undertake will be "quali-quantitative", in the sense that they will also address the *contents* of the documents as provided by the databases.

- First of all, the AFME's request specifically mentions fuel cells as objects of the study.
- Secondly, the word "fuel cell" is, according to experts in the field, a *label* designating a specific technical device[7]. This specificity of the "fuel cell" appellation makes it useless to reconstitute the object through classification codes.

• The nationality of a patent is not a piece of information readily available in the databases. We therefore referred to the country where the patent was filed in priority. As a result of the this approximation, our list will include the patents filed in priority in France by foreign firms, but will exclude the patents filed in priority in foreign countries by French firms. Thus, we are not analyzing the French fuel cell industry but rather France as a strategic place where firms choose to place their priorities in filing their fuel cell patents[8].

Before addressing the question of the scientific bases of fuel cells, we shall briefly present the results of our analysis of B, in order to be able to compare them to the results we shall obtain by studying publications.

FUEL CELL ACCORDING TO B: FROM THE DOORSTEPS OF THE MARKET TO THOSE OF THE LABORATORY

Without going into too much detail, we shall summarize the main conclusions of our analysis of patent documents. First of all, the annual counting of patents shows a filing rupture in 1975. From 1960 to 1987, B includes 200 documents. The six most productive years are: 1967 (15 documents), 1968 (26 documents), 1969 (32 documents), 1970 (20 documents), 1973 (12 documents), and 1975 (18 documents). These six years alone total 61 % of patent production (see figure 3-1).

7 In a way, the word "fuel cell" constitutes a *consecrated term*. However, this has no effect on the degree of stability of the technical device. In other words, even though technologists agree to unequivocally name the object of their research, they can still attribute very different technical and economic definitions to this object.

8 Nonetheless, this choice undoubtedly privileges French companies, when we take into account the preference firms have for filing patents in their country of origin (Schmoch et al. 1989).

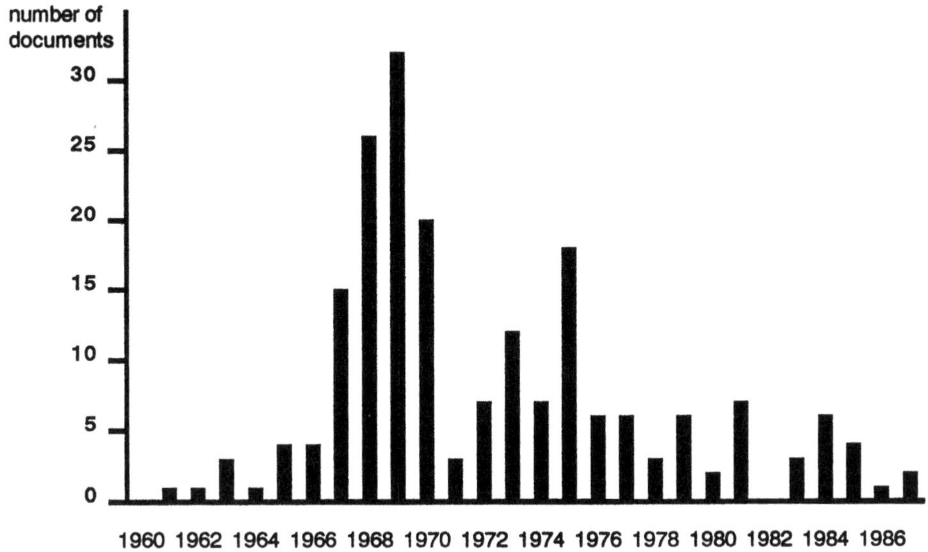

Figure 3-1: Number of patent documents listed by priority year[9]

Next, the profile of the filing firms changes around 1975. Before this year, the most active patent filers are large French companies: the Compagnie Générale d'Electricité (CGE), the Société de Constructions Mécaniques et Electriques ALSTHOM, and the Institut Français du Pétrole (IFP) together total nearly 63 % of patent production. After 1975, these three companies are only marginally present. However, foreign companies begin to appear (Occidental Chemical Corporation), and above all we see French companies that do research under contract, such as SORAPEC and the Société de Recherche Technique et Industrielle (SRTI) (cf. figures 3-2 and 3-3).

9 We would like to make a note about the way we counted these patents. Sometimes, the same patent has been subject to several priority applications. Thus, out of our 200 patents, 20 have more than one priority number. Nevertheless, we noticed that this does not have a large effect on the annual count. As an example, let us examine the year 1967. If we count using the oldest priority year as 1967, we find 15 patents. If we count using the most recent priority year as 1967, we find 14 documents. Therefore there is only a difference of 1 document. We found the same order of magnitude for the other years. Here, we have chosen to priviledge the first application. Thus, the histogram in figure 3-1 shows the number of patents according to the oldest priority year. In cases where the patent was submitted for application in several countries, we use the French priority dates.

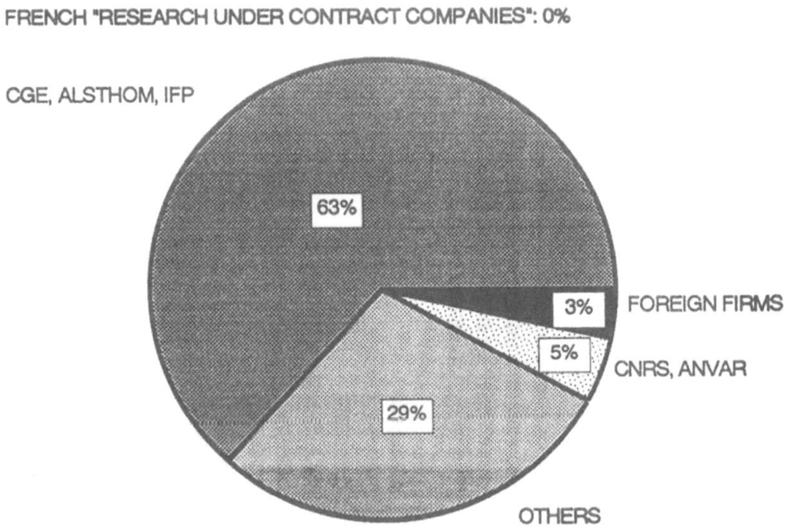

Figure 3-2: Percentage of patents filed by five types of firms before 1975

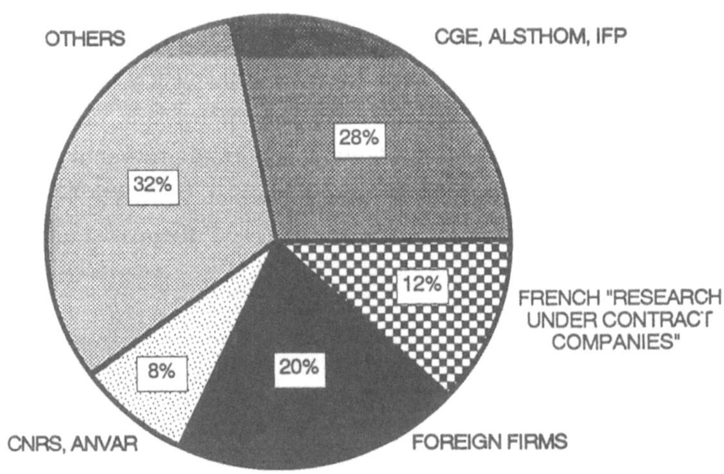

Figure 3-3: Percentage of patents filed by five types of firms after 1975

Finally, a detailed analysis of the titles and summaries of the patents[10] led us to conclude that the rupture of 1975 was accompanied by a change in the types of claims

10 By identifying the words appearing the most frequently in the titles, and analyzing the words associated by the patent documents.

made by inventors. Before this year, the patents are fairly *finalized*: inventors try to fix the various elements that make up fuel cells (electrodes, electrolytes, catalysts, etc), describe the satisfactory techno-economic performances of these cells, and predict promising commercial applications, especially for the electric vehicle. Towards the end of the 1970s, such hopes get dulled: patents begin to speak of unresolved technical and economic problems, and make suggestions to solve these problems. Fuel cells lose a bit of their "reality", and those inventors still interested in them fall back on objects, which though usable in fuel cells, could also fit into other technical devices (for example: reinforced fibres for the manufacturing of electrodes not only for fuel cells, but also for other electrochemcial generators).

In short, the analysis of B shows that fuel cells, as portrayed by patents filed in France, are practically at the doorstep of the marketplace in 1975; then, failing technical and economic trials, they become evanescent towards the end of the 1970s. Fuel cells are not an exception in this respect. For other cases of cyclic evolution see chapters 8 and 9 and in particular section 9.4. This conclusion shall be reinforced by the analysis of the articles published by the inventors. Beforehand, however, we shall describe the science designated by the articles appearing on the front page of the patents in B.

3.3 ARTICLES CITED ON THE FRONT PAGE OF THE PATENTS IN B: FUEL CELL RESEARCH DOES EXIST

STUDY OF THE ARTICLES CITED ON THE FRONT PAGE OF THE PATENTS: A LOCAL TEST

The hypothesis that the articles cited on the front page of a patent would be indicators of the scientific bases of the invention described by the document has been widely addressed by the literature on scientometrics (see, e. g. chapter 4). This hypothesis has encountered, and still does, various avatars. The heterogeneity of the results of different studies has initiated debates not only on practical questions, such as choosing a database, but also on the very problem of identifying and describing the relationships between science and technology. A sample of the conclusions drawn by certain scientometricians shows the hesitations generated by these questions. Seeing the small number of articles cited on the "front page" of patents, Collins & Wyatt (1988) question the competence of the patent examiners. They then remark that inventors have a much greater tendency to cite the scientific literature than the examiners. According to the authors, not only do the examiners cite few articles, but in addition, they may even cite these "badly"[11]! Certain scientometricians pick up the irregular nature of these front page citations and question the motivations that examiners might have to cite articles (van Vianen et al. 1989). According to the authors, the multiplicity of these intentions weakens the hypothesis which holds that these articles point to the scientific bases of the technologies described by the patents. Others, however, notice that the number of articles cited on the front page of patents has unceasingly increased in the past few years; they conclude that technologies have become increasingly "scientized". They thereby legitimate the role of these articles as indicators of the "science-based" nature of these technologies (Narin et al. 1989). The latest state of affairs is discussed in sections 4.2 and 4.3 in this volume pointing out that the mentioned authors used US patents only. For science citation in European Patent Offices things might be different.

Given these different comments, it seems that no global, unequivocal answer that would unite the scientometricians can be given to the question of whether or not the articles cited on the front page of patents trace links between different scientific and technical entities. Here, the individual case rules: certain examiners cite scientific articles often and consistently, thereby enabling scholars to circumscribe a "science" designated by these technologies (Narin et al. 1985); others only rely on science sporadically, making its proximity to technology indescribable. Analyzing citations on the front page of patents, then, is less a method leading systematically to science than a test at the end of which the patents analyzed do or do not mention science. This is the test to which we will put the patents in B.

ABOUT THE CITATIONS SHOWN BY WPI(L)

Before presenting the results of this test, it is appropriate to describe the WPI(L) database a bit further. First of all, this database only provides on-line citations as of 1978. This has two consequences:

- The citations made by patents filed before 1978 are not available;

---

11 By an analysis of the citation of articles on the front page, Collins & Wyatt arrive at the conclusion that the examiners are not very familiar with the scientific literature, and that they may only read the summaries of articles! Not satisfied by these front page citations, Collins & Wyatt analyzed the citations made by the inventors themselves. Unfortunately, at the present time, these citations are not available in computerized form. As a result, the generalization of their analysis is hard to conceive of.

- The citations received by the patents filed before 1978 are only partially available; indeed, only the citations received thanks to patents filed after 1978 are mentioned.

These constraints mean that a certain amount of information gets lost, which in turn means that it is difficult to perform the evaluation completely rigorously, unless one were to look for the citations pertaining to each patent of each family entered in WPI(L) in another database. We should also add that WPI(L) does not provide all the citations made by all the patents of a single family. It selects by privileging first the European judicial system, then the Americal judicial system:

- if the family contains a version filed in the European Patent Office, then WPI(L) only shows the citation made by this European patent;
- if the family does not contain a European version, but does contain an American one, then WPI(L) only shows the citations made by the American patent;
- if the family contains neither a European nor an American version, then WPI(L) does not show any citations.

Finally, we must note that WPI(L) only provides the numbers of articles cited by the patents. To gather the bibliographic data pertaining to these articles (such as the title or the authors), one must shift to the EPAT database. Thus, if the family does not contain a European version, but does contain an American one, then the bibliographic data pertaining to the articles cited by the American version, if any, are not available. We insist on these constraints in order to emphasize the extremely active role of databases in any scientometric exercise[12]: the results that we obtain from a citation study using WPI(L) can only be partial.

TOWARDS THE NETWORK OF ARTICLES LINKED TO B BY INTERMEDIARY CITATIONS

We would like to add to the articles cited by B:

- the articles cited by the patents that cite B, and
- the articles cited by the patents cited by B.

Indeed, these last two types of articles also designate research linked to B through other patents that surround it (see sketch in figure 3-4).

---

12 We have, in fact, addressed this question elswhere (Rabeharisoa 1990b). In this matter, we fully share the concern of certain scientometricians, who insist on describing what the databases can and cannot offer (Schmoch 1990).

PATENTS CITED BY B -------> ARTICLES CITED BY PATENTS CITED BY B

B -------> ARTICLES CITED BY B

PATENTS THAT CITE B ----> ARTICLES CITED BY PATENTS THAT CITE B

Figure 3-4: Articles cited by B and by its surrounding patents

After 1978, we found a total of 11 articles directly or indirectly cited by B. Of these 11 articles, only 6 are visible in EPAT. These articles talk about *technical research* on fuel cells, but they also discuss scientific themes connected to these cells, as the following titles show:

> Ross P.N. Jr. 1982,
> "Engineering analysis of an NH3-fuel cell system for vehicles", in Extended Abstracts.
>
> Tuller H.L. et al. 1975,
> "Doped ceria as a solid oxide electrolyte" in Journal of the Electrochemical Society.
>
> El Adham K. et al. 1983,
> "Grain boundary effect on ceria based solid solutions", in Solid State Ionics.

The names of these researchers and the words in the titles of their articles are as much markers of the techno-scientific network circumscribed by B (Callon 1990). Beyond noting the existence of fuel cell research, however, we can say nothing more about the contents of the science designated by B. Indeed, the small number of articles cited and the absence of data for the period preceding 1978 do not justify a more detailed study.

Thus, if the rare articles cited on the front page by the patents in B testify to an embryonic research network which is more or less directly interested in fuel cells, the small number of these articles leads us to think that B has few connections with this network. This does not mean, however, that B has no link with science. Indeed, as certain scientometricians (van Vianen et al. 1989) have forecasted, the articles published by the inventors also define science linked to the patents. As we shall see, these articles are less "complements" to the articles cited by the patents than indicators of another type of articulation between science and technology.

3.4 ARTICLES PUBLISHED BY INVENTORS: WHEN SCIENCE AND TECHNOLOGY INTERMINGLE

CONSTITUTION OF THE FILE A OF ARTICLES PUBLISHED BY INVENTORS

We choose to work with the PASCAL database. The choice of this database was guided by two factors:

- PASCAL includes a large number of journals, and in particular French scientific and technical journals; this allows us to make a *wide sweep*, since there is no a priori reason why the inventors in B would only publish in the *important* journals contained in SCISEARCH;
- PASCAL indexes each articles by a series of keywords, enabling us to perform co-word analysis.

In order to define the body A of articles published by the inventors, we adopted the following procedure:

1) For each firm that owns a patent belonging to B, we looked for the articles published by its inventors in the PASCAL database;
2) when we did not find any articles, we looked for the fuel cell patents filed by the firm in the CAS database[13]; we thus found other inventors in the same firm;
3) then we once more looked for the articles published by the additional inventors in the PASCAL database;
4) finally, let us note that in the case of the most active patent filers (CGE, ALSTHOM and IFP), we looked at their inventors present in both WPI(L) and CAS.

We wish to make a note about this delimitation of A. We could have "enriched" our population of inventors by extending our research to databases other than CAS. We did not judge such an extension necessary, however, in as much as passing through CAS enables us to find "inventor-authors" for most of the firms in B.

This definition of A thus leads us to a first conclusion: the fuel cell inventors in France, as we have selected them, also publish articles. In order to show the scientific or

---

13 We conducted the same interrogation of CAS as we did for WPI(L). Sometimes fuel cell patents appeared in CAS because they were considered as innovative in certain branches of chemistry but did not appear in WPI(L). This is why we extended our population of inventors to those visible in CAS.

technical knowledge produced by these inventors, we will now analyze a sub-set of A. In order to do so, we have circumscribed a sub-population of "inventor-authors", according to the following criteria:

- We were careful to retain inventors of large and small firms, both French and foreign, for the entire period 1960-1987;
- among the inventors we selected, certain disappear from the patent scene after 1975, while others only enter the scene as of this year. The majority, however, publishes articles throughout the entire period. By focussing on these inventors, we thus try to dispense with any a priori temporal correlation between patenting and article publication.

Our sub-population is composed of the following inventors:
Feuillade G. (CGE) - Jacquelin J. (CGE) - Hespel C. (ALSTHOM) - Bono P. (ALSTHOM) - Tissier A. (ALSTHOM) - Cheron J. (IFP) - Grehier A. (IFP) - Breele Y. (IFP) - Croset M. (THOMSON) - Chapiro A. (ANVAR) - Vignaud R. (Les Piles WONDER) - Laurent F.J. - Ruch J. - Brenet J. - Wiart A. (Jeumont-Schneider) - Blain J. (Terreau) - Pineri M. (CEA) - Bidan G. (CEA) - Genies E.M. (CEA) - Fauvel P. (Occidental Chemical Corporation/Fauvel) - Nenner T. (SRTI) - Roux M. (SRTI) - Bouthors P. (Régie Renault)[14].

PROFILES OF THE "INVENTOR-AUTHORS": MULTIPLE SKILLS AT THE SERVICE OF FUEL CELLS

We first listed, for each author, the keywords which appear most frequently in his articles. Completed by a reading of the titles of the articles, this first investigation put forward the following fact: Certain inventors are relatively specialized as authors of scientific articles. For example, Croset M., inventor for THOMSON, is a micro-electronician of thin layer dielectrics; Feuillade G., inventor for CGE, specializes in certain macromolecular membranes; Chapiro A., a CNRS researcher at the Laboratoire de chimie macromoléculaire sous rayonnement at Thiais, invests in polymerization techniques by radiochemical grafting (see figure 3-5).

---

14 The name of the inventor is followed, in parentheses, by the name of the company to which he belongs.

CROSET M. (THOMSON) thin layer silicium tantalum micro-electronic manufacturing gallium arsenate
FEUILLADE G. (CGE) electrochemical cell membrane polymer
CHAPIRO A. (Laboratoire de chimie macromoléculaire sous rayonnement - CNRS Thiais) gamma radiation kinetics radiochemical polymerization radiochemical grafting

Figure 3-5: List of keywords appearing most frequently in articles published by three "inventor-authors"

Next, some of these inventors have (or had) a university affiliation[15]. This is true for Roux M., an inventor for SRTI, who published articles as a researcher at the Departement des Etudes et Recherches en Technologie Spatiale of the ONERA. Blain J., inventor for Terreau, publishes as a researcher at the Institut de Génie des Matériaux of the CNR in Canada, and Brenet J. and Ruch J. are researchers at the Laboratoire d'Electrochimie of the Université Louis Pasteur in Strasbourg.

Based on these different profiles of "inventor-authors", we can advance the hypothesis that the field of fuel cells in France consititutes an intersection which attracts or requires varied skills[16]. This hypothesis will be consolidated during the study of the contents of the published articles. Indeed, we shall see that the "inventor-authors" contribute in many ways to the definition of fuel cells.

---

15 In PASCAL, only the affiliation of the first author is available. It is thus appropriate to pay special attention to joint publications.

16 This hypothesis is further confirmed by the multiplicity of technical themes addressed by the patents. We have fully developed this point in the report we submitted to the AFME (Rabeharisoa 1990a).

In order to describe the body of articles that they publish, we have proceeded to a study of their titles and summaries (search for the most frequent keywords, and analyses of the words associated by the documents, enabling us to enumerate and define the themes dealt with). From this study, it appears that the *place* occupied by the cells in their publications sees a sharp change in direction around 1976-1977, to such a degree that we could speak of two publication periods.

BEFORE 1976-77: TECHNO-ECONOMIC RESEARCH ON FUEL CELLS

Until the mid 1970s, fuel cells or their constitutive elements occupy a central place in most of the publications. As proof, we summon a few article titles and summaries. Feuillade G., researcher-inventor at the Laboratoire Marcoussis of the CGE, works on the perfecting of macromolecular membranes for solid batteries, one of the technical themes of the CGE patents:

> Feuillade G., 1973,
> "Elaboration de séparateurs ioniques à usage électrochimique" in Rapport DGRST - Action Concertée "Chimie macromoléculaire, Membrane".
> *Extracted from the abstract*: (...) préparation de membranes à base d'alcool polyvinylique, et étude comparative de l'influence des paramètres de préparation sur les propriétés finales (...)

It is interesting to note that the other articles of Feuillade, as well as those of other specialized authors, formulate problems close to those noted in certain patents filed before 1975:

> Feuillade G. 1975,
> "Rôle et utilisation des membranes en électrochimie et en électrotechnique" in Entropie.
> *Extracted from the abstract*: (...) contrôle et régulation des échanges de matières; applications: génération de courant, électrodialyse (...)

> Croset M. 1973,
> "Anodic oxidation of silicon in organic baths containing fluorine" in Journal of the Electrochemical Society.

> Brenet J. 1975,
> "Comportement électrochimique d'une électrode LA: : (2) NiO: : (4) en milieu alcalin" in Compte-rendus de l'Académie des Sciences.

These articles focus on the characterization of a few elements of fuel cells or on a few techniques necessary for their design. In a way, they prolong the *division* of the cell performed by certain patents filed before 1975. It is in such terms that we can speak of *technical research*: These articles show the state of knowledge concerning specific technical objects, and are published in *important* scientific journals: Journal of the Electrochemical Society[17], Entropie, etc. Furthermore, the authors cited above collaborate with *fundamental* researchers (Croset M. publishes with researchers at the Laboratoire de Physique des Solides of the Ecole Normale Supérieure), and are relatively *visible* in SCISEARCH (Feuillade G. is cited 52 times by both French and foreign electrochemists, and Croset M. participates in a *research front*: "Study of oxygen tracer diffusion using nuclear microanalysis"). In short, these authors are *specialists* who, until 1976-77, partly use their skills on certain elements of fuel cells. As for the other authors, they devote themselves to the techno-economic nature of certain cells, as indicated by the keywords appearing most frequently in their pre-1976 publications:

HESPEL C. (ALSTHOM) high power hydrazine fuel cell
JACQUELIN J. (CGE) air hydrogen fuel cell 500 W

Figure 3-6: List of most frequently appearing keywords in articles published before 1976 by two "inventor-authors"

The definition of possible markets for these cells occupies an important place, as in certain patents filed before 1975:

Hespel C. 1975,

"Piles à combustibles à forte puissance pour applications sous-marines" in International 9th

---

17 This journal is in SCISEARCH's panel.

Intersociety Energy Conversion.

Extracted from the abstract: (...) description des piles à combustible, à hydrazine comme combustible et peroxyde d'hydrogène comme réducteur; modules produits par ALSTHOM; performances (...)

Jacquelin J. 1974,

"Une pile à combustible hydrogène-air de 500 W avec reformage du méthanol" in 4th International Symposium on Fuel Cells.

Extracted from the abstract: (...) description détaillée de la pile et de ses accessoires mis au point par la CGE (...)

The IFP team is especially active in analyzing possible applications of fuel cells:

Breele Y., Cheron J., Grehier A. 1974,

"Problèmes liés à la commercialisation d'un véhicule urbain à pile à combustible hydrogène-air" in 4th International Symposium on Fuel Cells.

Extracted from the abstract: (...) pour atteindre un important marché, les deux principaux problèmes à résoudre sont le coût de l'unité de propulsion et l'utilisation de l'hydrogène à bord du véhicule; ces problèmes sont analysés en détail après une revue rapide d'une étude actuelle concernant une Renault 4L équipée d'une cellule hydrogène-air de 11 kW; conclusions favorables (...)

Breele Y., Cheron J., Degobert P., Grehier A. 1977,

"Les piles à combustibles. Les piles à hydrogène" in Revue Générale d'Electricité.

Abstract: Après avoir exposé les principes de la pile hydrogène-air et ses caractéristiques techniques de construction et de fonctionnement, on illustre les progrès accomplis dans ce domaine depuis une dizaine d'années par les réalisations de l'Institut Français du Pétrole: performances accrues des électrodes, progrès dans la construction des assemblages de piles, dans la construction des accessoires des batteries. On expose ensuite l'intérêt présenté par les piles à hydrogène: perspectives énergétiques de l'hydrogène avec l'avènement des centrales nucléaires, domaines d'application de ces types de piles comme groupes autonomes de fourniture d'énergie (capacité énergétique supérieure à 400 Wh/kg) et en traction automobile électrique.

In a way, these articles are techno-economic synthetic documents. Some of them are presented by their authors at international colloquia. Fuel cells are submitted to the economic question. Once more, we note that these articles are very close to certain patents from the 1970s, one concern of which was the commercial imperative. In particular, the electric vehicle, as a potential application, is widely mentioned: Its development is presented as a potential solution for a certain number of problems posed by transportation in urban areas.

Bouthors P. 1973,
"L'électricité dans les transports. Le véhicule électrique urbain. Stratégie de développement du petit véhicule urbain à la Régie Renault" in Revue Générale d'Electricité.

Breele Y. 1974 (avec Sale B.)
"Les nouveaux systèmes de propulsion envisageables pour les véhicules urbains" in Revue de l'Institut Français du Pétrole.
Abstract: On demande de plus en plus aux véhicules particuliers utilisés en zone urabaine de répondre à un certain nombre de contraintes (pollution, bruit, économie d'énergie). Examen des différents modes de propulsion pouvant être retenus en les classant en: systèmes thermodynamiques, systèmes électrochimiques. Réduction de la pollution des moteurs à explosion, recherche sur le moteur de Stirling. Problème de stockage de l'énergie électrique. Recherche dans le domaine des nouveaux générateurs électrochimiques: accumulateur sodium-soufre, pile zinc-air et pile hydrogène-air.

The structuring nature of the electric vehicle is further reinforced by a study of articles published between 1969 and 1975 in the Revue Générale d'Electricité. This technical journal is a kind of a *common denominator* of the 23 authors, since both specialists and non-specialists publish articles in it. In order to determine which themes were addressed by this journal in the early 1970s, we conducted a LEXIMAPPE[18] study of the publications therein. It appeared that "electric traction"[19], though not too central, is a particularly dense theme during this period.

What can we conclude about the articles published by the inventors before 1976-1977? First of all, we take note of two categories of publications:

- the first consists of articles published by authors who place their *specialized* knowledge at the service of certain fuel cell elements. These articles present the results of technical research on fuel cells in relatively highly *visible* journals;

---

18 LEXIMAPPE is a co-word analysis method that relies on calculating co-occurences of words in a group of documents. For more details, see Courtial 1989.

19 The words associated with this theme are: railway transportation - automobile - linear machine - high speed - <u>electric vehicle</u> - <u>electric motor</u> - <u>antipolluting motor</u> - design - functional characteristics (underlining ours).

- the second is made up of articles more oriented towards the uses of these cells. A non-negligible portion of these articles consists of papers given in colloquia devoted to fuel cells, thus making these cells exist as objects which have their own economic specificities.

Next, the concerns expressed by these articles are comparable to the claims made by patents filed before 1975. This is why we can easily say that the cells of this period are "house-cells": inventors propose the design of marketable cells. In order to do this, they undertake technical (or even techno-economic) research on these cells themselves.

Finally, the "inventor-authors" consider that fuel cells are fairly promising in terms of applications, especially with respect to the electric vehicle. We will see further on that, like the patents filed after 1975, the articles published after 1977 are a bit less enthusiastic!

AFTER 1977: A RETURN TO ELECTROCHEMICAL RESEARCH

Like the patents filed after 1975, the articles published after 1977 begin to be diversified. They include four types of concerns:

- Some articles are interested in other technical devices, such as heat pumps or systems for storing energy;
- others concentrate on more general electrochemical problems, such as that of equilibrium versus non-equilibrium;
- a few articles address very specific themes in physics or chemistry;
- others, finally, raise unresolved problems concering fuel cells.

AUTHORS "LET GO" OF THE FUEL-CELL-COMPACT OBJECT

Hespel C., inventor for ALSTHOM, moved from high power fuel cells for submarine applications to heat pumps. Jacquelin J. and Feuillade G., both inventors for CGE, in part use their skills investigating energy storage systems (cf. figure 3-7). Other technical problems and devices appear next to fuel cells that hold the attention of the inventors, as shown by the most frequent keywords in their publications after 1976:

HESPEL C. (ALSTHOM) floor device inverse Rankine cycle compression heat pump polytropic heat pump
JACQUELIN J. (CGE) electrolysis hydrogen production

Figure 3-7: List of the most frequent keywords appearing in the articles published after 1976 by two "inventor-authors"

Furthermore, those who speak about fuel cells focus on techno-scientific problems that get in the way of their industrialization:

> Bono P. 1977,
>
> "Les piles à combustibles. Pile à combustible méthanol-air à électrolyte tampon" in Revue Générale d'Electricité.
>
> *Extracted from the abstract*: (...) pour industrialiser la pile méthanol-air fonctionnant en milieu tampon, la puissance de 4,5 W/DM**2 est nécessaire; cette puissance n'a pas été atteinte à 9O**(O)C en tampon autocarburant, où les problèmes de diffusance et d'activité des électrodes sont importants; outre la sensibilité de l'anode à l'empoisonnement, l'activité de la cathode est faible; pour améliorer celle-ci, on doit utiliser des platinoïdes qui favorisent alors la perte en CH: :(3)OH par oxydation chimique et électrochimique; le problème essentiel de la pile CHOH en milieu tampon reste l'activité des électrodes qui n'a pu être améliorée à 150**(O)C dans les électrolytes concernés (...)

Thus, the enthusiasm of before 1977 gives way to a certain caution: Problems remain, at the level of both the electrodes and the electrolytes. It is time to draw up the balance sheets, and these take stock of technical and economic difficulties:

> Breele Y., Choffe B., Grehier A. 1980,
>
> "Incidence économique des piles à hydrogène" in Hydrogen as an energy vector - International seminar.
>
> *Abstract*: Trois filières sont surtout utilisées: la filière basse température à électrolyte basique, la

filière moyenne température à acide phosphorique et la filière haute température à carbonate fondu. L'analyse technique montre que les filières basique et carbonate offrent le plus d'avantages. Il faudra atteindre des objectifs contraingnants sur les prix (200 F/kW) et les durées de vie (20000 à 40000 heures).

There is thus no commercial miracle! In order to "hold" the industry and the market, the "inventor-authors" must "let go" of the cell-compact object to focus on pricing problems and immerse themselves in the *fundamental electrochemical mechanisms*. With respect to this last point, the articles published by the specialized authors whom we designated previously will also go in the direction of a displacement towards techno-scientific concerns.

SPECIALIZED ARTICLES: TOWARDS FUNDAMENTAL RESEARCH

As of 1976-77, the specialized authors invest themselves more exclusively in their favorite domains: Croset devotes himself to micro-electronic captor technology and to laser crystallization in thin layers of amorphous silicium. Blain invests himself resolutely in the study of chemical and thermal properties of aluminum-zinc alloy.

Those elsewhere who remain interested in fuel cells place themselves in the field of electrochemical research. This is the case of Brenet who, inspired by the work of the electrochemist P. van Rysselberghe, conducts research on:

"Equilibre et non-équilibre en électrochimie. Applications aux électrodes multiples et à un aspect macroscopique du mécanisme des processus de transfert d'électrons" in Journal de Chimie Physique Physicochimie Biologie, 1978.
Abstract: On reprend et développe les travaux de P. van Rysselberghe sur la notion d'équilibre et de non-équilibre en considérant le parallélisme entre des processus chimiques et électrochimiques. Cela démontre la nécessité absolue de traiter l'électrochimie à partir de la thermodynamique et de la notion de non-équilibre. Cela a conduit ensuite à envisager les caractéristiques des électrodes multiples au sens de E. Lange et les processus de transfert de plusieurs électrons sur les bases macroscopiques.

This is also the case for Feuillade G., who is doing a:

"Study of SO: electrocatalytic oxidation during the sulphur cycle" in Rapport CEE-Enregie, 1980.
Extracted from the abstract: (...) élucider l'étape électrochimique du cycle thermoélectrochimique de SO pour la production d'hydrogène (...)

and research on:

> "Le stockage chimique de l'énergie: IV - Utilisation de nouveaux vecteurs d'hydrogène dans les moteurs thermiques. Elaboration d'un programmeme prospectif général" in Entropie.
> Extracted from the abstract: (...) on examine en détail l'aptitude de l'ammoniac et du méthanol à l'utilisation en combustible (surtout moteurs) (...) il reste toutefois un effort important à fournir en dehors de la mécanique: recherche de catalyseurs favorisant la dissociation de l'ammoniac et l'élimination des imbrûlés, recherche d'additifs facilitant l'allumage et la combustion, choix de matériaux résistant à la corrosion (...)

In fact, these authors fall back on their specialties on the one hand, and on the other hand sometimes do research in electrochemistry; in the course of this research, the fuel cell gives way to a particular oxidation or reduction reaction or to a particular fuel. The results of our analysis of patent documents get reaffirmed: Towards the mid-1970s, the "inventor-authors" move from market concerns to techno-scientific investigations.

IN SUMMARY OF SECTION 3.4

Without returning to the details of the preceding analysis, we can speak of a certain *isomorphism* between the body B of patents and the group A of articles published by the inventors. Before 1975, the inventors tried to fix in their patents the techno-economic characteristics of fuel cells, especially by enumerating and describing different market configurations. Around the same time, these inventors published enthusiastic articles on the technical performances of these cells and noted promising commercial prospects. After 1976-77, the inventors publish articles which present a profile close to that of the patents filed after 1975: they take stock of the technical and economic obstacles to the industrialization of fuel cells, and list a few works of fundamental research to conduct in electrochemistry (for example: research on catalysts).

It would thus seem that the break of the end of the 1970s was accompanied by a displacement towards *science*, a link of the French fuel cell network in which, apparently, several holes remain to be filled.

3.5 CONCLUSION

Without returning to the points developed above, we would like, in conclusion, to present the main characteristics of the field of fuel cells in France, as shown by the patents, and to discuss the possible *generalization* of our qualification grid of links between science and technology.

We proceeded by following the articles designated by the patents. The first result of this course concerns the existence or non-existence of patents. Although obvious at first glance, this is not a trivial result. Indeed, the existence or non-existence of patents indicates a certain structuration of the field studied. Thus, the number of fuel cell patents filed in France indicates that these technical devices create an aggregation of interests of certain firms. In terms of patents, this translates into the fuel cell "label". Naturally, the latter can exist outside of patents. But this would not invalidate our exercise; rather, it would be a sign that fuel cells are also designed elsewhere.

The same comment applies to the articles designated by the patents: the existence or non-existence of these articles points to certain articulations between science and technology for the field studied. What are these articulations?

The first is the fact of article citations on the front page of patents. The main criticism directed against these by scientometricians concerns their hybrid nature: are these citations proof of the examiners' zeal, or are they markers of filiation between science and technology? As we have seen, the variety of answers shows the globally undecisive nature of this question. In a few words, the situation is as follows: It all depends on the domain chosen and on the database used! From our point of view, this is a recommendation rather than a handicap: These results are specific to each case studied. Thus the question of the relevance of the articles cited on the front page of patents as indicators of their science-based nature becomes a local question (for further support of this notion see sections 4.6 and 9.5).

The test of the body B of fuel cell patents filed in France provided us with the following results:

- The articles cited by B, the articles cited by the patents that cite B, and the articles cited by the patents cited by B show that technical research on fuel cells exists and designate research themes connected to these cells (for example, research on solid electrolytes).

- Because the WPI(L) database only provides a small number of citations, we were unable to conduct a precise analysis of the contents of the articles cited.

The main result of the *citation test* can therefore be summarized as the existence of research with respect to fuel cells. (For further applications of this type of analysis see chapter 4 and section 9.5).

The second articulation between science and technology results from articles published by the inventors themselves. These articles show that between 1960 and 1987, the field of fuel cells in France was the site of a simultaneous development of science and technology: the inventors built their cells and in the same move produced scientific, technical, and even economic knowledge about these cells. It is thus remarkable that in the heyday of fuel cells, when these were the object of enthusiastic market speculations, the articles published by the inventors were in part devoted to showing the techno-economic conditions favoring their development. Similarly, when fuel cells revealed themselves to be unreliable at the end of the 1970s (due to electrode poisoning, low life span, etc), the inventors took stock of the unresolved techno-scientific problems (such as: What catalysts to use? Or what are the advantages of methanol as a fuel?) This proximity, both temporal and thematic, of the patents filed and the article published by the inventors, shows us a dynamic of the development of fuel cells of France. We will qualify this dynamic as *learning by doing* in the strict sense of the term. Similar observations on cyclic progress of science and of technology are reported in chapters 8 and 9. In summary, we could say that companies involved did not programme science and technology as two different entities[20].

Such are the two articulations between science and technology that we found for the field of fuel cells in France. From our point of view, these two articulations neither confirm nor infirm each other. They designate two different linkage modes between science and technology. The articles cited by the patents mark the "loans" taken out by inventors from scientific research. As for the articles published by the inventors, they show the way in which R&D was conducted within the filing firms. Both make the field studied as a network, associating researchers, technologists, objects, and strategies in different ways.

---

20 This in no way predetermines their success or failure. What we can currently say about fuel cells in France is that, at present, they are undoing themselves as immediately marketable objects in order to exist as subjects of techno-scientific research.

Using these differences between *cited research* and *research produced* by inventors, we would like, in conclusion, to insist on the specific nature of each indicator of science and technology. Following certain authors (Grupp et al. 1989), we would say that an indicator derives its relevance from its ability to show certain peculiarities of the field studied. From our point of view, this definition of the effectiveness of an indicator offers at least two advantages. First, it constitutes a *criterion* for choosing methods for the scientometrician. If, for a particular domain, the patents all cite scientific articles in a relatively consistent manner, then these articles undoubtedly deserve special attention, since they constitute a *resource* visibly mobilized by the actors themselves. Secondly, it contributes to the establishment of tracing indicators adapted to each situation. Indeed, if the frontiers between science, technology and the market are fuzzy, they are mixed up in a way that is unique to each case. On the one hand, our fuel cell inventors publish scientific articles, but on the other, the Nippon Electric Company (NEC), for example, publishes very little in matters of radiotelecommunications[21] (Grupp et al. 1989). Thus, the relationships between science and technology get created in different ways in each case. Therefore, the indicators also take values according to the configuration of each domain.

REFERENCES

Bastide F., Callon M. and Courtial J.P., (1989), 'The Use of Review Articles in the Analysis of a Research Area', *Scientometrics*, 15, pp. 535-562.

Bijker W., (1987), 'The Social Construction of Bakelite: Toward a Theory of Invention', in Bijker W., Hughes T., Pinch T. (ed.), *The Social Construction of Technological Systems. New Directions in the Sociology and History of Technology*, MIT Press, Cambridge, Mass.

Callon M., (1986), 'Some Elements of a Sociology of Translation: Domestication of the Scallops and the Fishermen of St. Brieuc Bay', in Law J. (ed.), *Power, Action and Belief. A New Sociology of Knowledge?*, Routledge & Kegan Paul, London.

21 The reasons for which NEC does not publish very much are of little interest to us. Indeed, whether the NEC managers feel that their engineers are not paid to publish articles, or whether NEC has a secrecy policy with respect to its scientific discoveries in radiotelecommunications, the end result is the same: It is not through the rare publications of the inventors that we can describe the scientific bases of the inventions of this company. In order to do so, we would have to test other indicators (for example, a study of the papers presented by the engineering researchers of the company at colloquia on telecommunications).

Callon M., (1990), 'Les réseaux technico-économiques et irreversibilité' in Boyer R. (ed.), *Figures de l'irreversibilité en économie*, Economica, Paris.

Callon M., Laredo P., and Rabeharisoa V., (1990), 'Des instruments pour la gestion et l'évaluation des programmemes technologiques: le cas de l'AFME', in de Brandt J., Foray D. (ed.), *L'évaluation économique de la technologie*, Economica, Paris.

Collins P., and Wyatt S., (1988), 'Citations in Patents to the Basic Research Literature', *Research Policy*, 17, pp. 65-74.

Courtial J.P., (1989), 'Qualitative Models, Quantitative Tools and Network Analysis', *Scientometrics*, 15.

Dosi G., (1982), 'Technological Paradigms and Technological Trajectories', *Research Policy*, 11.

Grupp H., and Schwitalla B., (1989), 'Technometrics, Bibliometrics, Econometrics, and Patent Analysis: Towards a Correlated System of Science, Technology, and Innovation Indicators', in van Raan A.F.J., Nederhof A.J., Moed H.F. (eds.), *Science and Technology Indicators*, DSWO Press, Leiden.

Hyams M., (1985), 'Chemical Patents Information: The Challenge of Change', *J. Chem. Inf. Comput. Sci.*, 25, pp. 365-371.

Kuhn T.S., (1983), *La structure des révolutions scientifiques*, trad. Laure Meyer, Flammarion, Paris.

Law J., and Callon M., (1987), 'The Life and Death of an Aircraft: A Network Analysis of Technical Change', paper presented at the International Workshop on the Integration of Social and Historical Studies of Technology, Enschede.

Latour B., (1989), *La science en action*, La Découverte, Paris.

Narin F., and Noma E., (1985) 'Is Technology Becoming Science?', *Scientometrics*, 7, pp. 369-381.

Narin F., Rosen M., and Olivastro D., (1989), 'Patent Citation Analysis: New Validation Studies and Linkage Statistics', in van Raan A.F.J., Nederhof A.J., Moed H., (eds.), *Science and Technology Indicators*, DSWO Press, Leiden.

Pinch T.J., and Bijker W., (1984), 'The Social Construction of Facts and Artefacts: Or How the Sociology of Science adn the Sociology of Technology Might Benefit Each Other', *Social Studies of Science*, 14, pp. 399-411.

Rabeharisoa V., (1990a), 'Petit exercice de sciento-métrie: ce que chiffrent et racontent les brevets français sur les piles à combustibles', Rapport CSI-AFME, Paris.

Rabeharisoa V., (1990b), 'Pour une description des liens entre science et technique: des brevets aux articles scientifiques', in Vinck D. (ed.), *La gestion de la recherche: nouveaux problèmes, nouveaux outils*, de Boeck, Bruxelles.

Rosenberg N., (1982), *Inside the Black Box*, Cambridge University Press, Cambridge.

Schmoch U., and Grupp H., (1989), 'Patents Between Corporate Strategy and Technology Output: An Approach to the Synoptic Evaluation of US, European, and West German Patent Data', in van Raan A.F.J., Nederhof A.J., Moed F. (eds.), *Science and Technology Indicators*, DSWO Press, Leiden.

Schmoch U., (1990), *Wettbewerbsvorsprung durch Patentinformation*, Verlag TÜV Rheinland, Köln.

van Vianen B.G., Moed H.F., and van Raan A.F.J., (1989), 'The Assesment of the Science Base of Recent Technology', in van Raan A.F.J., Nederhof A.J., Moed H.F. (eds.), *Science and Technology Indicators*, DSWO Press, Leiden.

von Hippel E., (1988), *The Sources of Innovation*, Oxford University Press, New-York, Oxford.

4 PERCEPTIONS OF SCIENTIFICATION OF INNOVATION AS MEASURED BY REFERENCING BETWEEN PATENTS AND PAPERS: DYNAMICS IN SCIENCE-BASED FIELDS OF TECHNOLOGY

Hariolf Grupp and Ulrich Schmoch

4.1 INTRODUCTION AND STATE OF THE ART

The core of the relationship between science and technology in Europe can be traced back to Galileo Galilei (died 1642) or Leonardo da Vinci (died 1519). Thus progress in scientific knowledge to the heliocentric world system essentially advanced through improved Dutch telescope technology and excellent lens grinding and also conversely the then new scientific model of the cosmos could be used directly for maritime navigation. Science and technology appear to be closely interlinked. Owing to the connection - obvious even then - between technical change and economic prosperity, early innovations were in no doubt as the importance of science to economic and social advance was self.

In his book on "An Inquiry into the Nature and Causes of the Wealth of Nations" as far back as 1776, Adam Smith expressly stated that "each individual becomes more expert in his own peculiar branch (.....) and the quantity of science is considerably increased by it" (Smith 1776, p. 8). Karl Marx at various places in his opus points out that science is an important prerequisite for new machines, production methods or new technologies generally and hence also conditions economic growth and social wealth (Holland 1988, p. 137; similarly also Rosenberg 1976, p. 63). Take this verbatim quote, for example: "Mit der capitalistischen Production wird (...) der wissenschaftliche Factor zuerst mit Bewusstsein und auf einer Stufenleiter entwickelt, angewandt, und ins Leben gerufen auf einem Maßstab, von dem frühere Epochen keine Ahnung. (...) Erst in dieser Productionsweise stellen sich die praktischen Probleme dar, die nur wissenschaftlich gelöst werden können." (In capitalist production .... the science factor is first consciously developed stagewise, applied and called into being on a scale never dreamed of in earlier times (....). Only in this mode of production do practical problems arise that can only be solved scientifically.) (Marx circa 1862). According to this view, capitalist means of production appear to be both a special promoter of and also benefitiary from scientific advance.

Rosenberg (1974) and Freeman, in particular, cottoned on to the same idea in this century in which the discovery of certain quantum mechanical properties of solids resulted simultaneously in awarding the Nobel prize for physics whilst technological development yielded a micro-electronic unit, namely the transistor. Freeman headed the first section of his magnum opus "the Rise of Science-Related Technology" (Freeman 1982, p. VIII). Looking back at the most growth-intensive economic branches of the 20th century (chemistry, mineral oil, plastics) as well as on the development of atomic engineering and electronics, he recalls that industrial innovation activity is performed ever more professional and that technological development is increasingly dependent on scientific know-how. A modern perspective on the science base of innovation is also provided in chapter 1 of this volume. In German parlance, these phenomena in social sciences tend to come under the heading of "scientification" while German economic research largely disregards them. In this chapter, "scientification" is used only to characterize chronological changes.

It is important to differentiate between science intensity and intensity in research and development. According to the so-called "Frascati definition", "Research and experimental development (R&D) comprise creative work undertaken on a systematic basis in order to increase the stock of knowledge, including knowledge of man, culture and society, and the use of this stock of knowledge to devise new applications" (Freeman 1982, p. 225). R&D is therefore a generic term embodying scientific activities and also importantly something else (e.g., the construction of industrial prototypes and pilot plants).

The concept "R&D-intensive" is defined by expense parameters, financial or personal expenditures being used as yardsticks (among other sources most recently see Grupp and Legler 1990). An R&D-intensive area (technology, product, branch of the economy) is characterized by particularly high financial expenditure on research and development or by particularly high personnel deployment in this area who are involved, although not necessarily, in scientific activities. Gibbons & Johnston (1974, p. 231) based on limited interviews in the British economy were able to show that the frequency of resort to scientific literature does *not* correlate with R&D intensity in the branch of the economy.

The "science" or "natural science" concept is often poorly defined in empirical or statistical work. Criteria relating to the type of activity are confused with institutional definitions. Thus within the R&D activities area, the science concept is often equated with fundamental research (as opposed to industrial development) or - in the institutional definition - is understood to include activities in the academic area (as opposed to industrial R&D) or a combination of the two. On the other hand, West German statistics regard "science" as a generic term in which research and development merely represent a sub-area as it comprises scientific teaching and training (BMFT 1990, p. 331). According to this idea, the building of an industrial prototype, construction of a racing car say, would be one of many other scientific activities. With such a broad-based concept - however useful this may be for R&D statistics - it is hard to define clearly the conceptual link between science and technology. In the author's understanding science is a pool of formal knowledge represented by educated people - scientists - and mostly codified and disclosed in scientific publications (see chapter 1, section 1.2).

Even more difficult than defining an adequate concept is the evolution of a current and readily understandable quantifier. Also, so far there is no internationally standardized, universal definition for science involvement. Since this chapter cannot promote creation of a concept within a theoretical context and since sociological and economic schools in practice do not communicate with one another on the subject, science involvement will be defined pragmatically in terms of the set of indicators explained in section 4.2. The authors are well aware that this leaves much to be desired from the standpoint both of technical-historic, social-scientific and also innovation theory. They feel committed to the principle of "learning by doing". It is up to others to judge whether the aforementioned empirical findings can advance clarification and theoretical classification of adequate concepts.

Innovation research hitherto has essentially emphasized *scientific achievement* for industrial technology and seldom refers to the reverse relationship of the *industrialization of science*. However, the *interdependence* (Hack & Hack 1985) of the two aspects is clear from the latest apparatus equipping the scientific laboratory which is often much superior to the quality commercially available. Holland (1988, p. 170) has labelled this aspect as the dependence of scientific research of "modern research technology". From the Centre Européen pour La Recherche Nucléaire (CERN) there is a study which by investigating specially prepared research equipment quantifies the resulting economic spin-off on the equipment manufacturing industry and classifies the source of the "research technology" in terms of scientific subfields (Bianchi-Streit et al. 1984).

Institutional aspects play an important part in describing the interaction between "pure research" and industrial technology. Once again it was Freeman (1982) who pointed out that the extent of scientific involvement is dependent on the nature and interests of the bridging institutions and that an academic institution, in particular, even if it has technology transfer offices and the like, must be subject to other dictates (career orientation, opportunities for earning, opportunities for acquiring recognition, etc.) as for example the fundamental researcher in a major industrial laboratory (for example, the Nobel prize winning Bell Laboratories or the Swiss IBM research centres). Nelson (1988), Dosi (1988, p. 1137) and others include in the institutional picture the growing role played in the second half of this century by public or quasi-public R&D institutions even in the defence engineering sector which again take in other bridging functions between science and technology.

Pavitt (1984) presented a consequent operationalization of the theoretical relationships presented hitherto. He used many years of statistics on innovations in British industry to divide manufacturing industry into four groups, which in addition to conventional types (quotation-oriented sectors, specialised tenderers, scale-intensive sectors) for the first time postulate a formal science-based sector. In the science-based industrial sector, innovations are said to be directly linked to technical advances made possible by quantum leaps forward in scientific knowledge. Innovation-oriented activities of the relevant companies usually appear to be formalized in R&D laboratories and investments in the field of research and development appear to be disproportionately high. Innovative products from these science-based branches of industry are said frequently to be key requirements insofar as other industries are concerned. The companies are either very large or small to minuscule, highly-specialised newly formed companies directed by scientists (see also Dosi 1988, p. 1149).

The area of the economy cited largely based on British data material, classifies the following as being within the science-based sector:

- the electronics industry,
- large parts of organic chemistry,
- the pharmaceutical and bioengineering industry,
- aerospace, and
- defence engineering industry.

To the connoisseur of contemporary technology, the typology is plausible even without numerical substantiation. However, the trouble is that the typology is oriented to branches of the economy and - since on its own admission it refers to large concerns - it embraces technologically very heterogeneous structures. Nevertheless, two important deductions need to be retained that are also important for the present investigation: The industry-specific characteristics of science involvement are said to be "relatively stable timewise and between countries" (Dosi 1988, p. 1150; emphasis added). Thus only minor differences ought to be expected in the extent of the science involvement between individual economies and years (historic time periods excluded). (See section 4.3).

However, looking at things critically there are two aspects at issue: first of all, it cannot be assumed that Pavitt's four sectors are diametrically opposed to one another, i.e., that affiliation of a particular branch of the economy to one sector excludes the other features. Thus Gibbons & Johnston back in 1974 (p. 224) classified the "crafts" industry as the counterpart to science-based industry. On the other hand, nowadays at least owing to progressive "technological fusion" science and technological competence cross over branches of the economy. Thus Scherer (1982) evolved complex reciprocal relationships between branches of the economy and technological aspects empirically for the USA. He demonstrated convincingly that the R&D activity of industry is largely product-oriented and therefore technical change is product-oriented and not geared to a branch of industry. Analysing the R&D expenditures of various branches of industry for various product categories, Grupp (1991) concluded that in major OECD countries more than 50 per cent of all branches of the economy devoted less than 80 per cent of their R&D expenditure for the products they traditionally make. In the study of scientific involvement in technology, an industry-related concept therefore does not appear to be justifiable. Consequently, a technological sectoral model is employed in this chapter.

The methodology is further discussed in section 4.2. The results of the investigation are presented from three different aspects. The fundamental properties of science involvement in innovation expressed in terms of the indicators used here, are given in section 4.3.

The second main section of the results (section 4.4) provides a panoramic view of international development relative to scientific involvement. World-wide increased activity in science-based technology, increase in national activities in five selected countries and the activity of foreign companies and research institutions in West Germany are presented consecutively.

The third result-oriented component relates to the sectoral work breakdown in the former Federal Republic relative to science involvement (section 4.5). The contributions made by universities, public or semi-public research institutions and the West German industry are discussed in turn and analysed in regard to their sectoral interplay. Finally, specialization of the new federal states in East Germany in science-related technology is discussed. The conclusions of the investigation and proposals for further explorations and elucidations are summarized in section 4.6.

4.2 METHODOLOGY AND DATA

In view of the less-developed state of research, it is not surprising that there is hardly any formalised and quantified differential evidence of science involvement in innovation. The few exceptions rely on the fact that when patent applications are examined by officials at patent offices, there has to be a search for prior art. Invariably, it is described by earlier granted patents or by patent documents laid open for inspection. However, occasionally, patent examiners document the earlier status of science and technology by making references to scientific literature and not to patent literature. If the frequency of such non-patent literature references in patent documents is used as a measure of science involvement, a set of indicators can be devised for quantifying science involvement. Working on this assumption, Carpenter, Cooper and Narin (1980) drew up the first guides whereby US patents originating from certain "scientific" areas of technology, contain significantly more references to the prior state of the art which are not patent documents in themselves. In a subsequent investigation (Narin and Noma 1985), it was further specified in the bioengineering field that the time delay from references to scientific literature in patents is just as long as the standard citation interval within scientific publications.

Van Vianen, Moed & van Raan (1990) using the same indicator bases studied the chemistry field internationally and the national Dutch technology in breadth. They, too, relied on US patents and uncovered not only valuable results but also a series of inadequacies in the arrangement. Thus, in the US patents used, a very strong language barrier works to the disadvantage of all non-English literature. Coward and Franklin (1989) also tried to establish a particular relationship between scientific literature and patent literature in order to be able to deduce from it cross-overs between science and technology. In consequence the suggestion was made that the most productive processes for identifying potential profit yielding areas of scientific research for industrial technology involve defining fields in which the same researchers were able to furnish evidence particularly frequently, of both scientific publications *and* patents.

Method-wise it should be less controversial for patent literature to represent technology. There is comprehensive literature on the subject (e.g., Schmoch et al. 1988). Although not all technical inventions are patentable and not all patentable inventions are actually the subject of patent applications, undecided patentability, secrecy, and a low patent propensity play a small part nowadays. The remaining interfering factors can be corrected by quality indicators (op. cit., pp. 50 onwards). Allowing for certain peculiarities of the patent system, at least very good and virtually perfect random samples can be formed on technical development (op. cit., p. 53). With a few exceptions, large sections of the patent yield originate from industry. For the purposes of this investigation, patent literature therefore represents industrial technology sufficiently well.

The validity of citing non-patent literature in patents in regard to science involvement is more difficult to assess. It is questionable, in particular, in what circumstances the patent examiner bothers with *non-patent literature* (NPL) when establishing prior art. Usually, in their search report, they rely on patent literature (Greif 1982, pp. 54 onwards; Baré 1981; Claus & Higham 1982). In technical discussions with patent examiners, patent lawyers and other persons active in the patent field, the following six reasons were discovered for assignment of NPL reference:

1. Prior art in science and technology is still not covered by any one patent document. The patent examiner is compelled to substantiate novelty and the level of inventiveness of the patent application examined by reference to scientific publications.

2. The patent examiner does want to cite non-patentable research results (e.g., formulae, theories, discoveries, software developments, medical cures). He has to cite NPL documents because patent documents are not available.

3. The specialist area subject researched is developing so quickly that preceding patent documents from foreign patent offices are still not accessible to examiners owing to the publication periods (but only their own documents even prior to publication). If the inventor has published scientific papers on similar subjects in non-patent literature, then these documents can be cited.

4. A company dispenses with a patent application and publishes the result of the development in the company's own journal in order to safeguard the novelty claim vis-a-vis competitors even without a patent application. If this document is relevant to the examination process, the examiner will involve this document from non-patent literature.

5. The necessary references to Japanese language patent documents incomprehensible to most European and American examiners, may, via an English abstract service, feature in the search report in the form of an NPL reference whereas in reality they are patent citations.

6. Reference to very simple relationships that fall below the patentability threshold (e.g., how a screw is tightened or a mechanical clock wound) is essential. School books and encyclopaedias, therefore NPL sources, are mentioned.

Some of the six main motives listed above in fact reflect an intensive science involvement in the technical invention, others not so unambiguously. Thus, motives 5 and 6 are unimportant for science involvement. They must be relegated from the proposed indicator concept for measuring science involvement. Documents falling into the fourth category are inventions that, through management decision, fall outside the commercial interests of the producing company. NPL citations in this category also need to be excluded. The third motive pleads for the existence of a science connection (in short development cycles in micro-electronics, for instance). However, this is not absolutely necessary. Despite the unresolved situation, such NPL citations are permissible within the present investigation for establishing science involvement. The second motive reflects a science connection with a very high level of probability and the first motive reflects the classic situation. However, even here, care must be taken to ensure that not every journal citation is tantamount to a reference to fundamental research. According to a convention (Marquis & Allen 1966), a journal is only regarded as "scientific" if it is published by a scientific society. Journals from technical societies (such as the "IEEE Transactions on Electron Devices"; see Liebermann 1978, p. 7) on the other hand reflect the area of applied research. This case differentiation cannot be followed here, however, owing to poor operability, for even Gibbons & Johnston (1974, p. 228) assume that technical literature may contain basic research reports. In any case the method use in this chapter links only the codified parts of technology (Dasgupta & David 1991).

Having regard to these qualitative considerations, the measuring concept can be defined as follows:

i:	index for institutional demarcation (country, state, institutional groups),
j:	index for technological demarcation (technological sectors),
t:	time index (in the case in point, the five two-year periods 1979-80, 1981-82, 1983-84, 1985-86 and 1987-88),
$P_{ij}(t)$:	number of patent documents from institution (country) i in technical sector j and in year t.
NPL (P_{ij}, t):	number of NPL citations in patent literature from countries i in technical sector j and in year t.

The mean NPL indicator is then defined per formula (1) as:

$$NPLM_{ij}(t) = NPL\ (P_{ij}, t)/P_{ij}(t). \quad (1)$$

The NPLM indicator therefore gives the mean number of science references in patent literature for a particular year, a particular country and in a particular technical sector. If, for example, the search reports for these patents were to make specific reference to exactly one non-patent source, then the indicator would be given value 1. In reality, the NPLM values fluctuate between 0.1 and 2.0.

The basis hitherto has been that several references to non-patent literature in a research report indicate a stronger link to science than say only one such reference. This assumption is highly questionable. The patent office examiners must be meticulous and their search result must refer to the various patent claims individually. It can happen therefore that there is a central NPL citation for several or all patent claims and also the circumstance that several central claims need to be underpinned by various proofs from non-patent literature. In this case, the accumulation of NPL citations does not immediately point to a stronger science connection. Van Vianen et al. (1990) in their investigation of US patents discovered such a distorted distribution of the number of NPL citations per patent and concluded that the mean could scarcely be a representative figure for the number of such references in a particular random sample. These authors propose to introduce a second index evidencing distribution distortion. Their proposal specifically entails quoting the proportion of all patents with more than five NPL references in addition to the mean in order to identify a random sample. Lewison & Cunningham (1990, p. 107) suggest applying a mathematical conversion for the same situation before means are taken.

In the present paper, both proposals are not followed. The approach adopted is rather to regard an individual citation for science literature per patent as sufficient for the existence of a science connection and hence to ignore accumulations of NPL references. The indicator reflecting the proportion of all patent documents with at least one NPL reference is designated NPLQ (where Q denotes quota). The NPLQ index can in practice be assumed to be valued at between approximately 0.1 and 0.7 and in principle cannot exceed the value 1.

Further, a *weighted* index can be formed. It is defined by formula (2). The abbreviation W in this case stands for (mathematically) weighted mean.

(2) $$NPLW_i = \Sigma_j (NPLM_j P_{ij})/\Sigma_j P_{ij}.$$

The present investigations relate to a statistical total evaluation of all patent applications at the European Patent Office (EPO) since the priority year 1979. The *priority year* is the year for the first application to any patent office in the world in order to safeguard priority claims and does not necessarily mean that the year of application at EPO was 1979. The priority year therefore represents that year in which the invention materialised and will henceforth be termed (without patent law detail) the *year of invention*.

Comparison of EPO data with those from the American Patent and Trademark Office (USPTO) are extremely important for evaluating the interferences caused by differences in patent procedure. One of these differences lies in the fact that the USPTO does not recognize strict publication periods of 18 months and only publicizes patent grants and in so doing omits unsuccessful applications. In R&D-intensive areas of technology, the award process takes several years (Olivastro and Narin 1989, pp. 11 and 42). European data were compared to a parallel set of data for the USPTO for the 1975-1984 years of invention.

The European patent data are an excellent source for investigating references to science origins in patent documents. The European Patent Office route is certainly only one of several ways in which patent protection can be secured within any one European national territory. A European application is not necessarily available for inventions from, e.g., the Federal Republic of Germany. Owing to higher fees, it can be assumed that European documents are usually available for German inventions if the intention is to secure patent protection in three or more European contract countries (Cohausz 1978). European patent documents therefore take on the nature of foreign property rights in any case, *also* for Europeans. The single most active country is the US, not any European nation. Large economic enterprises and international concerns tend to be represented in European Patent Office data banks rather than medium sized companies, non profit making research institutes and universities or even independent inventors. For this reason, when investigating the sectoral distribution of a possible science involvement in Germany according to institutional groups, it was not worked with EPO data alone, but rather for this purpose combined documents from the German Patent Office (DPA) and EPO documents with FRG as the destination country. The same applies to investigation of the science link to East German technology. Owing to the economic and technological isolation of the former GDR, only comparatively few GDR inventions have been applied for abroad. West Germany was the preferred "foreign" country for the then GDR.

The data banks WPIL, EPAT and PATDPA used have not been set up for the purpose of statistical inquiries, but rather for patent lawyers, employees of patent offices and interested industries who wish to clarify and research individual cases. When conducting statistical inquiries, double counting, incomplete data and spelling mistakes have to be taken into account. The questions of reliability and in particular reproduceability which this raises call for a series of *standardisations* (for more details, see Grupp & Schmoch 1991). By personal inspection into random sets of patents a sufficiently consistent data set was arrived at.

Within the context of standardisation, it has been shown that motive 5 (concealed patents) is evident in each case and could be excluded from the count. Motive 4 (patent-like corporate publication) and 6 (references to circumstances below the patentability threshold) are not so clear-cut. The situation arises only rarely in the European patents system; corresponding citations have been excluded according to tendency and also where recognizable. However, this definition has been arrived at less satisfactorily. Citation motive 1 (reference to science), 2 (non-patentable scientific results) and 3 (quickly evolving high tech areas) were classified as valid in the sense of the concept for measuring science involvement; the corresponding citations have been included in the count. In most cases, data standardisation is reflected in minor corrections to raw data. The original and standardised indicators differ by no more than 1 per cent in the case of European countries. Annual fluctuations in data material are deduced to be of the same order of magnitude as the corrections mentioned. Admittedly, for the United States and Japan, the corrections for certain technical areas are considerable (up to 10 per cent).

A clearly classified observation structure is required showing the connection between science and technology. Since no generally valid classification of technology is available, in the field of patent analysis it is an important advantage for patent documents to be classified uniformly in terms of an internationally agreed, clear-cut, non-overlapping, and comprehensive classification matrix (the so called International Patent Classification; IPC). The matrix currently covers 58,400 individual elements and offers summaries at several hierarchical stages. Historically, these have emerged from grounds other than those with which one is concerned here and are not immediately suitable as the bases for a technological sectoral model.

As part of this investigation, a new technology classification has been evolved based on the IPC. Drawing on the technological know-how of specialists from the Fraunhofer society, individual patent symbols were arranged in 28 broad technological sectors representative and homogeneous in terms of modern technology (see table 4-1).

The intention was to divide technology into 28 areas that were as uniform as possible. In so doing, the IPC hierarchical structure was obtained with a view to putting each individual invention definitely into a precise technical field. Whenever the specialist considered it both possible and logical, an entire IPC class (two-digit hierarchical level) was assigned to a technological area. In cases where in the meantime technical development has differentiated and a coarse assignment of this kind appeared to be not upholdable, individual sub-classes (three-digit hierarchical level) were assigned to the various areas, one IPC class therefore being divided up into 2 or more areas. Even greater fractionation (main groups, sub groups) of IPC was avoided in order to obviate the blurring caused by erroneous or deviating use of IPC between various national patent offices. In particular, such problems of concordance between USPOC and IPC patent classification occurred with a rate of 12 per cent on patent class level, while at main group level there were 52 per cent misclassifications as defined by IPC (Stillger 1987, p. 261).

The FhG-ISI technology classification (table 4-1) - like any exhaustive classification - contains a few so-called residual categories covering patent documents which are too small in scope to constitute a technical area of their own or which cannot be classified in any other way on other grounds. For details of IPC subclasses related to the technology sectors see Grupp, Reiss, Schmoch (1990).

Residual categories therefore have to be interpreted with special care although the main areas dominate even there. At this point, it is worth mentioning particularly the

Table 4-1: Classification of technology by 28 sectors and related IPC (4th revision) codes (Source: FhG-ISI; see Grupp, Reiss & Schmoch 1990)

Code	Technical Area (keyword underlined)	
BA	Mining, Civil Engineering, Building Materials, Airconditioning, Waste Disposal, and related topics not included elsewhere	A47 (not J,L), A62, B09, B28, C02, C04, E01-21, F21, F24, F25D, F23 (not R)
CO	Coating, Crystal Growing	B32, C23, C25, C30
DA	Data Processing	G06
EL	Electronics, Electronic Components not included elsewhere	H01L, H05K
EN	Nuclear Technology, Electricity	G21, H02B, G, H, J, M, H05H
ET	Electrical Engineering, Electrical Machinery	A47L, J, H01B, C, F, G, H, J, K, M, R, T, H02K, N, P, H05B, C, F, G
GP	Bio- and Genetic Engineering, Genetics, Pharmacy	A61K, C07G, H, J, K, C12F, M, N, P, Q, R
HA	Handling, Conveyor Equipment, Robots	B25J, B65-67
IC	Inorganic Chemistry, Glass, Explosives	C01, C03, C05, C06, F42
IM	Image Transmission	H04N
IN	Instruments not included elsewhere, Controls	G05, G07-10, G12
IS	Information Storage	G11
LA	Lasers	H01S
MA	Mechanical Engineering, Machinery not included elesewhere, Armament	F15-17, F27, F41
ME	Biomedical Engineering, Medicine	A61 (not K)
MM	Material Processing, Machine Tools	B21-24, B25 (not J), B26-27, B30, C21-22
MS	Metrology, Sensors	G01, G04
MT	Engines, Motors, Turbines, Pumpes	F01-04, F22, F23R, F25B, F28
NA	Agriculture, Nutrition, Beverages, Tobacco	A01-24, C08B, C12C, G, H, J, L, C13
OC	Organic Chemistry, Petrochemistry	C07B, C, D, F, C10, C11
OP	Optics	G02, G03
PC	Polymer Materials, Polymer Chemistry	C08C, F, G, H, L
PE	Process Engineering, Separation, Mixing	B01-08, F25C, J, F26
PO	Polymer Processing, Synthetic Resins, Paints, and other Synthetic Substances not included elesewhere	B29, C08J, K, C09
PP	Paper & Printing	B31, B41-44, D21
TC	Telecommunications	H01P, Q, H03, H04 (not N)
TE	Textiles, Apparel, Leisure, Textile Machinery	A41-46, A63, B68, C14, D01-07
TR	Transport & Traffic	B60-64

building (BA), machinery (MA), electrical engineering (ET) and instruments (IN) area. Software is patentable only in conjunction with equipment and not on its own. Software therefore does not form a separate technological area, but is included in the relevant equipment section of inventions. Non-patentable areas are omitted from the classification; these are more especially medicinal cures and to some extent conventionally cultured breeds of animals and varieties of plants (genetic engineering processes and products are however contained in the GP sub-area).

The so-called 28-sector technology classification has been used as part of this investigation and tested and refined many times. However, its use has not been confined to the present investigation; in fact, in the meantime the model has already been applied successfully in over 5 other studies, some foreign. It can therefore be regarded as having been adequately validated, tried and tested (see chapter 9 of this volume, Grupp, Reiss & Schmoch 1990, Olivastro & Narin 1989, Grupp & Legler 1990, Schmoch & Kirsch 1990, Engelsman & van Raan 1990, and Team of Authors 1991).

The following definitions apply to identifying the *origins of technology*: the *national source* was the inventor's address (his postal address or the postcode for his place of residence). The place of residence of the inventor as a proxy for origin is based on the assumption that he is located in the vicinity of his workplace. If the location of the applying company were used as the bases, there would be substantial relocations particularly in the case of internationally interlinked concerns because the patent application may emanate from company headquarters. According to the approach adopted in this investigation, the technical development contribution made at IBM laboratories in Germany is regarded as one of German origin; the headquarters, shareholders and financial intricacies of IBM are ignored. If the inventors originate from more than two countries, the patent document has been counted fully for both countries respectively. Since such patent documents reflect trans-border cooperative efforts, which may be of particular importance to science involvement, fractionation would reduce the statistical importance of cooperative inventions within the data set of a country. Trans-border cooperative efforts are only discussed separately if at least one German partner was involved.

In the interests of data standardisation, where foreign cooperative endeavours are concerned, an important correction needs to be applied. Many large chemical concerns (Hoffmann la Roche, Sandoz, Ciba-Geigy) employing in their R&D laboratory German citizens are located in the industrial region around the Swiss border town of Basel. For tax and other reasons (welfare insurance, education system, foreigner laws) the residence of the German commuter is however usually in German border country that otherwise presents no R&D activities in the chemical sector. This commuting does not necessarily imply any international cooperation although the above-mentioned criterion is fulfilled. Consequently, such apparent cooperative situations were excluded as part of data standardisation.

Within the former Federal Republic, an *institutional sectoral model* is used which differentiates between three sectors: enterprises, public and semi-public research institutes plus universities. In so doing, definition of the academic sector is difficult, because in commercial property law, universities usually do not have status as legal persons. In the Federal Republic of Germany, matters of commercial property rights are regarded as the private concern of the teaching staff (clause 42 of the law on employee inventors; ArbEG). Consequently, the inventor himself and possible associates therefore feature as applicants for an invention. The title "Professor" was used in the inventor area of a patent document as a search criterion (for details, see Grupp & Schmoch 1991).

The institutional sector of the research institutes is demarkated in other ways. Since the research companies feature as legal persons and usually maintain their own patent department, assignment via the legal person is immediately possible. The following organisations were included in the institute sectors either alone or in conjunction with others:

- the Max-Planck Society, MPG (including their transfer company Garching Instrumente),
- Fraunhofer Society, FhG,
- all national research centres,
- four joint industrial research association institutes (AIF),
- the Heinrich-Hertz Institute for Telecommunications from the "blue list".

The other "blue list" institutes and other AIF institutes have applied for fewer than 10 patents respectively since 1968 and - in order to save on compilation costs - were discarded as being of minor interest.

Consequently, all public and semi-public research centres within the Federal Republic of Germany have been included with the exception of certain so-called departmental research centres, which evidence such low patent activity that an enquiry on a name basis would prove to be statistically non-relevant.

With a demarcation of this kind, certain double allocations do occur, perhaps if a person with a "professor" title is represented in patent law by one of the institutes. This double assignment is however desirable because usually such people are both active university lecturers (and recruit their junior staff from universities) and have a business relationship with the respective institute. All patent documents from the Federal Republic of Germany not allocated to the university sector and not to the institute sector were assigned to the enterprise sector. The so-called individual inventors (inventors who are themselves patentees minus professors) have been left to the industrial sector although there are a few inventors from the hobby area and therefore outside the commercial context. The reason for this was that, in individual cases, many owners of medium-sized companies are indistinguishable from individual inventors and hence private persons (Täger 1989).

A parallel investigation relied on US patent documents (Olivastro & Narin 1989). The 28 sector technology model used was identical, likewise the timing of the inventions to the respective invention year. The findings in regard to the number of NPL citations and hence the science involvement are therefore bound to lead to similar results unless the interfering influences from the American or European patent system are so serious that they would lead to different estimates in regard to science involvement. In this case, they were indissociable from the specific patent system and assessment of science connection by the patent method is said to be system-dependent. However, such is not the case: there is a marked correlation between both patent systems in regard to the NPLM indicator and hence the science connection. Regression analysis between both distributions shows a very high positive correlation at the 0.1 per cent significance level ($R^2 = 0.61$). *Consequently, in the following analysis European patent data only have been used because the index concept clearly cannot be patent system dependent insofar as science involvement is concerned.*

Further two steps to warrant the validity of the set of indicators are the following:

- Comparison of the features of various alternatively possible indicators,
- expert assessment of 30 randomly selected patent documents and the relevant NPL references by specialists in the field of laser medicine.

The measurement concept and data availability permit the formation of two different indicators for measuring scientific involvement: the mean frequency of NPL references per patent document (NPLM indicator) or the percentage of patent documents with at least one NPL reference (NPLQ indicator). Each suffer from various drawbacks: The taking of means is questionable because citation frequency distribution is skewed and certain NPL references have to be excluded by extrapolation for inapplicability reasons. The NPLA indicator is free from these problems but is tantamount to a yes-no decision and assesses patent documents with cumulative NPL citations in the same way as those that may perhaps contain only one such citation. An empirical check was therefore conducted to see whether the assessment of the science relationship was dependent on the indicator chosen.

There was very good concordance in the series and in indicator values between both methods of measurement. All correlation coefficients carry a probability of error of well below 0.1 per cent. Such being the case it is immaterial which indicator is used. The result in practice will be unaffected by the choice made. However, it was evident that the NPLM indicator exhibits substantially greater variance and hence *measurements of science connection via mean NPL citation levels should be preferable.*

Specialists can assess reliably selected areas of technology only namely those in which they are competent. For this reason and in view of the immense complement of data, validation of the measurement concept was carried out by expert assessment only in the laser medicine field. Laser medicine is at the interface between the two sectors medicine and laser technology (see table 4-1).

From this technical area, 30 EPO patent documents were selected randomly of which 10 were characterised by the fact that they contained no NPL references, 10 more displayed hardly one NPL reference in their research report and the remaining 10 research reports were typified by the fact that they listed more than three NPL citations. These patent applications were assessed from various aspects in full text version by a subject specialist from the Fraunhofer Society (for details see Noyons et al. 1991).

Overall, the random sample from laser medicine clearly showed that the field is actually being evaluated in a quasi-scientific way. Specialists recognized this in the content of the patent documents and the inventors themselves behave 50 per cent as productive scientists. Practically half of the inventors (and of these over 80 per cent of authors) were not from the industrial sector. There is therefore clear evidence that the measurement concept using NPL indicators is a valid way for establishing science affiliation. It is naturally debatable as to whether the field of laser medicine can be extrapolated to technology as a whole. Conversely, however, such a time consuming validation by individual consideration in the grand style would be impractical.

4.3 FUNDAMENTAL CHARACTERISTICS OF SCIENCE INVOLVEMENT IN TECHNOLOGY

Nowadays, technical change is occurring internationally, especially in quasi-scientific fields of application. National differences in science dependence within a defined sub-area of technology are in fact unlikely. Yet papers have been published that assert precisely this and in particular that Europe is lagging behind the USA and now even Japan in the science and technology stakes. On many occasions, the spectre has been raised of technological shortcomings, this time in connection with science dependence, therefore in relation to the academic infrastructure of a country and its propensity for involvement in industrial technology development.

Thus, finally in 1989 based on a progressive series of investigations, Narin, Roosen & Olivastro (1989, p. 41) assert that a hierarchy of science dependence exists between American, European and Japanese US patents in the field of electronics. Thus, measured on the NPL indicator based on US patent data, American electronics appears to be substantially more science-intensive than West European, which appears to be clearly surpassed in turn by the extent of science involvement in Japan.

It therefore needs to be tested statistically whether the variance in science involvement between technical areas is as large as the variance within a technical area between countries. It appears that the variance in terms of technical areas is substantial, i.e., one to two orders of magnitude above the variance in country groups per technology area. *The science connection strongly differentiates between technical areas and yet tenuously between countries.* It is an internal feature of technology; no gaps exist between economies.

An entirely different approach to investigation of a possible technological deficiency in science involvement between international and German technology, for example, involves comparing the national indicators for science involvement in the Federal Republic of Germany with international ones. Figure 4-1 shows this comparison for the period 1985-86. The figure also witnesses an excellent correlation between the science involvement for Germany and that measured internationally (quadratic correlation coefficient of R^2=0.91 with an error probability of well below 0.1 per cent).

From the examples presented, it is again apparent that science involvement is an inherent characteristic of technical fields and that hardly any differences exist between countries. Science involvement certainly differs considerably between individual technical areas. Technological deficiencies between countries cannot be established in general and except for one or two individual cases, it does not matter whether the indicators are country-related or measured internationally.

The connections presented are in agreement with science and innovation theory but allegedly in contention with the most recently published results (Narin et al. 1989, p. 41). The apparent discrepancies fade away however, if it is remembered that the papers cited relate to US data. Research literature from the USPTO was fully automated in the 80s thereby considerably simplifying systematic access to non-patent literature ever since. This has probably prompted the growing tendency among US patent examiners

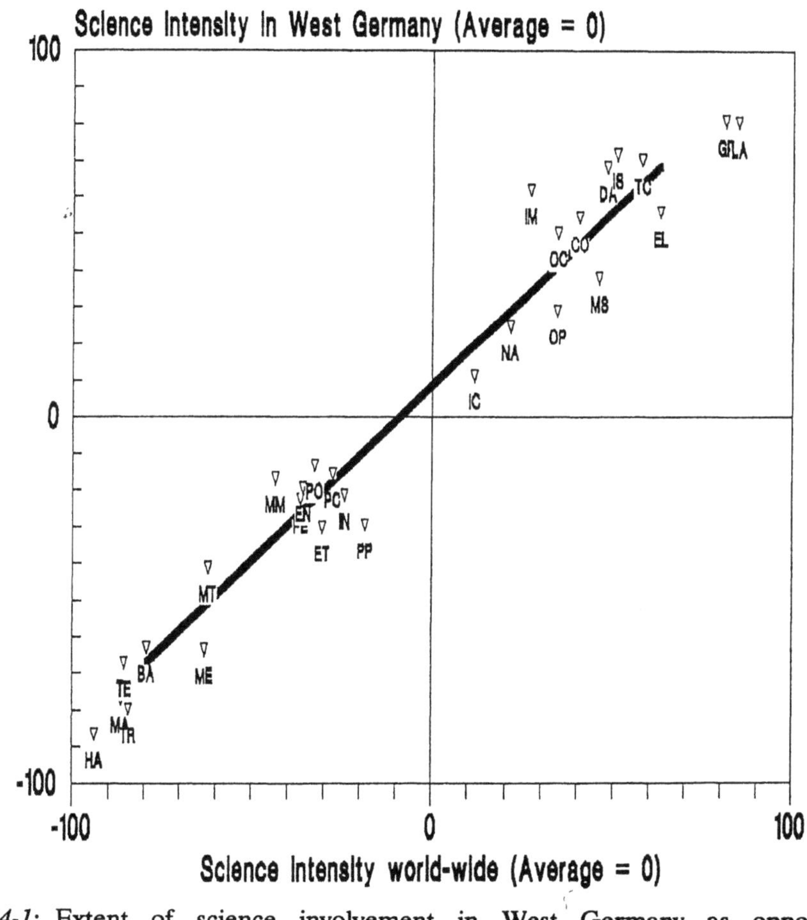

Figure 4-1: Extent of science involvement in West Germany as opposed to international technology based on 28 sub-areas over the years 1985-86 (measured on the NPLM indicator based on EPO data; sector codes see table 4-1).

to resort to non-patent literature. The electronisation of the American Patent Office ought at first sight to be neutral on a country by country basis. Unlike the European and Japanese context, the US patent system certainly does display the peculiarity that first publication occurs concomitantly with the grant and hence at an unpredictable time. According to the respective priority date, domestic patent applications are published on average one year ahead of foreign ones. Since from the source quoted, the time allocation was based on the award and hence publication date, the state of the art in regard to domestic US patents has been examined at a later point in time more advantageous to electronic research than the corresponding foreign papers for the same year of publication. On the other hand, if US data relate to priority year, then the greater differences in science involvement between countries disappear as suggested by the group of authors mentioned earlier (Olivastro & Narin, 1989, pp. 8 and 15). The alleged contradiction concerning country dependent science involvement is, thus, directly attributable to administrative patent law formalities.

It should also be noted that all non-patent literature for determining the science relationship was involved in the investigation mentioned. According to the approach adopted in this study, differing motivations of patent examiners are at issue and not all references to non-patent literature are indicative of a science connection (for details, see section 4- 2). In order to exclude irrelevant references, this data inquiry was subject to standardisation. Discrepancies in evaluation particularly on electronics may very well originate from the fact that in the United States references to company journals frequently occur (the IBM Disclosure Bulletin, for instance), that do hardly correspond to any science connection.

Similar references to European or Japanese company journals are unlikely owing to the fact that American examiners are probably not in a position to understand the corresponding languages of the country and the sources required are not available in their card index. This argument too confirms that the above-mentioned independence of the science connection from national technology also applies there despite the contradicting results published. In the meantime, Olivastro & Narin (1989, p. 23) subscribe to this finding when they note that "the number of references given to science dependends on the field itself (....). Neither the national origin of the inventions, nor the extent to which a country patents in a technological field, determine the number of references going to scientific articles".

Next to the issue of country gaps is the question of ongoing "scientification". The electronisation of the American Patent Office has most probably facilitated the technical access to scientific literature and thus leads to a rise in the overall number of references to non-patent literature. Even if all the said citations are justifiable, the deduction cannot be made that there are more references to science today for it is conceivable that such citations were also required previously but with limited working capacity and unsystematic documentation they simply could not be identified. Since easier access and genuine intensification of scientific involvement are therefore difficult to differentiate, for some time it has been assumed that it was the science connection which had increased substantially in the 80s (Narin et al. 1989, p. 41). The increase in the United States was apparently more marked than in Europe and Japan. The European Patent Office, founded in 1978, right from the outset was equipped in a more modern way than the large national patent offices of the post war period. Consequently, this office should not suffer from the phenomenon of increased citations through improved access. It is therefore interesting to check the time constant for science involvement as presented in EPO data.

From the standpoint of innovation research one should not assume that a technical field can change substantially in only 10 years that a marked increase of science-dependence is detectable. Scientification occurs in the context of new scientific paradigms and therefore at decade intervals. Even from the practical technology policy standpoint, the research system and the prevailing development approach therein is known to change only extremely slowly and state interventions both in the medium and short term in particular often remain ineffectual. Accordingly, there is practical value in ascertaining whether changes in the scientification of technology are feasible over short periods.

The international and West German indices (NPLM indicator) have therefore been investigated by regression analysis to see whether a significant change has occurred since 1979. Summarising for technology as a whole, the indicator for Germany for scientification has remained constant. Internationally, a vague and slight rise is noticeable (for an error probability of approximately 10 per cent). Apparently, despite there being a faint impression of country-specific differences, internationally a slight increase in scientification occurs, whilst in Germany there is no increase whatsoever.

If the individual technical fields themselves are considered then, in the main, the indices are found not to have changed over the last 10 years. Today, most sub-areas have the same high or low scientification as they had in the course of the 70s. This means that based on a generous 10 per cent error probability by reference German records still 23 out of 28 technical sectors, worldwide 17 out of 28 sectors have not undergone further scientification. The exceptions to this rule of thumb are summarized in table 4-2. For all that, poorly significant sub-areas have been included without which the figure for "scientificated" technical areas would have worked out even smaller.

Table 4-2 compares the chronological change in scientification based on worldwide and West German trends. No concordance between international and West German development is noticeable for above average science-based technical sectors. Over here, increased links with science apply only to electronics, while international metrology & and sensor systems plus genetic engineering & pharmaceuticals have developed with the corresponding dynamism. As far as optics, surface engineering (coating) and nutrition technology are concerned, there is only a poorly significant increase. Conversely, in the area of sub-average science-based technology sectors, international and German trends are proceeding largely in parallel. Low level scientification is occurring in machine tools, polymer processing (resins) and in engines & turbines. (International) mechanical engineering and (German) handling technology are close and therefore comparable sectors. The scientification of traffic technology which involves electronisation is worth noting.

Table 4-2: Chronological changes in science involvement between 1979 and 1986 (significance level for an error probability of less than 5 per cent; an asterisk denotes slightly significant at less than 10 per cent)

	International	West Germany
Above average Science-based	• Metrology & Sensors • Genetics & Pharmacy • Optics *) • Coating *) • Nutrition *)	• Electronics
Not or less science-based	• Machine tools • Mechanical machinery • Resins • Printing • Engines • Traffic	• Machine tools • Handling • Resins • Engines

The technical areas mentioned in which, diverging from the general trend, new or other dependencies on science research are emerging, are inherently interesting and merit more accurate analysis. To do this, however, there would need to be more intensive disaggregation, because technical conglomerates such as communications technology or sensor systems cannot be interpreted logically at this technologically unsophisticated level of analysis. Most probably the increased science involvement relates only to sub-areas in these sectors. A detailed in-depth investigation is however neither possible nor even aimed for within the present terms of reference of this paper.

An apparent increase in scientification in the 80s, as substantiated by US data, is certainly attributable to administrative procedures at the USPTO and in the European patent system only faintly perceptible and then in isolated technical areas. It cannot be excluded that older effects of scientification before 1979 could have manifested themselves in the late US patent grant data up to well into the 80s.

The common supposition of further scientification, therefore ever stronger scientific involvement in technical development which according to the above research results over the last decade at any rate needs to be modified considerably, also conflicts with the fact that by the end of the 80s some of the technical sectors appear to be less science-dependent than 10 years earlier. This partial regression of scientification is not conspicuous (the error probabilities are around 10 per cent or even higher). Nevertheless, for the sake of completeness, mention should be made of the technical areas which are tending to "descientify": worldwide and in Germany, these are electricity & nuclear engineering, (traditional) instrument making, data processing and information storage. Only according to the German indicator, the areas of textiles, medical engineering, polymer chemistry and sensor technology also fall into this category.

Also, from the innovation theory standpoint, it might be expected that apart from the scientification of certain technical areas, these, after maturity then start to "descientify", namely independence is attained within industrial development without the need for acquisition of any new scientific knowledge. Lieberman (1978, p. 13) argues that the link between the scientific and technological sector timewise is like a series of overlapping waves. Over a period of time, close connections form only to vanish again gradually. He was able to show by reference to two empirical case studies in the field of electronics that the electronics sub-areas he studied (microwave diodes and metal oxide semiconductor components) between 1965 and 1975 have severed their links with science. The sum total for technology as a whole is therefore obtained not so much in the form of increased science links over a period of time, but in details. Rather at the level below the 28 technical sectors defined in this paper, furtherances and recessions of technical development recur in relation to the science connection.

So which of these areas are in fact science-dependent? Failing a theoretically justifiable cut-off threshold, involvement of all indicator values lying above the average for technology as a whole should be involved in identifying science-intensive areas of technology. In so doing, there is an opportunity to standardise absolute indicator values at their respective average for technology as a whole so that *above average and below average science-intensive areas* can be clearly identified.

According to figure 4-1, the science connection is at its most pronounced in genetic engineering & pharmaceuticals on the one hand and laser technology on the other. These two sub-areas are *markedly science-dependent.* These are followed - a considerable way behind - by a series of technological areas belonging more or less to the broad area of micro-electronics. These are telecommunications, electronic components, information storage and data processing, image transmission as well as sensor technology. These *above average science-dependent technical areas* also include optics, surface technology, organic and also - just - inorganic chemistry and food technology.

Although, in theory, it is impossible to cite any threshold value between science-dependent and non science-dependent technology and although in figure 4-1 merely the average value for technology as a whole has been selected as the boundary line, very surprisingly the science relationship is not distributed over all possible indicator values rather there is a turning point below inorganic chemistry (IC).

Classified as slightly *sub-average (or not very) science-dependent* are 8 further fields, namely the field of polymer chemistry and resins, instrument building, printing technology, both electrical engineering and also electricity & nuclear engineering plus machine tools and process technology. After that, once again there is a slight turning point in the series so that the 7 remaining areas engines & turbines, biomedicine, building, textile technology, mechanical engineering, transport & traffic plus handling have to be qualified as science-independent.

The split between above and below average science-dependence plus the further inherent differentiation for strong and weak involvement is shown in another way in figure 4-2. Structurally, it is a cartographic representation of the link between technical areas. Sub-areas with many cross-references (multiple classifications) are depicted more closely as such without overlaps. For a clarification of the cartographic representation, see Engelsman & van Raan (1991) and section 11.4 in this volume.

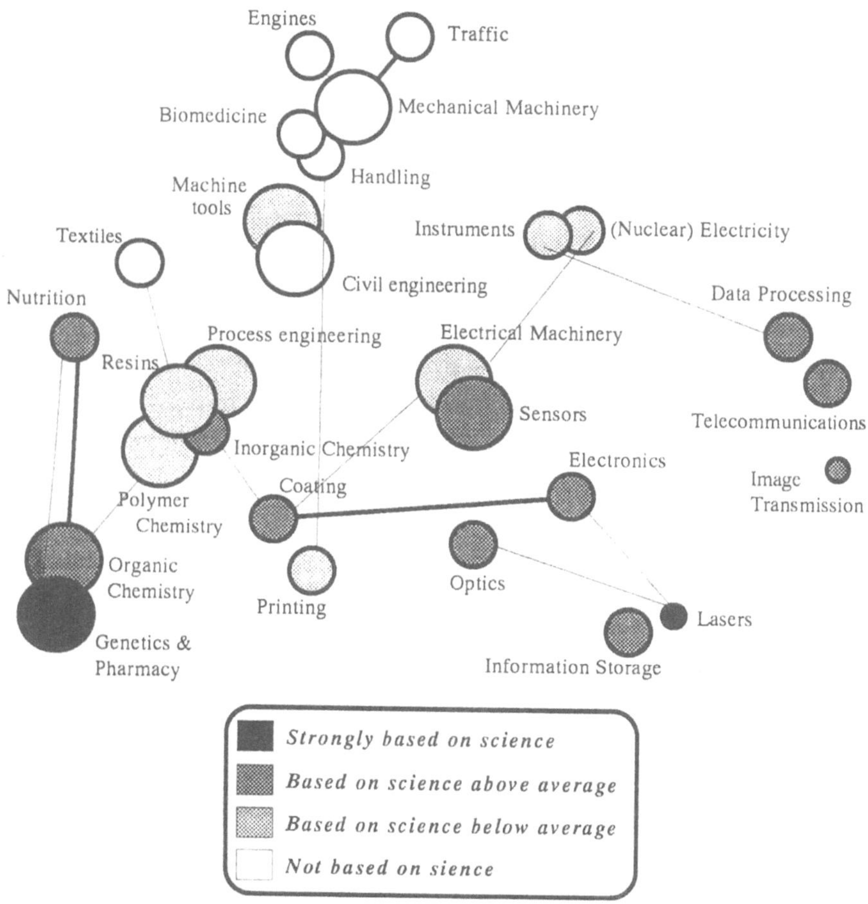

Figure 4-2: Cartographical depiction of the structure of technology in 28 areas and their science relationship (for clarification of the figure, compare section 11.4; the lines between technology fields denote closeness which is difficult to visualize in this two-dimensional figure otherwise).

Figure 4-2 very clearly shows that the science relationship is concentrated on the one hand in the larger area of micro- electronics assisted information technology and also in organic chemistry and its most recently impacts in genetic engineering and now also in food technology. From the cluster around more traditional chemistry, inorganic chemistry is an outrider owing essentially to developments in modern materials including high-temperature superconductive materials. The large area of mechanical engineering with its many offshoots can be classified as not dependent or barely dependent on advances in science.

There are, yet, some considerations to be taken into account: Technical inventions are allocated using the patent classification very precisely in relation to their narrow contribution to the furtherance of technology so that roughly all component developments, for the electronisation of the motor vehicle, say, are arranged in the immediately "relevant" area (in the case of the example) within electronics, but not within traffic. According to the technological sectoral model, biomedicine also has its mechanical aspects; pharmaceuticals, on the other hand, are categorised with genetic engineering in figure 4-2. To this extent, one can actually talk about a rough split in the science relationship between electronics and mechanical engineering, with chemistry somewhere in the middle and with its organic section (in the broadest sense) science-dependent, but its inorganic section (in the broadest sense) less affected by the latest scientific knowledge.

The above division into more or less science-dependent sub-areas of technology can now be compared with statements made in other publications. Essentially this refers to the investigation by Pavitt (1984) who arranged branches of the economy and therefore not technical sectors according to their science involvement. According to Pavitt, electrical engineering and electronics plus chemistry are science-based (op. cit., p. 260). In the area of electronics, this can be confirmed unrestrictedly by this newly available data material while the rather traditional areas of electrical engineering including energy generation, according to the inquiries in this chapter, fall outside the science-intensive sector. Chemistry, bio-engineering, pharmaceuticals and organic chemistry are included in the data presented here and evaluated identically by Pavitt and can thus be graded as quasi-scientific in any case. Resins, polymer raw materials, chemical process technology and other areas however are not noticeably particularly science-dependent, at least by presentday standards. Differences between the evaluation presented here and those of Pavitt derive essentially from the fact that industrial sectors cannot be so finely differentiated as technical sectors. Multi-national concerns, particularly, with widely divergent business activities are technologically heterogeneous. Consequently, assessment of the science relationship in terms of branches of the economy is bound to be less convincing than one in terms of technical disciplines.

A more obvious contradiction with Pavitt is obtained in the field of *food technology* which the author cited grades as being non science-dependent, whereas the latest findings state otherwise. It is pointless agonising over whether this is the outcome of the current scientification of food technology which seven years ago was not yet perceptible or whether evaluation of the "food and nutrition" branch of industry leads to conclusions other than the technical definitions. This is one of the most remarkable results of this study particularly, since food technology is one of the few areas in which there was statistically significant evidence during the 80s of increased links to science (compare table 4-2).

Techniques for the manufacture of foods have a thousand year long tradition behind them (bread making, beer brewing, wine production). A science-based food technology has, however, existed only since the work by Justus von Liebig and Louis Pasteur in the middle of the last century. The food branch still tends to be a very low research and development disburser at least within the German economy (East and West) which is pronouncedly small or medium scale. Owing to the structure of industry, the high costs of basic research and additional measures required for links to scientific development - that clearly dynamically precedes it - one is tempted to assume that this technical area is undergoing dramatic restructuration for which the branch at any rate appears to be inadequately prepared. This state of affairs has become the object of other inquiries (Team of Authors 1991).

4.4 INTERNATIONAL DEVELOPMENT PANORAMA

The basic features of the science connection with technology as presented in the previous section, provide scant evidence to support the theory of the scientification of technology and therefore an ever increasing dependence of technology on science as time progresses. A completely different (and slightly confusing!) matter is whether the increase in invention activity in science-intensive areas over the last 10 years was greater than in the less or completely non science-dependent. A further consideration is whether the R&D centres in a country are focussing more on science-dependent and in others more on science-independent sub-areas of technology.

Figure 4-3 substantiates the special dynamic development of science-intensive technical areas without any doubt: *the more closely a technical development area is interlinked with scientific knowledge, the stronger its growth in the 80s* (this applies for a quadratic regression coefficient of 0.42 and an error probability of less than 0.1 per cent). Correlations however fail to reveal any reasons why the reverse relationship can also be propounded: *The technical sub-areas with the greatest growth in the 80s are more science-dependent than the others.*

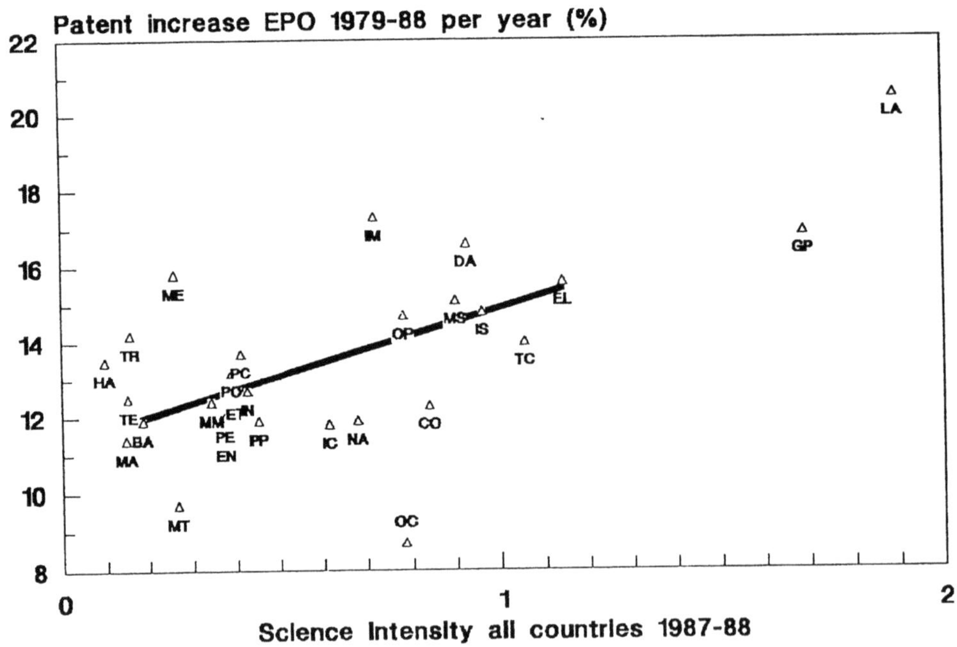

Figure 4-3: World-wide increase in invention activity as a function of science involvement (mean annual increases 1979-88 measured as a percentage of the 87-88 patent applications; the alphabetical list with the identifications for the technical areas is located in table 4-1).

Patentable technical developments in areas with especially high science involvement, such as laser technology for example, are showing an annual growth rate in excess of 20 per cent per annum; genetic engineering and pharmaceuticals are registering a growth of almost 17 per cent per annum. However, the fact should not be overlooked that the advent of patents at the European Patent Office is bound to register universally high growth rates owing to the increasing attractiveness of this patent property rights area; viewed in this light, the growths say in mechanical engineering (with a low science involvement) at approximately 12 per cent are still remarkable. These values are certainly noticeably lower than the growth rates for the more science-dependent development lines. The ten areas with the greatest dynamic development are the following:

- Laser technology (20.5%)
- Image transmission (17.3%),
- Genetic engineering & pharmaceuticals (16.9%),
- Data processing (16.6%),
- Biomedicine (15.8%),
- Electronic components (15.6%),
- Metrology & sensors (15.1%),
- Information storage (14.8%),
- Optics (14.7%), and
- Traffic technology (14.2%).

In the face of such growth rates, it is thoroughly surprising that an area such as telecommunications (TC; growth rate 14%) is perceived as being only moderately dynamic. This should give pause for thought to those strategists who entertained strong hopes, with tentative liberalisation within the European telecommunications market in the mid 80s, of accelerated advances in technical development. Organic chemistry (OC) in particular appears to be stagnating; however this may have something to do with the fact that company R&D centres hitherto active in this area are switching their strategic objective to genetic engineering and hence the traces of structural change are obvious in this field.

The driving forces behind these dynamics need not necessarily be evenly distributed over all participating countries. Lately, on the scene in technology policy there has been much talk about an ever-widening technological gap between Europe on the one hand and Japan and United States on the other. It is therefore helpful to competitive and technology policy debates, to analyse the dynamics in international development at least for major OECD national economies.

Even where all technical sectors in all countries are rated roughly equal in terms of science dependence, it still cannot be said that the national development activities in individual countries are concentrated *to the same degree* on the more science-dependent areas of technology. In section 4.1 a quantifier has been introduced in order to describe this state of affairs (formula (2)), which weighs the actual invention activities with their science index and thus supplies information about *weighted activity in science-intensive technology.* Figure 4-4 gives the weighted activities in the 1987-88 invention years for several countries in comparison to the world average.

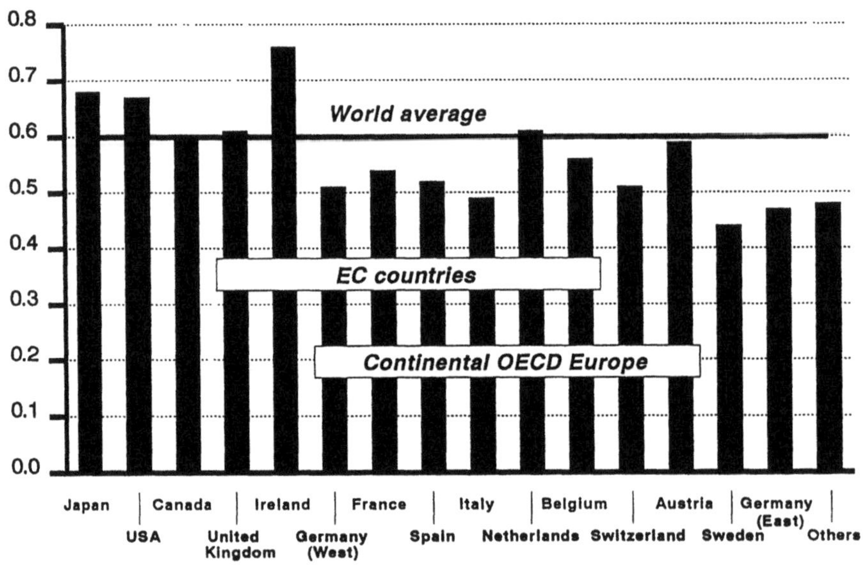

Figure 4-4: Weighted activity in science-intensive technology per country 1987-88 and in comparison to the world average.

The United States and Japan appear to be stronger in science-related areas of technology than the Europeans - with the exception of Great Britain and the Netherlands which are world-average rated. The two Germanies, France and other countries (in terms of numbers, the following determining the average of this group: Italy, Scandinavia, the two Alpine Republics, Belgium, and Canada) have corresponding below average specialisation in science-dependent areas of technology.

Figure 4-4 highlights this situation but says nothing about whether the momentary strategic direction of the countries considered has a stable pattern over the years.

Consequently, later on there is a country by country analysis of development trends over the last 10 years.

Further information is important when assessing national trends. Not every economy pursues a broad-based course of technical development; this is an issue of size. Specialisation in a few selected areas in technical development for medium-sized countries on the one hand is dictated by restricted R&D budgets, whilst in international competition from the point of view of revealed comparative advantage it is profitable insofar as in selected technological niches small and medium sized enterprises can stand up to the commercial Goliaths. The deviation of technological specialisation from the world average generally increases with decreasing resources potentially allocable by national economies to R&D (Grupp & Legler 1990).

Technology-wise, Great Britain and the United States are the most widely diversified of the countries considered. In 3rd place comes the Federal Republic of Germany. Other countries - all with smaller R&D resources - are intensively pursuing a niche strategy. This is certainly true of France who among the three major European countries is adopting a much more selective approach. Japan is certainly more highly specialised than the small countries in selected technical areas, although measured in terms of R&D resources it is the second largest OECD country. This should not be taken as a criticism but merely as a statement of fact - all the economic competitive indicators point to an extremely successful Japanese strategy (cf. Grupp & Legler 1990). Because of these distinctions in subsequent discussion of national trends, the corresponding breadth of national specialisation is a factor to be taken into consideration.

Since 1979 the *United States of America* has clearly had above-average activity in science-intensive technology. Up to the mid-80s, the involvement of American technology in fundamental science appears to have further intensified, while the present position reflects a clear cut tendency towards the world average. This strong position in science-based technology is in no way attributable to a niche strategy directed at the correspondingly strongly linked sub-areas of technology. The USA have the most broadly-based activity in all areas - even in those that are not very science-dependent.

The importance of the United States in science-based sectors is undeniable - there are even many qualitative indications to support this. This country has for many decades has been the prime mover for fast introduction of new radical knowledge into the industrial innovation scene. Primarily in conjunction with data processing and genetic engineering, the USA, its universities and companies are constantly being applauded for their pioneering efforts. Such qualitative evaluations can therefore be fully confirmed by the present quantitative instrumentarium.

The same position as for the USA is also evident in *Japan*, namely a propensity for science-intensive technology above the world average. Unlike America, its upward tread away from the world average appears to have increased - not decreased - if appearances are not deceptive. Despite the similarity in Japanese and American leadership, the driving forces are fundamentally different. Japan is active in a much more selective way in chosen technical areas. Overall Japanese technology ostensibly owes its extreme science-dependence not to a broad foundation but to concentration of its endeavours in the area of electronics in the widest sense. To this extent, outwardly identical science-intensity indeces need to be evaluated totally differently from the situation with the United States.

Initial observations show that over the last 15 years, Japan has not metamorphosed with an across-the-board technological power (Grupp & Legler 1990, pp. 72 onwards). This is not intended as a criticism: whether Japan as a technological late starter in the post-war period has chosen the "right" mix between breadth and depth must remain open to debate. Perhaps the key for competitive success in future - since science-dependent technology seems to be financially more expensive than conventional technology - will lie in the selective commissioning of innovation-oriented activities, based on division of labour. Such considerations can only be touched on here; they cannot be deduced from observation of dynamic development over the last decade. Conventional, technological policy and structural peculiarities in Japan tend to play an important part in assessment of future development which cannot be discussed comprehensively at this point.

In the cross comparison of nations (figure 4-4) the (former) *Federal Republic of Germany* appears to be less oriented towards science-based sectors. Such was not always the case. Germany has moved away from the world trend. Over the last 10 years, in redirecting its innovation-oriented activities to fields requiring massive science support, the Federal Republic has not advanced to the same extent as expected from international development (sustained by the United States and Japan, see above). At the beginning of the decade, the Federal Republic was averagely science-dependent; by the end of the decade it is stronger in the less science-dependent fields (compare figure 4-10 in the next section). The tendency of the Federal Republic towards less science-dependent technology is the result of sound anchorage in machine construction (with all the modern trappings), motor vehicle construction and conventional chemistry.

Specialisation in invention activity in the Federal Republic is almost as broadly based as in the USA; in any case, it is clearly more intensely breadth-oriented than Japan. The comparatively low science involvement is therefore not the result of deliberate avoidance of science-dependent technical areas. It originates rather from a broad orientation of development activities not specifically towards science-based technology.

This recognizable gravitation towards the less science-based technical sectors also derives from another consideration. The more strongly the company and research centres in the Federal Republic are devoted to one area, the weaker its growth turns out to be (or vice versa). *The Federal Republic is inclined towards engagement in the less dynamic technical areas (quadratic correlation coefficient 0.16 for an error of probability of less than 5 per cent).* This statistical connection is the key to understanding the less pronounced science-involvement in the Federal Republic. As it devotes itself to technical advance in the less dynamic fields which in turn are also less involved with science (see figure 4-4), a corresponding dissociation is bound to occur. This characterisation of Federal German technical development cannot be negatively rated from data submissions alone. Given that engagement in science-dependent technical developments ties up larger research capacities, it can be assumed to be comparatively expensive.

Major R&D participators in the Federal Republic of Germany possibly avoid such costs and adopt the necessary new technology in some way other than by their own endeavours because all the technological competition indicators for the Federal Republic show excellent performance on world markets (cf. recently Grupp & Legler 1990). She has registered exceptional economic success in R&D-intensive areas. The advancement of broad, medium science-dependent technologies has been her own particular path to trade success.

Certainly, present-day competitor successes on international markets for technological goods relate to work done in previous years. To this extent, the *present-day level* should not be viewed critically only because of lesser participation in science-dependent technology, but rather in terms of unfavourable time trend. The critical trend may only be reflected in future market results and then - because it has been spotted late - be difficult to rectify. At the micro level of individual companies, the growing tendency to discard internal R&D in science-dependent technology primarily in the field of electronics, means that supply, licensing, purchasing and joint enterprises will become increasingly important. This is what is happening in group strategies abroad instead of internal science links. Whether this specifically - say for the customer-specific manufacture of micro-electronic components - would be a viable avenue for German industry, is often disputed as is also the trendy movement towards the less dynamic areas of technology which means that industrial property rights are being taken out for the less dynamic product markets of the future which will probably result in a correspondingly smaller market potential.

France, like the Federal Republic of Germany has a below average profile in relation to the world average science involvement. The gap between national and international development certainly does not appear to be so wide; France is following the international lead half-way between Germany and the trend. This development has occurred for a clear-cut limitation on certain sub-areas of technological development. In comparison to other European countries, development of industrial technology is not so broadly based. On the other hand, the selectivity of French technology development is not so great as in Japan; here, too, France is steering a middle course.

Technological activities in *Great Britain* in regard to science over the ten year time period considered follow the world average exactly. Within Europe, the United Kingdom apparently raises the issue of duality of this commitment, namely economic and organisational endeavours aimed at closely allying scientific fundamental research and industrial development on the one hand and the difficult achievement of market-appropriate, profitable innovative advances on the other. Great Britain's world-average commitment to a good structure in the field of scientific research is largely achieved. For the British position, a prominent research level in organic chemistry, pharmaceuticals and peripheral areas whilst keeping pace with sub-areas of electronics is likewise characteristic. Within Europe, the British technological activities in science-related fields are well above continental Europe.

This can also be thoroughly disadvantageous for a medium-sized industrial country if it fails to equip rival companies equally well with qualified personnel and other resources so that they can break through in all production areas. In fact, it has the appearance that at least if the international technological competitive position of Great Britain is studied (as in Grupp & Legler 1990), that this breadth is unacceptable on the market in world economy terms and hence the approach adopted is to be regarded as critical. For technological cooperation in Europe, the strong links between British technology and science surpassed only by the United States and Japan nevertheless represent a handsome dowry.

With increasing *internationalisation of important technological markets,* foreign patent rights are generally gaining in importance - not only in the Federal Republic where strong concern was expressed. Foreigners are also forging ahead on American, other European and even on the Japanese market. This characterises only world-wide cooperation in research-intensive areas at the end of the 20th century and cannot be perceived as negatively as some individual voices in the technological policy debate would have us believe. So the high proportion of foreigners on the patent rights market of the Federal Republic of Germany also merely mirror generally increased international competition and in the final analysis represent the high regard in which technology is placed by the German home market. Based on data material now available, at most there may be concern that the technological dominance of foreign companies in key areas of the anticipated information society, for instance in the field of data processing, information storage, or image transmission, is so great. Figure 4-5 shows that sub-areas with high foreign dominance are also markedly dependent on science. It is therefore not hard to predict that closure of the corresponding gaps will be neither quick nor easy. Without adequate links to relevant basic research on the one hand and without high standards of development on the science side on the other the race for these areas of industrial development can hardly be successful.

It is a common prejudice of certain policy monitors that the West German research system is very self preoccupied and has not reacted flexibly enough to international innovations. Other observers are on the other hand convinced that it is strongly dependent to an unacceptable level, on foreign development endeavours - at least in selected strategic areas. It was *not* checked within this investigation what amount of foreign cooperation would be appropriate. However, it is possible to supply quantitative data as to whether foreign cooperation is directed intensively towards science-dependent - and hence long term strategic - or other areas.

The following considerations certainly only deal with those types of cooperation in the field of research and development that occur in patents. This is only a fraction, possibly a small at that, of all cooperative endeavours. In this treatise, formalized and to a lesser extent informal relationships have been covered. One example: a joint patent application by two companies or the materialisation of inventors from an institution other than that of the applicant document a legally binding cooperative undertaking entered into previously by all parties in concert. Such cooperations lay the foundations for the subsequent analysis. However, in addition, there is a less obvious practice called "implicit patent pooling" which exists in certain branches of industry and in the case of technologically competing companies, via secret agreements, avoids mutual legal actions for patent infringements (Nelson 1988, p. 318).

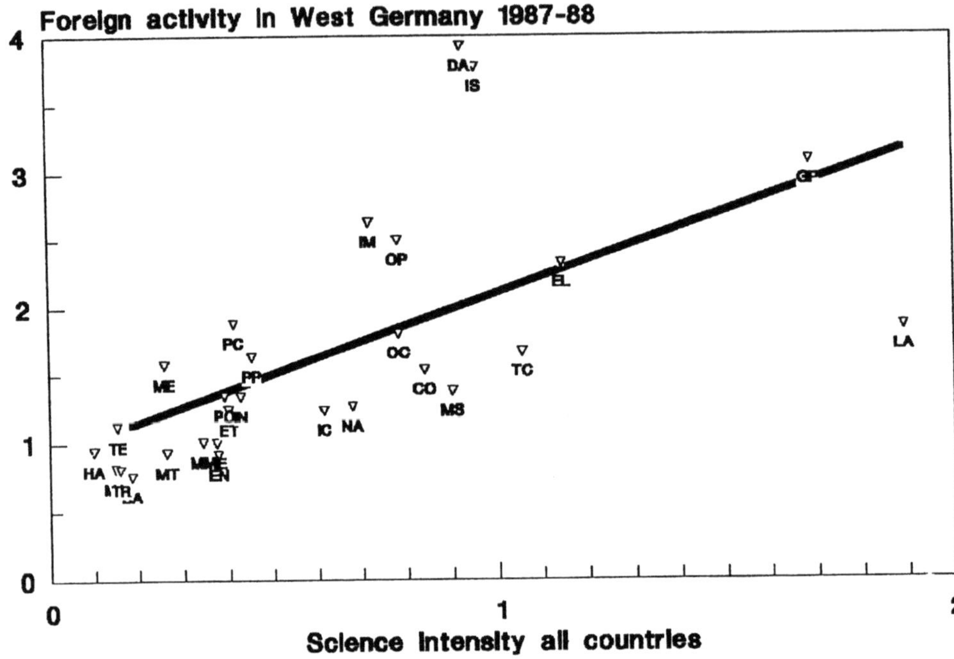

Figure 4-5: Innovation-oriented activities of foreigners in West Germany 1987-88 and their science involvement (German activity is equal to one. The alphabetical list with the identifications for the technical areas is located in table 4-1).

Besides this, there are entirely different forms of cooperation that certainly do not occur in patent documents. Hagedoorn & Schakenraad (1990, p. 27) in a more comprehensive study of the whole multiplicity certainly only cite four recurrent motives as the basis for cooperation between companies:

- the chance to locate and supply to new markets,
- shortening of the innovation process,
- technological competence of the partner and
- the prospect of being able better to identify and use technical facilities jointly.

Financial bottlenecks often cited in literature as the reason for cooperation, according to information supplied by the said authors appear to be comparatively unimportant. So it actually seems as though direct international cooperative efforts tend to evolve for strategic and technological rather than other reasons. Consequently, the assumption not confirmable here and perhaps not wide of the mark is that a study of cooperative behaviour via patents documents the main motives for cooperation over technological strategy even if numerically only a small proportion of the many cooperative relationships is available on record. The following explanations support this limiting assumption.

From the data used, an admittedly weak dual dependence on science and overseas can be detected. West German R&D participants cooperate to the extent of 5 to 10 per cent with foreign individuals or facilities and this covers practically all fields of technology. Cooperation in the field of chemistry, even genetic engineering, is higher, yet - and this is remarkable - in most micro-electronic sub-areas even in laser technology, it is lower. In the field of micro-electronics, where foreign patent rights on the territory of the Federal Republic are strongly dominant, it is found that this is not accompanied by a marked strategy for allying foreign competitors.

Cooperative behaviour in micro-electronics might have been expected, from all the assurances about a European concerted action, to be definitely higher in this field. This is all the tougher because in the Federal Republic there is a backlog as she has had no marked strength in this area and also therefore, because it is reckoned to be amongst the most dynamic sub-areas of technology. Amongst the cooperating partners of West German companies, the German speaking zones predominate as before. At the top there are Switzerland and Austria, also Sweden in selected fields. Put another way, the high cooperation potential in other countries is nowhere near exhausted. It would be very interesting to investigate what reasons specifically are important for intensive cooperative behaviour with German-speaking countries which are not EEC members. Answers to this area of inquiry cannot be given within the terms of reference of this paper.

4.5 SECTORAL TABLEAU OF SCIENCE INVOLVEMENT IN GERMANY

It is by no means the prime concern of universities of all kinds to conduct research and development to achieve patent-relevant results. The tasks of teaching and fundamental research tend to be paramount. Nevertheless, nigh on 2,000 patent applications from university staff are submitted per annum, corresponding to over 4 per cent of all West German national applications. How does one arrive at this figure? In the Federal Republic of Germany, by reference to 1987, universities generated approximately 13 per cent of the national R&D (BMFT 1990). Discounting non patent-relevant sciences, i.e., social sciences and humanities, law, the arts and other areas, even so roughly 7 per cent of patent relevant R&D can be attributed to the universities (Irvine et al. 1990). Against this yardstick, the patent contribution of university lecturers at 4 per cent is exceptionally high, because in the natural and engineering sciences also broad areas of fundamental research are not patentable. *West German universities are important centres for technology production.*

Nelson (1988) who studied the United States in this regard, even goes so far as to insist that since the 19th century, universities have allegedly become an even more important part of the "capitalist engine". They are a depository and a treasure trove both for public sciences and also for technological know-how. Apparently in the United States, university science and engineering has grown alongside science-based industries. Thus, chemistry became an academic field of activity roughly at the same time as chemists began to play a part in industry. Academic research and teaching in the field of electrical engineering appear to have emerged precisely at the same time as the electrical machine industry grew in the United States. Nelson (op. cit., p. 320) however also comments critically, that there are findings to suggest that academic research in chemistry and electrical engineering has ceased to be an important resource for technological knowledge in industry. At the moment, academic biology and computer science are tending to be the major university partners for cooperation over newly developed ideas and processes in industry.

If this line of thought were extrapolated for the Federal Republic of Germany, then it would have to be checked to what extent universities direct their research activities to particularly science-dependent fields. For if it is true that in the course of the decade ever more academic research areas have been imparting fundamental knowledge into industry, then the present-day science-dependent areas would coincide with the present-day prevailing activity areas across all technical sectors.

Figure 4-6 clarifies this relationship by reference to the sectoral model of technology used here. Calculated ideal values for the activity profile of technically-oriented university research deduced from science involvement are quoted in descending series. The actual percentages of patent applications from university professors are contrasted against all West German applications. The ideal distribution was estimated by regression analysis. It suggests that the actual patent activity within universities correlates with science involvement to an exceptionally high level (error probability less than 0.1 per cent).

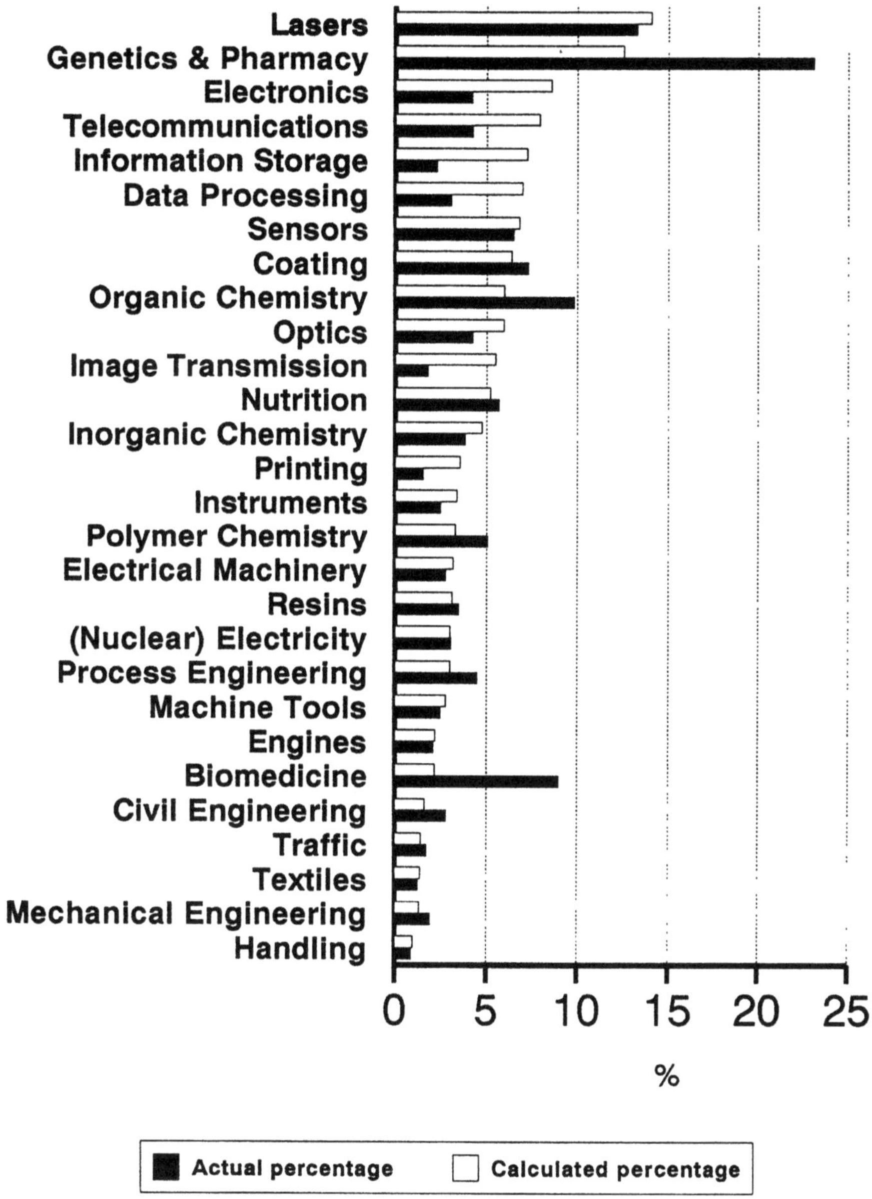

Figure 4-6: Sectoral German participation of universities in science-dependent technology (as a percentage of all West German national patent applications between 1987-88).

However figure 4-6 also details deviations from the ideal value: in the quasi-medicinal areas (genetic engineering & pharmaceuticals plus biomedical engineering) and in organic chemistry, university activities are more markedly defined than the value anticipated from science involvement suggests. In the overall area of micro-electronics and information research (electronics, telecommunications, information storage, data processing, image transmission) on the other hand they are below average. Participation in the field of laser technology, also in sensor technology, surface technology and in the nutritional field is ideal.

This corresponds to a qualitative assessment by experts who - cum grano salis - note that the German chemical industry knows precisely which chemical faculties are important for it, while the electrical engineering industry bumbles around between physical and electrical engineering faculties and those for information science and applied mathematics. The argument of those involved applies not only to research but also to recruitment of qualified young talent plus vice-versa the importation of the most modern course content by industrial researchers into the academic teaching environment which according to this estimate, is more sharply defined for chemistry than for electrical engineering. Liebermann (1978) alludes to the importance of the bridging function between respectively a "group of industrial scientists" and the "university authors of technical publications", whereby scientific knowledge from very fundamental research is seldom transferred straight into technical applications, but rather by intercession of the groups of people mentioned (op. cit., pp. 12 onwards).

If thought is given to the present situation in the Federal Republic, then the defective links between industry and universities in the micro-electronics field appears to go some way towards explaining the corresponding deficit. This is also evident from the fact that when it comes to the new initiatives in this area (JESSI et al.) apart from industry not so much universities but other public research institutes play a part. If the state of affairs supported here by only some quantitative observations is true, then a structural and institutional problem appears to be present which undoubtedly cannot be eliminated in the short term.

The public research institutes sector amongst the multitude of institutions that in principle could be included, comprises only those with continuous and non-evanescent patent activity (see section 4.2). The patent activity of the research institutes altogether amounts to some 500 patent applications per annum or a quarter of the university sector. This is not surprising for the R&D budget of institutes is smaller than those of the universities (approximately 10 per cent of the national budget in West Germany), and the research profile also comprises many unpatentable areas (social sciences and humanities, high energy physics, museums and many others). The scope of fundamental research according to respective institution is as high as at the universities. Nevertheless, the universities are regarded as being comparatively patent-intensive. The focal points of innovation-oriented activities of institutes can be deduced from figure 4-7.

Figure 4-7, as in the case of universities, shows national patent percentages arranged in descending order plus the scope for science involvement per sub-area. The calculated ideal values have been estimated by regression analysis and ought to be obtainable in reality if the group of institutes were to direct their patent-oriented R&D activities precisely according to the importance of science involvement. This is inconceivable even in extreme cases, because individual institutions within the public institute sector take on completely different projects. Some institutes have a degree of liberty within the research sphere, whilst others have their R&D activity more or less circumscribed by overall state control (to some extent, state or commission-oriented tasks as in space or nuclear research, for example). Overall, the general bias of institutes towards science-intensive technology generation is as high as at the universities (correlation having an error probability of less than 0.1 per cent). In figure 4-7, activities in laser technology are particularly prominent. In the case of the less science-dependent technical areas, the field of electricity is noteworthy, as it takes in nuclear engineering as a whole according to the definition of this technical area. By this the commission-oriented activity of the two nuclear research centres is visible.

As in the university sector, once again considerable restraint in the field of micro-electronics and information technology is apparent (see, for example, the large discrepancy between actual percentages and calculated ideal values in the fields of telecommunications, information storage, data processing and image transmission). Only in the field of electronic components i.e., the as yet non system-specific section of information technology, do the institutes take on their anticipated role.

The criticism may be advanced that other than in the university sector, it is not logical to put diverse institutions such as the Max-Planck Society, Fraunhofer Society and the nuclear research centres together as an institutional sector. This could be countered by the statement that under the terms of reference of the present general inquiry it could not be divided arbitrarily into technical sectors and also into institutional sectors. Therefore, a microanalysis is presented subsequently purely for the five largest patent producers in the Federal Republic.

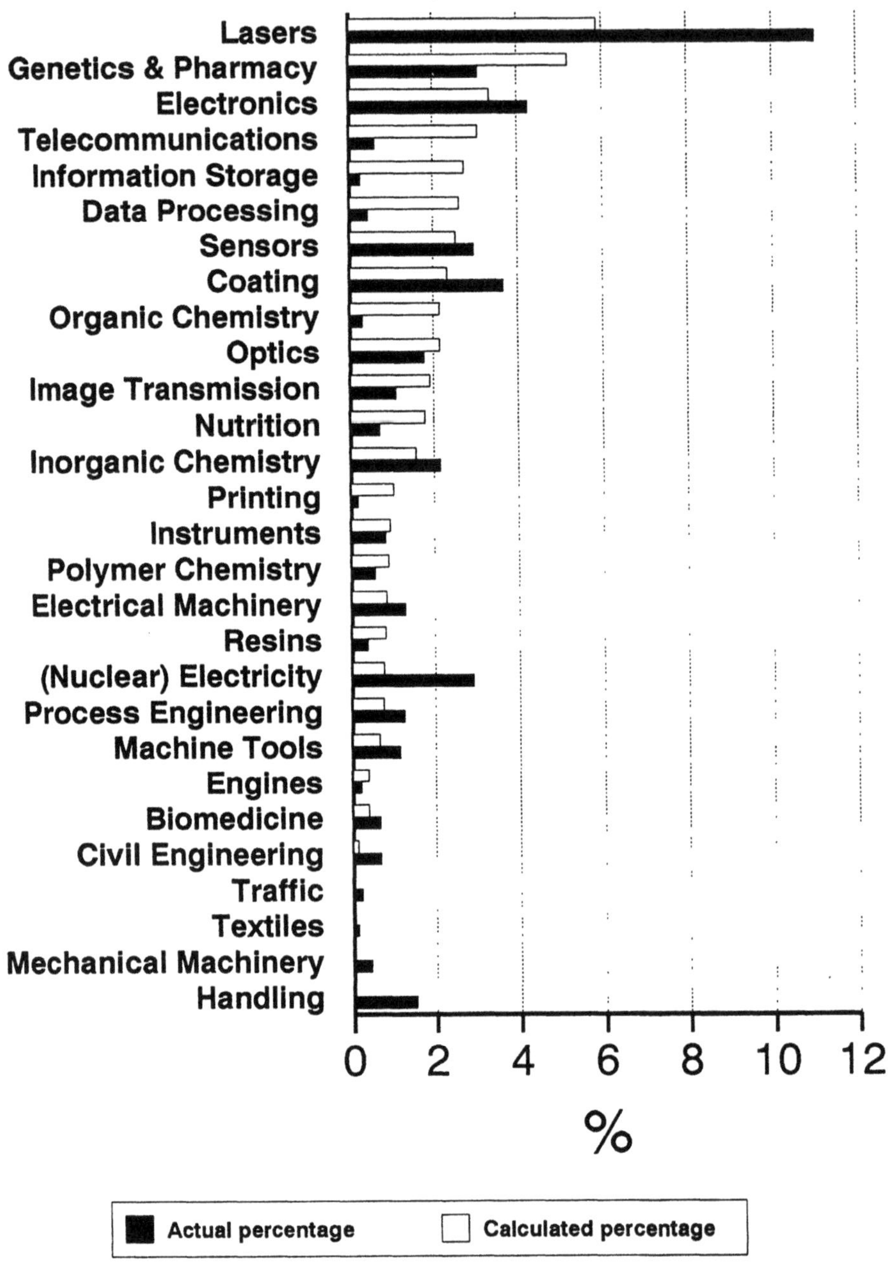

Figure 4-7: Sectoral German participation of institutes in science-dependent technology (as a percentage of all West German patent applications in the year 1987-88)

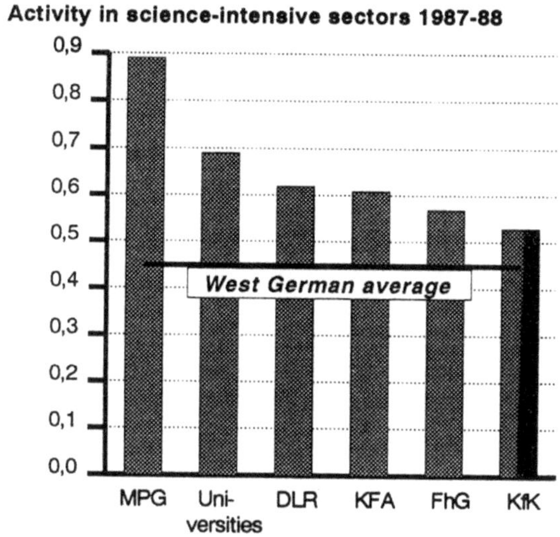

Figure 4-8: Science involvement in technology generated by individual institutes (1985-86)

Figure 4-8 compares the five largest institutes and all universities in terms of their weighted science involvement. The selected institutes and research organisations *without exception* are above the German average defined in the main by the enterprises. Peak science involvement rating goes to the Max-Planck organisation which registers a level even higher than that of the universities. This ties in with the marked activities in the fields of lasers, electronics, sensor technology, and genetic engineering, which without exception are science-based.

Somewhat below the level of the universities come the DLR (German Aerospace Research Establishment) and KFA (Nuclear Research Centre, Jülich). In the case of the DLR, this has less to do with the commitment to aerospace and rather more to that of sensor technology, laser technology, and satellite communications. KFA engages in the science-dependent areas of bio and genetic engineering plus surface technology. FhG (Fraunhofer Gesellschaft) like KfK (Nuclear Research Centre, Karlsruhe) are similarly above-averagely science-intensive but do not rank as high as other institutes. In the case of FhG, science-intensive activities in the field of electronics and science-impoverished projects in the field of mechanical engineering (handling, machine tools) plus materials are more or less evenly balanced. In this connection, it should certainly be noted that robotics, a major FhG focal point, is classified in its entirety in the science-impoverished area of handling. Even KfK displays a marked research component in the field of mechanical engineering which according to the data presented here is not regarded as science-intensive.

Most patent-oriented activities are ascribable to business enterprises. This is not unusual for major industrial countries (Grupp & Schnoering 1991, chapter 11). Overall, German companies are responsible for almost 95 per cent of patent-oriented activity. Their innovation-oriented behaviour leaves its mark on Germany. Only where universities and institutes are particularly strong (laser and genetic engineering) can the latter achieve worthwhile percentages in terms of patent events in addition to industry.

How can the sectoral interplay in the Western German Federal states be assessed?

It has already become clear in section 4.1, that drawing of boundaries in a highly differentiated research system such as that in the former Federal Republic of Germany is not easy and also that no straightforward arrangement can be applied. Even if Lundgreen et al. (1986, pp. 17 onwards) and Hack (1988, pp. 64 onwards) create an academic, state and industrial research "triangle", it is impossible to state that parts of "state" research cannot also be academic as Weingart & Winterhagen (1984, p. 58) find for MPG, which they assign to academic research. According to this latter viewpoint, FhG does not come under state research but industrial research. The West German Ministry for Research and Technology (BMFT) on the other hand includes under the heading "extra-university research" all state research facilities and the non profit making institutes so financed (BMFT 1990). This study pursues the idea of a triangle and calls the three sectors (universities, institutes and business enterprises) in terms of their main supporters in order to avoid labels such as "academic" or "publicly financed".

Since universities and institutes are more than averagely active in science-based technology (to the extent they produce mainly patentable results), industry must be somewhat more strongly anchored in the low-science areas. The science involvement figures do not reveal anything about the pursuit of opposed R&D aims by the sponsors of the corresponding facilities. It was already rather obvious that the targets assigned according to institutional sector and within the sector of the institute deviate very markedly from one another. It is therefore worthwhile distributing the targets between individual R&D sponsors and looking for similarities or mutual complementarity. Figure 4-9 attempts to present the result of a number of mutual regression analyses on patent profiles in terms of 28 technical areas. In so doing, only the five major research organisations and research centres could be taken into account as the others produce only few statistically non-exploitable patent applications (at any rate relative to a technologically differentiated model such as the one here).

Structurally, figure 4-9 is a cartographic representation of the "closeness" between institutions. Two institutional sectors with highly correlated activities (shares of patents per any one technology sector) are depicted more closely as those with no or inverse correlation. The two axes in figure 4-9 do not have a natural meaning as the multi-dimensional scaling technique used here produces arbitrary axes. However, it happened to be the case that academia turned out to be "West" and industry R&D is "East". Roughly, the different size of the R&D expenditures per institution is represented by smaller and larger boxes.

It is obvious from figure 4-9, that many areas in the map are empty, i. e., the contrasted institutions present activity profiles that do not correlate altogether. This supports the view that the Federal German R&D system is highly differentiated and characterised by a multitude of independently acting R&D supporters. Frequently, the activity targets are largely self-determined, whilst in other cases, they do follow the overall state projects (say in the field of nuclear or space research). Industrial companies place their emphases according to economic dictates and in regard to future markets targeted. If the multitude of unoccupied spots on the R&D landscape is anything to go by, there are marked similarities and likewise marked differences.

The Max Planck organisation and universities present a closely-correlated activity profile. This is another reflection of the generally established fact that MPG research is largely identical to that of the universities. However, it is interesting that this state of affairs also applies to the (small) area of particular research activity devoted to technical development. The two nuclear research centres also present a highly concordant profile. The business sector, on the other hand, has no similarity with the other R&D sponsors in terms of the structure of its R&D activity. Even in the case of the FhG, which amongst the institutes works most intensively with the enterprises, there is no complete coordination between the respective working areas. This is immediately understandable bearing in mind the fact that companies permanently run their own R&D departments in specialised fields, and are not overall channelled into contract research by external agencies. Business enterprise specialisation runs absolutely counter to the directions taken by the universities and the Max-Planck organisation. In the field of patent-relevant R&D activities, the academic can be distinguished very clearly from the industrial sector. This suggests that cooperation has marked complementary traits. Space and aviation R&D (DLR) is remote from most other portfolios of activity.

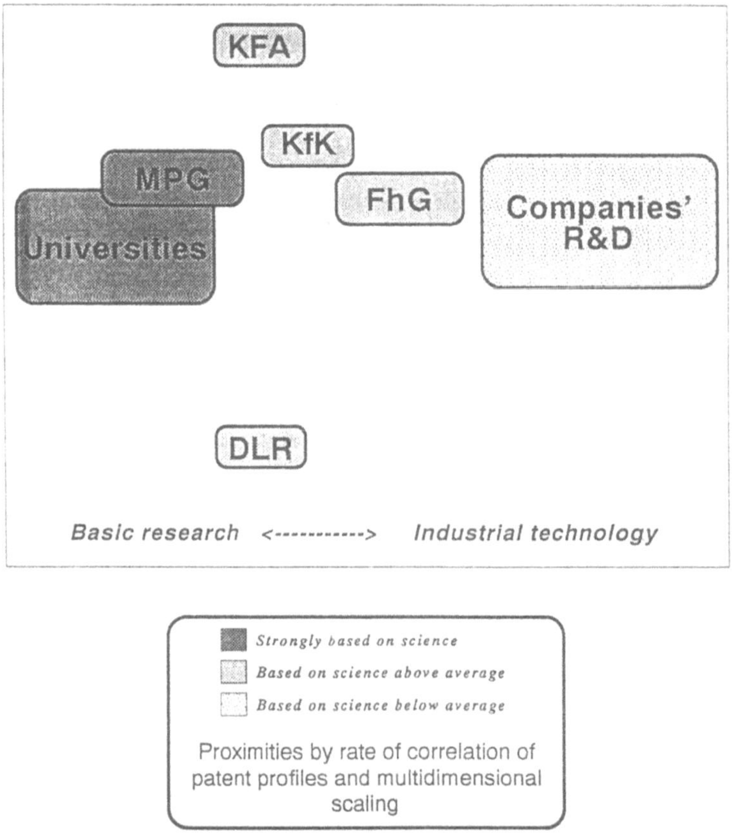

Figure 4-9: Correlation between the activity targets of various R&D sectors in the Federal Republic of Germany between 1985-86 in regard to their technical innovations (patent applications) according to correlation coefficients by multidimensional scaling.

As far as patent law in the *GDR* to date is concerned, other relationships apply that are determined by socialist ownership attitudes. Consequently, national patent applications at the former Office for Inventions and Patents of the GDR (AfEP) can hardly be compared with those in the West. The main difference lies in the fact that the conventional patent as conceived in the West has had hardly any importance other than the so called "economic patent". Economic patents have accordingly been regarded as a feature of socialism. Inventions were supposed to be used by society generally, because the competitive ethic was alien to the socialist culture and a fundamental allying of interests exists between the various social groups and strata (Albrecht et al. 1991, p. 4). The economic patent has brought the GDR company no additional profit expectations. Furthermore, the company only had very limited decisionary competence, since the objectives of production were predetermined by the drawing up of state plans. The strategic interests of the company were therefore distorted (op. cit.), and it cannot be assumed that the scope, specialist distribution and commercial importance of patent activity in the GDR could be compared to the corresponding values for West Germany.

Consequently, another approach was adopted for this section of the inquiry: patent activity abroad in Western Europe was used to reflect the science involvement of GDR technology. At the German and European patent office, for example, in 1988 there were around 600 patent applications from the GDR. Western Europe and the Federal Republic in particular belonged to the preferred export market for the GDR outside the Council for Mutual Economic Aid (CMEA). As might be supposed, only a tenth of the patent applications for West Germany were also recorded at the USPTO. The basis for dealing with GDR-invented West European patent applications is an expedient for circumventing all patent law peculiarities and contriving direct comparability to the rest of this inquiry.

Figure 4-10 compares the change in patent activity with time in science-intensive technology in the old and new Federal states of Germany. The world average is likewise given for the purposes of comparison. In the nations cross-referencing, the former GDR seems less oriented towards science-based sectors. This was consistently so over the period considered. The deviation from world level has remained roughly constant, i.e., over the last 10 years, the GDR has reoriented its innovative activities to some degree towards the science-intensive as can also be noted with the world average - but only to the lower level respectively. This is interesting in comparison to West Germany, because the East-West gulf in this area has therefore been narrowed accordingly. The former Federal Republic (see section 4-4) has tended to dissociate itself from the world trend and has progressively lowered the science basis which originally ran at world level. It is interesting to compare this with the GDR which in world economy terms had to operate under entirely different conditions.

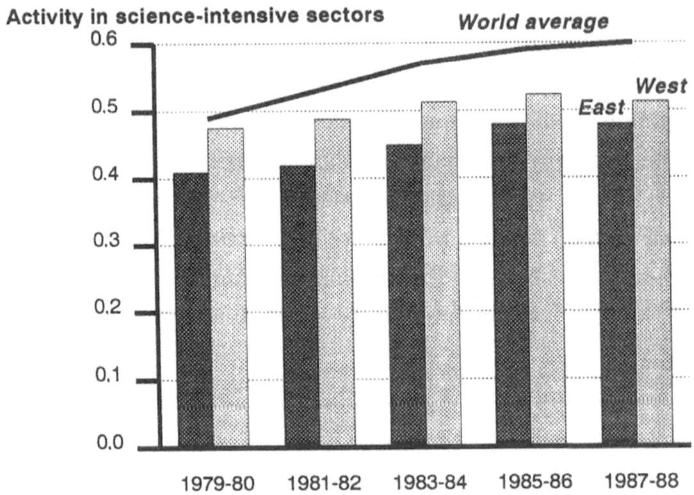

Figure 4-10: Weighted activity in science-intensive technology in the former GDR from 1979 to 1988 in comparison to West Germany.

In figure 4-11, the specialisation of Western and Eastern Federal states for the years 1987-88 are compared. In so doing, the so called specialisation indicator (Grupp 1991) is used which expresses the respective positive and negative deviations from world average (in terms of positive or negative values). The technical areas are arranged in figure 4-11 in terms of decreasing science intensity and therefore precisely as in figures 4-6 and 4-7. Figure 4-11 very clearly shows that the GDR exhibits less marked R&D activities in science-intensive areas. One exception is the field of sensor technology. Here, the GDR displays above-average activity and at the same time this is an area with a high science involvement. The situation is similar only for the areas of surface technology, optics, and food technology.

Other areas with above average patent activities such as printing technology, process technology, textiles, machine tools, and handling point to a rather low science-dependence. In all areas with extremely high values for science involvement such as data processing, telecommunications technology, image transmission, and electronics, below-average R&D activities are recorded.

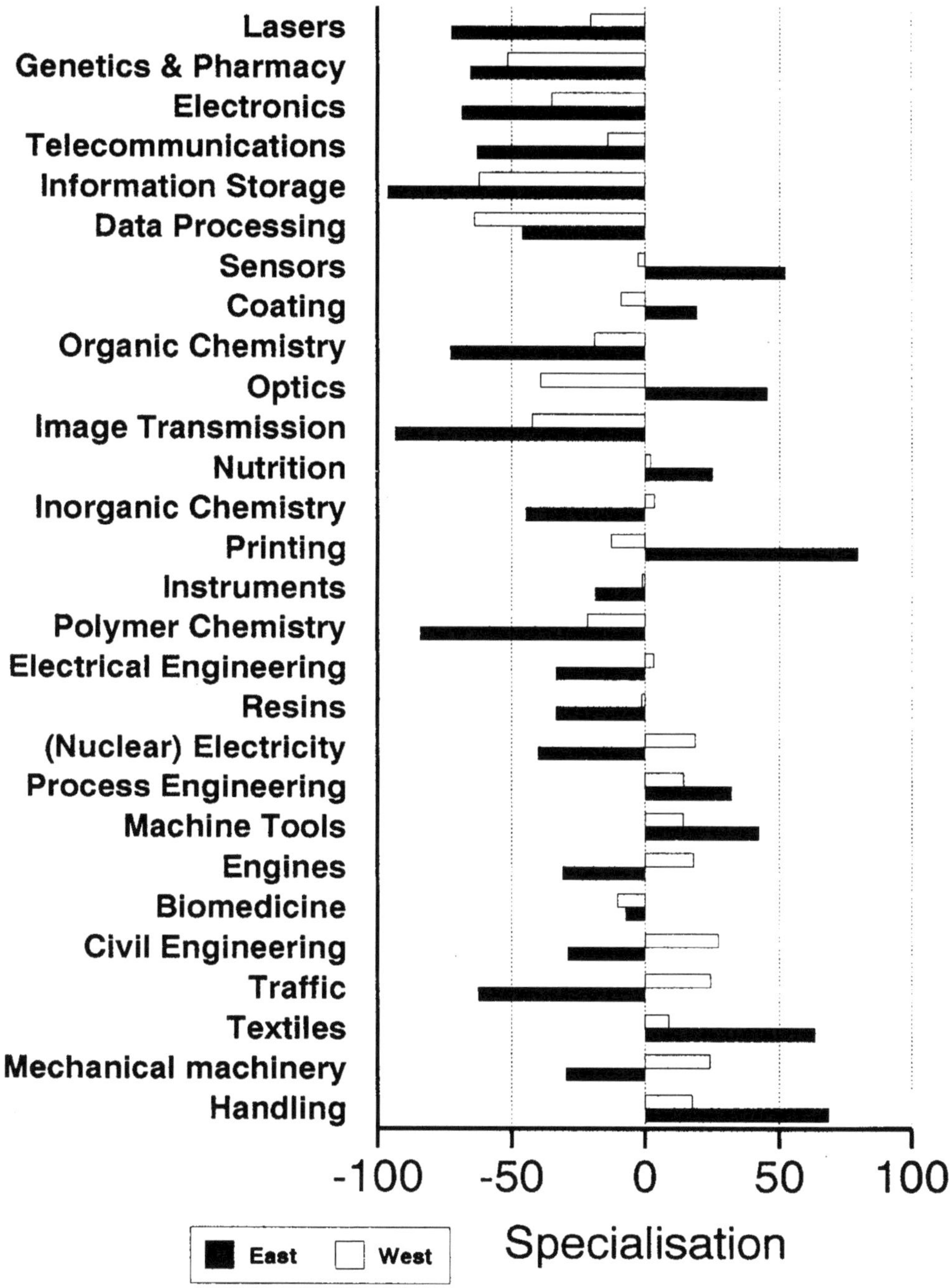

Figure 4-11: Comparison of the specialisation of the old and new Federal states of Germany in technology (standard deviation of patent activities from the world trend for 1987-88)

Despite many differences say in the area of traffic technology or in building plus optics, striking areas of agreement (this is true of the entire range of micro-electronics) are evident between Western and Eastern Federal states. Actually, there is a good correlation between both specialisation profiles in technology (for an error probability of less than 5 per cent). Despite totally different economic circumstances for 40 years (Grupp & Legler 1990), *in technology there are obvious still broad areas of coincidence between West and East Germany.*

The rather less evident science involvement of GDR technology arises in just the same way as in the former Federal Republic of Germany. It cannot be termed conspicuous in the international comparison. In addition to the decidedly quasi-scientific R&D systems of USA and Great Britain and in the case of the substantial research services of Japan in more recent time, they can be termed as is the case for continental Europe as thoroughly typical. France and other European countries are at a similar level. In the case of the expanded and GDR size-related decidedly large science operation, it is however surprising that technical development is unable to derive more benefit from its assets and has remained comparatively science-impoverished.

4.6 CONCLUSIONS

In this inquiry, patentable technology as a whole has been divided into 28 areas across the board. In so doing, certain compromises have had to be made. From the results presented, it was obvious at several points that such a classification was bound to be too coarse for many interesting items. Some of the similarities mentioned between countries or between German sectors would disappear if technology could have been more finely classified. Other observations would however then no longer be discernible, because the statistical quality of the then smaller figures is no longer good enough to detect significant trends.

A particularly enduring repercussion of the classification model relates to definition of science dependence itself and therefore to the main variable investigated here. The 28 technical areas are all far from homogeneous. Science-dependent and low-science sub-areas are mixed together and offset one another. Figure 4-12 tries to provide a silhouette for the science involvement schematically by introducing the indices of two research specialities: laser medicine and polyimides. From this, it is apparent that the science involvement index is dependent on the extent to which technology aggregates. Both special fields are highly science-dependent but belong to broader sectors with little influence from basic science.

In general, the sector index for science involvement is average if the sector divides into low science and high science halves. The science-based fraction in this technology sector will not be reflected in the mean index for science involvement, so that one is inclined to conclude that another sector would be more science-intensive just because the latter was demarcated in such a way that it contained *only* science-intensive specialities.

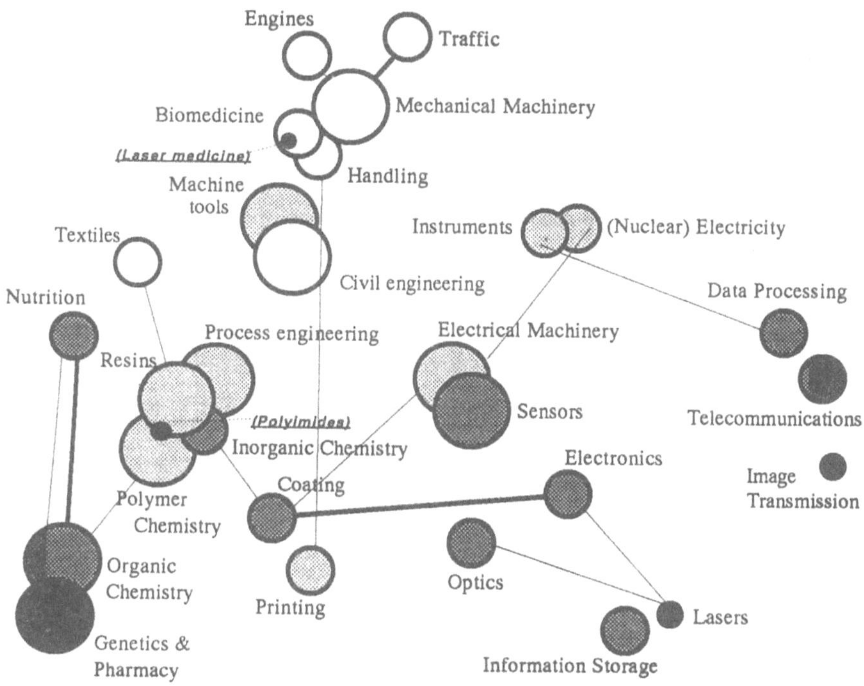

Figure 4-12: Silhouettes for the function of mean science involvement in the technology classification in the case of laser medicine (within the biomedial sector) and polyimides (within two chemical sectors). The remainder of the figure correspondents to figure 4-2.

The example in figure 4-12 has not been selected entirely arbitrarily; certainly, it was not possible within the terms of reference of this pioneering inquiry to go into the science-intensive "pcaks" systematically in all the 28 sectors. In the case of laser medicine and polyimides however it could be shown that the silhouette drawn fits in principle (cf. chapter 9).

The relatively high aggregation level needs to be treated with care when drawing conclusions in *research policy*. In the investigation, not individual, clearly definable technical developments have been identified, but only technical sectors. For all that, even at this level there is a need to take action in some cases: the food and nutrition industry would hardly be in a position with its present compliment of R&D resources to keep pace with the scientification of its technological bases. Even for the entire area of information technology and micro-electronics, it is true that substantial industrial R&D capacities in Germany need to link up even better with the public research system in order to manage future knowledge-intensive projects. Even in this broad area, it would be a valuable task, for the future, to "survey" selected technical development lines finely in order to obtain more specific ideas about the silhouetted structure of such areas in regard to science-involvement. Such a quest even in modern information technology would probably be of greater practical importance to research policy.

Also, based on the present investigations, certain recommendations can be addressed to *industrial research management*. To the enterprises, the search for particularly science-dependent sub-areas in the entrepreneurial businesses is much more important than in the area of state R&D politics. For the company, it is becoming increasingly true that links to science need to be mastered in areas outside the traditional businesses. The question of industry's ability to link up directly with science-intensive areas is thus unavoidable. The internal R&D structures in business enterprise will also be affected by the dynamic importance of science-based technology. The liaison capability already mentioned should not be apportioned arbitrarily to business areas whose development projects must conform to the immediate needs of current markets. Liaison capability is often required in areas in which the company still has no market share. Therefore, central R&D departments must concern themselves with the problem and the opportunities for science-intensive technology.

*

Financial support by the German Federal Ministry for Research and Technology (BMFT) for the research work underlying this chapter is gratefully acknowledged.

REFERENCES

Albrecht, E., Dohnert, O., Schneider, M. and Bourcevet, H. (1991), 'DDR-Forschung im internationalen Vergleich unter Zugrundelegung der Patentstatistik', Study for the Science Counsil, Fachbereich Angewandte Informatik und Wissenschaftsforschung, Humboldt-Universität, Berlin.

Baré, R. (1981), 'Results of a Statistical Study of the References Cited in the Search Reports Established by the EPO (January 1981)', *World Patent Information*, 3, pp. 56-60.

Bianchi-Streit, M., Blackburn, N., Budde, R., Reitz, H., Sagnell, B., Schmied, H. and Schorr, B. (1984), 'Economic Utility Resulting from CERN Contracts', CERN 84-14, CERN, Geneva.

BMFT (ed.) (1990), 'Faktenbericht 1990 zum Bundesbericht Forschung', BMFT, Bonn.

Carpenter, M. P., Cooper, M. and Narin, F. (1980), 'Linkage between Basic Research Literature and Patents', *Research Management*, 13, pp. 30-35.

Claus, P. and Higham, P. A. (1982), 'Study of Citations Given in Search Reports of International Patent Applications Published under the Patent Cooperation Treaty', *World Patent Information*, 4, pp. 105-109.

Cohausz, H. B. (1978), 'Das Europäische Patent, Die europäische und internationale Anmeldung', Erweiterter Sonderdruck aus *VDI-Nachrichten*, Nr. 8,9,10+11, Düsseldorf.

Coward, H. R. and Franklin, J. J. (1989) 'Identifying the Science-Technology Interface: Matching Patent Data to a Bibliometric Model', *Science, Technology, & Human Values*, 14, pp. 50-77.

Dasgupta, P.S. and David, P.A. (1991),'The New Economics of Science-Resource Allocation and the Institutions of Science', Center for Economic Policy Research Working Paper, Stanford University, Stanford.

Dosi, G. (1988), 'Sources, procedures and microeconomic effects of innovation', *Journal of Economic Literature*, 26, pp. 1120-71.

Engelsman, E. C. and van Raan, A. F. J. (1991), 'Mapping of Technology', Report CWTS-91-01, Centre for Science and Technology Studies (CWTS), Leiden.

Engelsmann, E. C. and van Raan, A. F. J. (1990), 'The Netherlands in modern technology: A patent-based assessment', Report to Dutch Ministry for Economic Affairs, Den Haag.

Freeman, Ch. (1982), '*The Economics of Industrial Innovation*, 2nd edition, Pinter Publishers, London.

Gibbons, M. and Johnston, R. (1974), 'The roles of science in technological innovations', *Research Policy*, 3, pp. 220-242.

Greif, S. (1982),
'Angebot und Nachfrage nach Patentinformationen - Die Informationsfunktion von Patenten', Göttingen.

Grupp, H. (1991), 'The measurement of technical performance I. Framework of R&D intensity, technometric, and patent indicators', to be published.

Grupp, H. and Legler, H. (1990),
'Innovationspotential und Hochtechnologie - Technologische Position Deutschlands im internationalen Wettbewerb - Aktualisierung 1989/90', Report to BMFT, Fraunhofer Institute for Systems and Innovations Research, Niedersächsisches Institut für Wirtschaftsforschung and Gesellschaft für Wirtschaftsförderung und Marktplanung mbH, FhG-ISI, Karlsruhe and Hanover.

Grupp, H. and Schmoch, U. (1991),
'Wissenschaftsbindung der Technik', Final Report to BMFT, FhG-ISI, Karlsruhe.

Grupp, H. and Schnöring, Th. (eds.) (1990 and 1991),
Forschung und Entwicklung für die Telekommunikation - Internationaler Vergleich mit zehn Ländern, Vol. I and II, Springer Publishers, Berlin, Heidelberg, New York.

Grupp, H., Reiß, Th. and Schmoch, U. (1990)
'Knowledge Interface of Technology and of Science: Developing Tools for Strategic R&D Management', Report to Stiftung Volkswagenwerk, FhG-ISI, Karlsruhe.

Hack, L. (1988),
Vor Vollendung der Tatsachen. Die Rolle von Wissenschaft und Technologie in der dritten Phase der Industriellen Revolution, Fischer, Frankfurt.

Hack, L. and Hack, I. (1985), *Die Wirklichkeit, die Wissen schafft. Zum wechselseitigen Begründungsverhältnis von 'Verwissenschaftlichung der Industrie' und 'Industrialisierung der Wissenschaft'*, Campus, Frankfurt.

Hagedoorn, J. and Schakenraad, J. (1990),
'Inter-firm partnerships and co-operative strategies in core technologies', in: Freeman, Soete (eds.), *New Explorations in the Economics of Technical Change*, Pinter Publishers, London.

Holland, D. (1988),
'Wechselwirkung von ökonomischer Dynamik und Theoriendynamik - Dargestellt am Beispiel der Entwicklung und Anwendung hochpolymerer Materialien', Dissertation B (Habilitationsschrift, Humboldt-Universität, Berlin.

Irvine, J., Martin, B. R. and Isard, Ph. A. (1990),
Investing in the Future, Edward Elgar Publishers, Aldershof.

Lewison, G. and Cunningham, P. (1989), 'The Use of Bibliometrics in the Evaluation of Community Biotechnology Research Programmes', in: van Raan, Nederhof, Moed (1989), pp. 99-127.

Liebermann, M. B. (1978),
'A Literature Citation Study of Science-Technology Coupling in Electronics', *Proceedings of the IEEE*, 66, pp. 5-13.

Lundgreen, P., Horn, B., Krohn, W., Küppers, G. and Paslack, R. (1986),
Staatliche Forschung in Deutschland 1870-1980, Campus, Frankfurt.

Marquis, D. G. and Allen, T. J. (1966), 'Communication patterns in applied technology', *Amer. Psychologist*, 21, pp. 1051-1060.

Marx, K. (circa 1862), 'Zur Kritik der politischen Ökonomie', MEGA edition II. 3.6 from 1982, Berlin.

Narin, F. and Noma, E. (1985), 'Is Technology Becoming Science?', *Scientometrics*, 7, pp. 369-381.

Narin, F., Rosen, M. and Olivastro, D. (1989), 'Patent Citation Analysis: New Validation Studies and Linkage Statistics', in: van Raan, Nederhof, Moed (1989), pp. 35-47.

Nelson, R. (1988), 'Institutions supporting technical change in the United States', in: Dosi, Freeman, Nelson, Silverberg, Soete (eds.), *Technical Change and Economic Theory*, Pinter Publishers, London.

Noyons, E. C. M., van Raan, A. F. J., Grupp, H. and Schmoch U., (1992), 'The Science and Technology Interface: Inventor-Author Relations in Laser Medicine Research', to be published.

Olivastro, D. and Narin, F. (1989), *'Definition of Fields of Close Cooperation between Science and Technology'*, Report submitted to BMFT, CHI Research / Computer Horizons, Inc., Haddon Heights.

Pavitt, K. (1984), 'Sectoral patterns of technical change: towards a taxonomy and a theory', *Research Policy*, 13, pp. 343-373.

van Raan, A. F. J., Nederhof, A. J. and Moed, H. F. (1989), *Science and Technology Indicators*, DSWO Press, Leiden.

Rosenberg, N. (1976), 'Marx as a Student of Technology', *Monthly Review: an indepentend Socialist magazine*, 28, pp. 56-77.

Rosenberg, N. (1982), *Inside the Black Box: Technology and Economics*, Cambrigde University Press, Cambridge.

Scherer, F. M. (1982), 'Inter-Industry technology flows in the United States', *Research Policy*, 11, pp. 227-245.

Schmoch, U. and Kirsch, N. (1990), 'Analyse der Raumfahrtforschung und ihrer technischen Ausstrahlungseffekte mit Hilfe von Patentindikatoren', Report to DARA, FhG-ISI, Karlsruhe.

Schmoch, U., Grupp, H., Mannsbart, W. and Schwitalla, B. (1988), *Technikprognosen mit Patentindikatoren*, Verlag TÜV Rheinland, Cologne.

Smith, A. (1776), *An Inquiry into the Nature and Causes of the Wealth of Nations*, Dent edition from 1910.

Stillger, J. (1987),
'Eigenschaften von Offline-Datenbanken mit Patentdokumenten', *Nachr. Dok.*, 38, pp. 257-262.

Team of Authors (1991),
'Erster Technikreport', Study for the Office of Technology Assessment of the German Federal Parliament (TAB), FhG-ISI, Karlsruhe.

Täger, U. Ch. (1989),
'Probleme des deutschen Patentwesens im Hinblick auf die Innovationsaktivitäten der Wirtschaft', Final Report to BMWi, ifo, Munich.

van Vianen, B. G., Moed, H. F. and van Raan, A. F. J. (1990),
'An Exploration of the Science Base of Recent Technology', *Research Policy*, 19, pp. 61-81.

Weingart, P. and Winterhager, M. (1984), *Die Vermessung der Forschung*, Campus, Frankfurt.

PART II

CHANGES IN NATIONAL SCIENCE AND INNOVATION SYSTEMS

5 LINKING SCIENCE AND INNOVATION IN EAST GERMANY - A FIASCO?

Edo Albrecht

5.1 A MACROVIEW ON R&D IN THE FORMER GERMAN DEMOCRATIC REPUBLIC

In chapter 1 some remarks on symptoms of crisis and stagnation in the development of the world economy in the seventies can be found. In the 1970s the conditions of economic growth changed also in the socialist countries. The economic growth was in the 60s and 70s mainly achieved in an extensive way, i.e. the expenditure of resources increased more rapidly than the national income (Albrecht 1988). The cardinal problem was to find a way to turn to a funds-saving type of economic growth, which basically means to increase the national income while the expenditure of resources is being reduced.

The most important precondition for that was the development and use of new technologies. But the necessity to switch over to a new technological and economic paradigm was realized by most of the socialist countries either to late or it was not realized at all.

This is actually amazing. The Marxist political economy gives a largely consistent explanation of the phenomenon of long waves. In addition to that there were other starting points which were not used: Malthus who in the last century formulated the law of the decreasing produce of soil was contradicted from the Marxist point of view in stating that this law was valid only partially, i.e. on the condition that the methods of production do not change. If, however, new production methods were introduced then it would be possible to increase the produce of the soil. Freeman (1984) draws attention to the principal theoretical content of this conclusion.

In the GDR there was at least a first sign for the realization of new production methods. With regard to economic policy, the following demands have been made:

- Production has to grow while the consumption of energy and material absolutely decreases. The reduction of the specific consumption of energy and material must be faster than production is growing.
- Labour productivity has to increase more rapidly than production. The growth of labour productivity has increasingly to result in an absolute saving of labour.
- Labour productivity has to increase more rapidly than the value of fixed assets.
- Profits from exports of new products and of those with an improved quality have to grow more rapidly than the expenditure on science and technology.

Striving for these economic demands, it was pointed out that economic growth could be ensured in the long run only if new technologies were introduced on a large scale, i.e. high and key technologies were wanted. And, indeed, central state planning orientated towards

- intensified application of modern computer technology, particularly for the computer-aided designing and projecting as well as computer-aided manufacturing (CAD/CAM) and the automation of production, information and service processes including routine processes, in management and administration, particularly by the large-scale introduction of personal and office computers, automated word processing and transmission, exemplary solutions to local networks and industrial visual display texts as well as by the creation of the software required,
- development of universally automated production lines for a flexible production by a broad application of micro-electronics, robotics, and complex rationalisation as well as the provision of highly productive machines and equipment, highly effective technologies for machining and processing as well as of a productive measurement technology and testing technique for increasing labour productivity in all branches of the national economy on the basis of a fundamental increase in the technological level,
- increased utilization of biotechnology, particularly of genetic engineering, immune, enzyme and cell culture technologies as well as of the bioreactor and processing techniques for the low-cost production of high-quality products of the pharmaceutical industry as well as of the food industry competitive on the world market, for the development of an effective merchandizing, for the development of effective techniques in the vegetal and animal production as well as for the production of fodder rich in protein and the development of energy sources (Directives 1986, p. 22 and pp.32).

These demands and state planing objectives have had a far-reaching influence on the link of science and industry and especially on the organization of R&D (see below).

The scientific potential of the GDR in terms of institutionalisation is made up of the following four major sectors:
a) Research and development centres attached to companies;
b) Government-affiliated scientific units;
c) Institutes of the Academy of Sciences;
d) Universities and other schools of higher education.

If the definitions of the so-called "Frascati Manual" are obeyed, then the former GDR spending on research and development amounted to 9,672 million Marks in 1989, among them 7,438 million in industry, 1,664 million in non-university institutions, and 571 million Marks in universities and colleges. The 1989 figure represented an increase by 27.1 percent on the 1981 total of 7,608 million Marks. The R&D outlays of the company sector (1987) was spent primarily by three industries, electronics (30.4 percent), mechanical machinery (13.2 percent), and the chemical industry (9.0 percent), summing up to about half the amount of all industries (Stifterverband 1990).

Centres for research and technology were set up at selected corporations for the purpose of orientation to high-tech. R&D spending (teaching excluded) has been relatively low at universities and other schools of higher education, which, however, should not be taken as a measure of their importance. The employment figure in this sector amounts to 43,000, including about 7,000 professors and lecturers.

5.2 LINKS BETWEEN ACADEMIC AND INDUSTRIAL RESEARCH: A MICROVIEW

A wide variety of new links between academic and industrial research has come into being since the early eighties. Sciento-methodological centres, technological units, centres for research and application, methodologico-diagnostic centres as well as university-industry complexes are some of them. Technological units, for example, were established for the purpose of research and development up to workshop maturity and low-quantity production. There had been numerous joint R&D teams of academic and industrial professionals.

A new mode of funding was introduced in 1985, with R&D no longer being exclusively financed from the national budget but substantial areas were supported by industries from now on. A list of selected university-linked technological units and some university-industry complexes is given in the tables 5-1 and 5-2.

Table 5-1: Technological units at the institutions of higher education

Institutions of higher eduction	Topics of the technological units
Technical University Chemnitz	Micro-electronics
School of Engineering Ilmenau	Manufacturing of electronic and precision instruments
College for Transport and Communication Dresden	Process control and simulation technology
Technical University Dresden	Assembly technique in micro-electronics
Friedrich-Schiller University Jena	Thin-film technology and special components
Ernst-Moritz-Arndt University Greifswald	Fluorescent substances
Humboldt University Berlin	Optoelectronic components
Martin-Luther University Halle-Wittenberg	Biotechnology
University Rostock	Ship's elasticity ship's vibrations
Mining Academy Freiberg	Hot rolling technology (high-speed rolling)
School of Engineering Cottbus	Building technology (pre-fabrication)
School of Engineering Köthen	Drinks and fermentation technology
School of Engineering Zwickau	Mechanical engineering and motor vehicle construction
School of Engineering Dresden	Medical engineering

Table 5-2: Selected university-industry complexes

Institutions of higher education	Partner in industry
Humboldt University Berlin	Television Electronics
Technical University Dresden	Combine Progress Agricultural Machinery
Martin-Luther University Halle-Wittenberg	Chemical Combine Bitterfeld
University Rostock	Combine Shipbuilding
Technical University Chemnitz	Combine Textima Combine for Machine Tools "Fritz Heckert"
School of Engineering Leipzig	Combine Polygraph
Technical University Magdeburg	Combine of Heavy Engineering
School of Engineering Zwickau	Sachsenring Car Factory

With a view to illustrating the problem situation, let us have a closer look at the research tasks in the Departments of Physics, Mathematics, and Electronics at Humboldt University, East Berlin, with particular reference being made to micro-electronics and opto-electronics. This status analysis is based on perusal of annual reports compiled at the above Departments in the context of research activities. Judgement and qualitative appraisal have been deliberataly avoided. It has rather been the author's intention to demonstrate by example the nature of the present problem situation.

Research have been primarily focussed on the following projects of micro-electronics and opto-electronics within the scope of the Department of Physics:

- HgCdTe material research;
- Light wave conductors for telecommunications;

- Opto-electronic infrared components;
- Very large-scale integrated systems (VLSI);
- Photo-biophysics and molecular biophysics;
- Semiconductor microstructures.

The following topics were part of the Techno-Scientific State Plan:

- HgCdTe material research;
- Semiconductor microstructures;
- Interaction of ion-beam-induced defects;
- Laser diode spectrometers;
- NIR detector II;
- Studio-scale production of photodiodes.

More than 50 per cent of the research capacity were financed from WFB (Werk für Fernsehelektronik Berlin). The technological centre of opto-electronics succeeded in elevating selected developments to the level of studio-scale production. Up to 45 persons of the enterprise for opto-electronic circuits were temporarily involved in the programmes of the Department of Physics.

Although no commitment had been taken by industry of funding exploratory basic research, the Department of Physics succeeded in making substantial contributions to progress in semiconductor theory which met with international recognition. Major achievements were scored, for example, in the context of charge carrier transport, and quantum-related Hall effect as well as field dependence of generation and recombination processes.

The need for a wide-ranging scope of research has been emphasised in progress reports issued by the Department of Mathematics. Responsibilities should be more specifically defined for major lines of research, including algebra and geometry as well as algebraic geometry, theory of numbers, differential geometry, global analyses, integral and operator equations, parametric and stochastic optimisation, mathematical statistics, fundamentals of informations as well as principles and applications of distributed computer systems. Practice-related research accounted for some 55 % of all activities and was increasingly reflected in software development. Work was primarily oriented to following priorities:

- Design of integrated circuits;
- Development of software for a portable Pascal compiler;
- Digital switching equipment (studies into theoretical principles, underlying operation and dependability of digital switsching systems).

In the design of integrated circuits, great attention was given to the enhancement of algorithms and implementations for computer-assisted layout design on the basis of standard cells. This provided a basis for development of various systems, such as HULDA (Humboldt Uni-Layout Representation), a design system for control circuits layouts, as well as RELACS (Recursive Layout Computing System), a geometrical design system for recursively structured control circuits. While a computer laboratory had been established, attention was drawn in the related progress reports to the prevailing shortage of high-quality computers.

Patents did not play any role in the scientific results of the Humboldt mathematicians, though certain approaches had been made to providing software licences. A programme package under the label of "Petrinetzmaschine" was purchased by Siemens.

Research conducted at the Department of Electronics, in full keeping with its predetermined profile, was primarily geared to the development of technological systems. Technological innovations were successfully implemented in two major fields of the Department: light conductor signal transmission and flexible automation systems, with selective emphasis on priorities, namely micro-electronics, opto-electronics, industrial robots, and computer systems.

Work is now being done, among others, on the following topics:

- Image processing (design and testing of configuration of an image processing module system for industrial applications);
- AN connection line (development of LAN pin control and basic software for local network)
- AD/CAM systems (development of a strategy for computer-assisted processes for manufacture of radio-operated systems for traffic communication);
- Component technology and cycle I opto-electronics (technology-related material research on interaction between process-induced micro-effects, on the one hand, and component parameters, on the other);
- Optical broad-band communications (studies into systems for optical broad-band communication, using integrated optical circuits);

- Cycle II opto-electronics and component technology (investigations on light guide structural technologies for chip and module assembly as well as quality assurance).

Work on those topics was conducted in close cooperation with the research division of the "Kombinat Nachrichtenelektronik" (Communication Electronics Corporation) and of the WFB (Werk für Fernsehelektronik Berlin).

Research activities in that Department were linked in a very particular way to the general conditions of economic development. Here are some of the problems to which references was made in the reports:

- Insufficient funding of exploratory basic research;
- Absence of a strategic concept for product development;
- Inadequate quality of research equipment, with only 34 percent of fixed assets needed for defined projects being available in 1986;
- Insufficient availability of highly integrated control circuits;
- Excessive waiting periods in placing orders for modules and instruments.

5.3 DISCUSSION AND ASSESSMENT OF SCIENCE-RELATED INNOVATION IN EAST GERMANY

Was the linking of science and innovation in the GDR a fiasco? Section 5.2 gives the impression of a rather close institutional coupling of industry and academia much better than in some Western countries (see e.g., chapter 6). Yet, the answer to the question is: Yes, it was a failure. The State Plan objectives were not sufficiently accomplished, nor were new technologies devised to desirable dimensions or rendered economically effective. The gap was not reduced to industrially advanced countries. On the contrary, is was widened further. The pattern of foreign trade underwent sizeable deterioration in the last years prior to unification. It should be borne in mind, in that context, that foreign trade contributed 45 per cent to the national income of the GDR.

That failure was attributable to various reasons. While a comprehensive analysis still has to be made, the following aspects can be enumerated already now, as seen from the position of the GDR:

- The goals formulated above were tackled under conditions of isolation or self-made isolation from international division of labour. Attainment of economic and scientific self-sufficiency, autarcy, actually was one of the major political premises for the development of new technologies. Numerous developments, consequently, were simply reconstituted with more or less strongly pronounced time lags.

- Economic spin-off was not obtained from new technologies, since the Council for Mutual Economic Aid (CMEA) failed to practise mutually adjusted innovation strategies and was even less successful in setting up a common market.

- Adequate investment capital was not available due to persistent instability of growth.

- The mechanisms of administrative state planning continuously failed in linking up with the new demands. They proved inadequate for coming to grips with the complexity of innovation processes. There was a general trend towards deceleration of circulation and, consequently, towards extension of planning and decision-making cycles.

All this had a particular impact upon science. It, basically, took two years from the date of order, to obtain fixed assets for a certain project. Financial funds alone did not help; one had to be listed on the national balance of resources. Domestic development of necessary hardware proved to be expensive beyond any proportionality. The problem situation typical of the development and introduction of new technologies may be appropriately outlined by the example of micro-electronics.

What standards had been achieved in reality? The point must be made today that realistic analyses had been available even before the change (Marschall & Reichel 1989).

- The component and modules industry was behind contemporary technological progress by one generation in development and by two generations in production, that is between five and eight years. The backlog in application was even more severe. Prototypes of new control circuits were usually not available in the GDR until commercial production of the same models had already been started in industrially advanced countries.
- The cost implied in one transistor function in the GDR used to be one decimal power above the international level.
- Manufacturing lot sizes of control circuits were fare below those in industrially advanced countries.

The 1989 plan, for example, provided for an output of 7.8 millions of 64 kbit memory modules. Another 100,000 256 kbit memories and 20,000 1 Mbit memories were planned for the same year. These were dimensions of pilot or mini-scale production. The contribution made by the GDR to worldwide electronic production, in spite of tremendous efforts, declined from 0.8 percent in 1979 to 0.4 percent in 1989.

A comparison of these data with international figures is likely to illustrate the inferior position of the GDR component and modules industry and to offer an explanation for its non-competitive prices.

Even more serious, the effects expected from application were not accomplished in the computer, telecommunication, mechanical engineering, and consumer goods industries. The investment potential of the GDR proved insufficient for concomitant production and large-scale application of electronic components. The paradoxy of the situation actually was related to the fact that against the background of given economic mechanisms GDR corporations tended towards behaving like closed systems. Inter-company R&D cooperation in the forefield of innovations was not sufficiently encouraged. Developments of modules and systems in many cases were not streamlined to each other.

Many of the expectations hinged on new forms of cooperation between science and industry (see section 5.2) had not been met. Some of them had come into being under pressures for reaching a certain critical potential magnitude. Universities and colleges were quite often forced to undertake activities which should have been part of industrial research. Basically new technological solutions were very rarely achieved under such conditions. If real innovations were successfully accomplished at research level, there was usually no potential for translation into practice with good economy. Nevertheless, global and undifferentiated judgement would be totally wrong.

We have at our universities and other academic institutions doubtless numerous highly competent scientists who are properly familiar with the latest developments in scientific basic research and are absolutely capable of producing results of high international standards. They have acquired skills of working under difficult conditions and have accumulated wide-ranging experience in transfer of knowledge and technologies between the academic sector and industries. Inadequate standards of research at acadamic institutions in East Germany have occasionally been the subject of pretty one-sided discussion in the West German public. The spezific political and economic conditions under which research activities had to be conducted in recent years quite often have been left unconsidered, in that context. I rather feel that it would be a more productive approach to the problem, if we tried to define what contributions might be made by East German institutions to a joint all-German research landscape and how obstacles to effectiveness might be overcome and what might be done to take advantage of new opportunities.

There can be no doubt that restrictions on international cooperation inhibited effectiveness, with the latter being additionally impaired by low-quality research equipment as well as by the need for excessive spending on the development of scientific instruments and high-purity materials. Many goals in applied research were oriented to catching up with highest world standards, that is to achieve something that had already been achieved at international level. These goals, of course, cannot be upheld for the future.

Nevertheless, really new findings were made at top level in some areas of basic research. Yet, most of them were not translated into technological progress and economic benefit. An answer should now be found to the question which efforts can and should be continued with a promise of success. The experience obtained in the past from university-industry cooperation can be of lasting value. It can be utilised in the future in setting up technology parks and innovation centres. A remarkable initiative has been taken in this situation by a group called "Arbeitsgemeinschaft deutscher Technologie- und Gründerzentren e.V. (ADT)" trying to establish technology centres in East Germany (Däumichen et al. 1990). Economic upswing cannot be ensured for this region unless innovation-oriented, intelligence-intensive industries are preserved and stimulated. Yet, under such circumstances, there will also be a need for an innovation-conducive infrastructure in the form of technology centres or technology parks. Along with reintroduction of administrative units (Federal States or Länder), brand-new importance will be assumed by aspects relating to regional distribution of innovation-oriented and techno-scientific potentials. According to ADT thinking, innovation centres should be set up with a view to ensuring no-delay evaluation and translation into practice of research results in the context of state-of-art technologies and to initiating small-series production. Small innovation-oriented firms might provide an appropriate environment for academic institutions and should be encouraged. So far 17 technological centres have been planned or are being established. More or less clear-cut concepts have been worked out for Dresden and Jena.

In Dresden the emphasis will be laid on the following fields of technology: construction, environment, biomedical engineering, electronics, including communication engineering, process engineering for mechanical engineering, food processing technologies, etc. Preparations are now being under way for a technological park in Jena which include Carl Zeiss Jena, Jenaer Glasswerke, Friedrich-Schiller University as well as the latter's departments for optics, industrial chemistry, specialised components, coating technologies, etc. Also linked to it will be some institutes of the Academy of Sciences, such as the Physico-Technological Institute, the Central Institute of Microbiology, and the Central Institute of Physics of the Earth.

REFERENCES

Albrecht, E. (1988), 'Dialektik des ökonomischen Wachstums', *Wissenschaftliche Zeitschrift der Humboldt-Universität*, Reihe Mathematik/Naturwissenschaften, 37, pp. 5-17.

Däumichen, K., Ekerhardt, I. and Irmer, H. (1990), 'Technologietransfer und Technika in der DDR', ADT-Focus, Weidler Buchverlag, Berlin.

Directives (1990), 'Directives of the XIth SED Congress on the Five-year Plan for the Advancement of GDR National Economy', 1986 to 1990, Berlin, Edited 1990.

Freeman, Ch. (1984), 'Prometheus Unbound', *Futures*, 16, pp. 494-507.

Marschall, Reichel (1989), 'Reproduktionstheoretische Untersuchungen zur Entwicklung, Produktion und Anwendung der Mikroelektronik in der DDR bis zum Jahr 2005', Zentralinstitut für Wirtschaftswissenschaften der AdW und Hochschule für Ökonomie, Berlin.

Stifterverband (ed.) (1990), 'Forschung und Entwicklung in der DDR. Daten aus der Wissenschaftsstatistik 1971 bis 1989', SV-Wissenschaftsstatistik GmbH im Stifterverband für die Deutsche Wissenschaft, Essen.

6 LINKING SCIENCE AND INNOVATION IN JAPAN - A SUCCESS?

Shinichi Yamamoto

6.1 THE ROLE OF UNIVERSITIES IN JAPAN

The university sytem in Japan has played an important role in the development of science and technology since the beginning of the modern education system more than one hundred years ago, when Japan established a new political framework, called the Meiji Restoration. The new Meiji government, at that time, planned to establish higher educational institutes as the chief means of training leaders to meet the demands of the new age and for the urgent task of assimilating Western arts and sciences. Tokyo University, renamed Imperial University in 1886, was set up as the top institute for the country. It should be remarked that engineering research was included as one of the five schools within the university system. (The Imperial University consisted of five schools i.e. School of Law, School of Medicine, School of Engineering, School of Art and School of Science. These sequence of schools still remains today at the University of Tokyo although it does not show the difference of each school's prestige. The basic facts about Japanese university history can be obtained from Monbusho (1980) and other sources.)

In the early stage, however, the Tokyo University played a minor role in creative research except for introducing Western advanced science and technology. Many teachers were invited from Europe and America to train students in science and technology because there were few Japanese faculties who could do so. Gradually, foreign faculties were replaced by Japanese according to the increase of staff who studied abroad and according to the development of modern and formal education at the elementary and secondary level in Japan.

Not until World War I, when Japan reached an industrialized stage and must develop her own science and technology for a while, was creative research needed and academic research regraded as a important function in Japanese universities (Nakayama 1982, p. 6).

The university system expanded around this period. The Kyoto Imperial University was established in 1897. Soon after Kyoto, Tohoku (1907), Kyushu (1911), Hokkaido (1918),Osaka (1931) and Nagoya (1939) followed the imperial universities. Kyoto, for example, was the most active rival to Tokyo and was known for its liberal minded research activities, Tohoku aimed at being a research university and soon became famous through Dr. Kotaro Honda and the Metal Material Institute which was attached to the university (Nakayama, op. cit., p. 5). Outside of the university system, the Institute of Physical and Chemical Research (RIKEN) was established as the first non-profit research institute in 1917.

In addition to these prestigious institutes, there were various type of higher education institutes, both public and private, but they were not allowed to be called universities until the 1918 reform. After that, the number of Japanese universities and students grew steadily, partly because graduates of science and technology were needed by the government and industry and partly because the people's demand for higher education was very strong.

Since the latter part of Meiji (1868 - 1912), the people's demand for higher education had been increasing because graduation from a university guaranteed a student a better job and higher salary. The expansion of higher education opportunities and the reformation of the university entrance examination sytem were among the most important government educational policies (Kuroha 1984, pp. 8-16).

The more science and technology developed, the more universities were expected to be the centre of academic research and before World War II universites did have the prestige. The growing student population and the popularization of university education caused the function of the university system to change. One question was whether or not the integration of research and education was possible.

One solution was to expand graduate school education and another was to build research institutes attached to universities. Graduate school education in Japan was not so successful before War II because at that time degrees were granted mainly by submitting theses to universities instead of studying at graduate schools. In Japan, from the beginning, this Ronbun-doctoral degree system worked very well. The latter option of establishing university-attached institutions was successful. The university-attached institutes enjoyed the full prestige of the university and also were free from the teaching load (Nakayama 1982, p. 4).

After World War II reform, the university system in Japan changed, but the change was not as drastic as that for elementary and secondary education. As for the academic research system, the university system survived without changing its main framework.

During the 50s and 60s, although the main concern of government policy for higher education was to recover what was lost during the War and was to respond to the rapid increase of demand for education opportunities by people, the academic research system within the university sector also expanded.

Today, scientific or academic research is promoted mainly at the universities and their related institutes. Outside the university system, there are government research institutes under the jurisdiction of Monbusho (Ministry of Education, Science and Culture), such as the National Institute for Educational Research and research institutes set up by private juridical persons supervised by Monbusho. Besides the research institutes under the auspices of Monbusho, there are mission-oriented research institutes set up by other government ministries and agencies or their special juridical persons, as well as private corporate research insitutes to improve the products. The number of researchers and the expenditure for research by sector are shown in tables 6-1 and 6-2. Besides the research activities, universities play an important role in supplying researchers to industry and government research institutes.

Table 6-1: Number of researchers by sector and by field of science (full-time) in 1989 (Source: Monbusho)

	Total	Universities	Research Institutes	Industry
Total	535000	201000	40000	294000
Natural Sciences	86 %	68 %	91 %	99 %
Humanities and Social Sciences	8	23	6	1
Others	6	9	3	

Table 6-2: Research expenditures by sector and by type of activity (million yen) in 1988 (Source: STA, White Paper on Science and Technology; natural sciences only)

	Total	Universities	Research Institutes	Industry
Total	9675548	1179266	1276964	7219318
Basic Research	13 %	53 %	15 %	7 %
Applied Research	24	39	25	22
Development	62	9	60	72

6.2 SEPARATION OF RESEARCH FROM EDUCATION

In the university sector, officially speaking, education and research are lintegrated at the undergraduate and graduate levels, but in reality the former became more popular due to the rapid expansion of higher education opportunity. The too much popular undergraduate education has had little to do with advanced levels of research and most undergraduate students also have had little interest in the traditional training of the university curriculum. There was a gap between education and research even before World War II, but it became especially serious after the 60s. The main concern of most undergraduate students is to get good jobs in big companies. They may get them not because they study hard during thier undergraduate studies but because they succeed in passing the entrance examinations of the universities. So, the main function of undergraduate education has become more to select human resources by difficult entrance examinations. A rough scheme of the truly different research functions is given in table 6-3.

Table 6-3: Expansion and separation of university functions

Sub-sectors of universities	Purpose	Function
Undergraduate	Education	Selection of human resources
Graduate School	Research and education	Supply of researchers
Research Institute attached to university	Research	Performance of research (education loard free)

On the other hand, the latter plays an important role in training researchers. After World War II, universities which have graduate schools have been increasing year by year. The number of universities and the number of students are shown in tables 6-4 and 6-5. This was supported by both the academic community and industry. The academic community within universities welcomed the expansion of graduate schools because if a university attached a graduate school (after the War, not many national universities had graduate schools), the government supplied much more research funds than it did when there was no graduate school. The university would get more funds if it had doctoral degree courses.[1] In addition to funds, graduate students were needed as "research man power" in science and engineering laboratories because there were very few post-doctoral researchers. The post-doctoral fellowship system was not prepared enough to pay their salary.[2]

Table 6-4: Number of universities (Source: Monbusho, Basic Survey on School Education)

	Total	G	National	G	Public	G	Private	G
1970	382	180(92)	75	59(25)	33	19(7)	274	102(60)
1980	446	257(169)	93	77(39)	34	21(14)	319	159(116)
1989	499	303(215)	96	94(59)	39	23(17)	364	186(139)
G: Universities with graduate schools (number with doctoral courses in brackets)								

It was mainly the engineering field, however, that was welcomed by industry, because, at first, industry needed well-qualified engineers but not narrow-minded researchers. In other areas, science and humanities, the function of graduate schools remained as training for researchers who wanted to enter the university community as full-time research assistants (Joshu) immediately after graduation.

Graduate schools have traditionally been structured along the lines corresponding to undergraduate courses, enabling university to be managed as one united body. However, in response to the recent scientific development and social needs,

1 Faculties at a university with doctoral degree course usually get general research fund about 100 to 120 per cent more than faculties with no graduate school do. Master's degree course is between undergraduate and doctoral degree course.

2 Post-doctoral fellowship problems have been often discussed. The Final Recommendation by the National Council for Education Reform (1987) may be the latest example.

postgraduate courses which are not based on specific undergraduate departments and new types of universities which consist of graduate schools are begining to emerge.[3]

Table 6-5: Number of students by field (Source: Monbusho, Basic Survey on School Education)

		Total	Humanities	Social sciences	Science	Engineering	Others
70	Undergraduates	1344358	170907	562162	42071	283674	285544
	Graduates	27714	5157	4607	2983	10251	4716
	Doctoral	13243	1876	1727	2263	2356	5021
80	Undergraduates	1741504	239990	704737	54579	337767	404431
	Graduates	35781	5469	4050	3741	14864	7657
	Doctoral	18211	2860	2430	2589	2358	7974
89	Undergraduates	1929137	290387	759636	63997	379405	435712
	Graduates	58228	5926	5749	6185	26777	13591
	Doctoral	27035	3459	2852	2962	3859	13903

Another expansion of the university research function are the inter-university research institutes. From the early stages of the university system, various research institutes have been set up and attached to universities, in order to promote research activities in specialized fields of science as necessary. Especially during the Taisho period (1912-25) and up to the end of the first decade of Showa (1926-35), quite a few such institutes were set up in the national universities, which became centres of excellence in the fields concerned, and thereby pushed forward the Japanese standard in those fields.

In 1953, a new type to institute was established. It was new in that, although it was attached to a certain national university, it was open to all university researchers in Japan, and the trend has continued since then. Another break-through was made in 1971 when an institute system independent of specific universities, so called National Inter-University Research Institute, was established, which is open for nation-wide joint use. Since then, emphasis has been given to setting up this type of institute.

In order to meet expectations and demands, Monbusho has been striving to promote scientific research activities through various measures. Basically, it is trying to

3 The reform of graduate education system has been promoted by Monbusho since the enactment of Graduate School Establishment Standard in 1974.

strengthen the reseach function of universities. And the idea of detaching the research endeavour from undergraduate education was introduced. One reason is the expansion of the new type of graduate schools, as is pointed out above, and another is the establishment of research institutions within universities or inter-university research institutes.

Inter-university research institutes are not attached to any particular universities and are open for the joint use to all university researchers. As the national centres of research activity, they have been established in fields of science where there are clear needs for either the use of large-scale research facilities and equipments or the systematic collection of data, or for team research with the participation of a large community of scientists.

Each institute is playing a significant role in promoting its specific field of science. There are 63 research institutes attached to national universities to promote research in particular fields. Emphasis has been placed on the promotion of cooperative research and joint use of research facilities by all university researchers. Of the 63 institutes, 14 are available for joint use. In addition to the institutes mentioned, there are some 355 research centres or facilities attached to the national universities, some of which are for inter-departmental joint use or open to all researchers.

Table 6-6: Number of institutions as of fiscal year 1990 (Source: Monbusho)

	Natural Sciences Science and engineering	Medicine	Others	Sub-total	Humanities, social sciences and others	Total
Inter-University Institute	8	1	1	10	5	15
Institute attached to a National University	33	15	5	53	10	63
Joint use	12	1		13	1	14
Others	21	14	5	40	9	49
Total	41	16	5	63	15	78

Many private and public universities have their own research institutes, some of them have a long history and carry out distinguished research activities. They differ greatly in terms of the number of researchers, research facilities, etc., and two-thirds of them are in the fields of humanities and social sciences. The members of these various institutes are given in table 6-6.

6.3 OPENING UNIVERSITIES TO INDUSTRY AND THE GENERAL PUBLIC

Recently, research activity in the university sector seems to be behind that of the industrial sector, as far as research expenditures and the number of researchers are concerned (compare table 6-7). The demand for more funds and more researchers into university research is strongly supported by the academic community. There are several difficult problems to solve which are main results of the research funding system and the employment system of researchers etc., combined with the belief of academic freedom or university autonomy supported by academics. The university sector should pay more attention to the outside environment and its changing situation, and this is an important academic science policy agenda.

Table 6-7: Research expenditures and number of researchers in Japanese universities (natural science only) as a percentage of Japan total (Source: STA, White Paper on Science and Technology)

	expenditure	researchers
1970	18.2 %	32.1 %
1975	19.7 %	32.1 %
1980	17.6 %	33.3 %
1985	13.2 %	31.0 %
1987	13.4 %	29.0 %

CLOSED CIRCUIT OF THE POLICY FORMULATION SYSTEM

In formulating policies for the promotion of scientific research, Monbusho needs to consult the Science Council which is set up "to examine important matters on the field of science as requested by the Minister of Education, Science and Culture, and to give advice and suggestions to the Minister". The Council submitted in 1973 its first major report on "Basic Policies for the Promotion of Science". Since then, studies on the measures to develop certain priority areas of science such as space science, nuclear fusion, life science and ocean science and on the ideal map of the scientific research system have been conducted so as to form Monbusho's policy decisions. This is an example of Council-oriented public policy.

In this sense, the intiatives are held by the Council or the academic community which sends members to the Council. The number of its full members is limited to 30 and most of them come from the university academic community and very rarely from industry[4]. In addition to full members, there are some 1,200 ad-hoc members attached for special tasks such as the review of research proposals for funding. This is partly because there are very few administrators within the bureaucracy (Monbusho) who have research experience or some expertise, and partly because the academic community is separated from the industrial sector or political influence.

In the report submitted in 1984, due attention was given to the new circumstances surrounding scientific research and remarks adressed the scientific research response to social needs. The report outlined issues the Council thought were important to improving Japan's scientific research capabilities, and described the measures which might be employed to address the issues. The report has important implications for Japanese science policy (for a summary of the 1984 report see Monbusho 1988, p. 5).

Along the direction which was made by the Council, industry-university cooperation policy has been implemented.

PROBLEMS RELATED TO RESEARCH FUNDS AND RESEARCHERS' TREATMENT

Research in the university has been supported primarily by the following funds:

- General and basic funds to the researchers allocated according to a standard formula;
- Specific funds to be allocated for the specific research organization, facilities and equipment, research activities etc., and
- Research grants to be awarded directly to researchers or groups of researchers with excellent research projects.

Apart from the funds or financial support from central or local governments, the university has its own internal and external sources of income such as tuition fees,

4 Of 27 full members in 1990, only 3 are from the non-academic community.

contract research fees and donations. Of these funds, the increase of main funds is shown in table 6-8. As shown, the universities are getting more money from industry.

One of the problems is the rigid system of research funding. General funds are provided to each university by Monbusho but after being provided to the university, the funds are usually divided equally by the number of faculties or research laboratories. Each faculty can only get a small amount of money and it is not enough to buy large equipment. This equal distribution custom is good for a minimum guarantee of free and open choice for the researchers but is no great incentive for a better quality of research.

Table 6-8: Comparison of major research funds in million yen; indexes in brackets (Source: Monbusho)

	General funds for national universities based on the number of faculties	Kakenhi	Donations from private companies for national universities
1980	89.3 (100)	32.5 (100)	10.2 (100)
1985	98.8 (111)	42.0 (129)	22.4 (220)
1990	117.2 (131)	55.8 (172)	39.2 (384)

Although every university faculty seems to prefer research activities to education, in reality, not so many faculties conduct enough research activities. This is partly because there are not many universities which can be called research universities and the main mission of the faculties in "non-research universities" is education in the result, and partly because they do not lose their tenure positions, once they succeed to get it, if they do not conduct research. The general research fund system, combined with the guarantee of civil servant status (in case of national and public universities), helps them survive. But in some cases, for example if he or she wants to get a tenure position at a research university as an assistant professor or wants to be a researcher with good reputation, the faculty's preference to research must be very strong because faculties are evaluated almost only by the output of the research activities. So even an assittant professor at a junior college thinks more of research and his or her belonging to academic associations (Yamamoto).

To stimulate agressive and excellent research by researchers, Monbusho makes another type of research support available in the form of research grants called "Kakenhi" (scientific research grant). The grants are given directly to researchers or groups of researchers working at universities and research institutes. Research grants are awarded on the basis of merit after proposals are examined by specialist committees establised within the Science Council's Research Grant Committee (peer review system). Recipients of the grants are expected to make important contributions to the progress of science. The amount of 58,800 million yen is budgeted for 1990.

The share of Kakenhi in university research funds has been increasing in recent years, although the increasing rate is not as high as that of industry's donations. Several problems regarding Kakenhi are (a) it does not include salary for research staff, (b) usually, it does not include the expense of traveling abroad and so on.

How a researcher is treated is also a problem. In Japan, research in science and engineering is usually done by a group in the same laboratory directed by a professor and graduate students are regarded as the research manpower, because there is not enough support for post-doctoral fellowships. Professors can use graduate students without paying them anything but the students must pay their tuition fees, because students are generally regarded to help research activities as a part of their educational activities. Improvement of the financial condition of young researchers is needed (Amagi et al. 1989, pp. 90-93).

Today many undergraduate students prefer the financial business to the manufacturing industry. The most important reason for this is the large salary difference between the two sectors. And this tendency may became more or less popular among graduate students, which will have a big impact on the recruitment of future researchers.

UNIVERSITY-INDUSTRY COOPERATION

In recent years, industry and other sectors of society have expressed increasingly diverse expectations for scientific research activities in universities. Among many reasons, getting research funds from industry is the main reason for promoting cooperation on the university side. For industry, especially smaller companies, getting the advanced knowledge and technology from university labs is a big reason.

Several measures either directly or through the Japan Society for the Promotion of Science have been taken by Monbusho, in order to meet the need for increased cooperation between the universities, especially national universities and industry.

According to Monbusho's explanation, the following measures have been taken (Monbusho 1988, pp. 90-93).

JOINT RESEARCH WITH INDUSTRY

This system was created in 1983 to encourage researchers at national universities to conduct research in cooperation with researchers in private enterprises for outstanding results. Both universities and industry have shown growing interest in this system and the number of joint projects under this system has been increasing year by year. This joint research is actively carried out especially in materials development, electronics, biotechnology etc.

When a joint research project has led to a joint invention, the application for the patent is made jointly by the university and the other body concerned. In principle, the patent is held by both parties, and the company or other body enjoys priority for the license for the approved patent for a set period (not more than seven years). The budget for this joint research in 1990 is 3,800 million yen.

CONTRACT RESEARCH

This system enables national universities to carry out, on a contract basis, the research commissioned by industrial firms, government institutes and other external bodies. The costs of such research are naturally born by the external bodies.

National universities accept 2,000 projects a year under this system. A truster can enjoy license priority for a set period (not more than 7 years) of the government-owned patent resulting form contract research. Such patents may even be shared with external bodies concerned under the provisions of the Research Exchange Promotion Law. The budget for this is 7,100 million yen in 1990.

CONTRACT RESEARCHERS

The contract researcher system provides researchers and engineers in industry with the opportunity to conduct research in national universities at the graduate level. More than 1,200 contract researchers are implementing research projects at national universities under this system. The budget for contract researchers is 300 million yen in 1990.

DONATIONS

National universities are authorized to receive donations form outside organs, such as private companies, for the purpose of encouraging scientific research in the universities. These donations are utilized flexibly for research and educational activities which are in line with the objectives of the donors. The amount of donations in 1990 is 29,200 million yen.

CENTRES FOR COOPERATIVE RESEARCH

To promote full-scale cooperation through joint research, contract research etc. with the private sector, Monbusho has since 1987 endeavoured to establish centres for cooperative research at national universities. There are 18 centres in 1990.

JOINT RESEARCH BY MONBUSHO GRANTS

Joint research under the category of "Developmental Scientific Research" of the research grants (developmental and applied with a high potential for practical use), between university and industry is being encouraged. The budget for these grants is 4,900 million yen.

JSPS PROGRAMME

The Japan Society for Promotion of Science (JSPS) sets up an Advisory Committee on University Industry Research Activities consisting of eminent members from academia as well as industry to study the fields of research which are likely to develop by university-industry cooperation. JSPS also provides a forum for promoting cooperation by organizing Industry University Cooperative Research Committees on specific themes of technological improtance. 40 such committees are active at present.

6.4 POPULARIZATION OF GRADUATE SCHOOL EDUCATION

Although the share of research activities in terms of research expenditure and numbers of researchers is decreasing in the university sector, universities are becoming more important to train future researchers. After World War II, graduate schools have been expanding their capacity and now popularization of graduate education is occurring, especially in the fields of natural science and engineering. Table 6-9 shows that in becoming a specialist like a scientist or a university researcher, a master's or doctoral degree is necessary.

Table 6-9: Scientists and engineers, and the level of education (in brackets in per cent) (Source: Monbusho, Basic Survey on School Education)

		Non-University scientists	University researchers	Engineers and technologists
1970	Bachelor	1555 (84)	909 (35)	46138 (93)
	Master	186 (10)	608 (23)	3340 (07)
	Doctor	102 (06)	1106 (42)	225 (00)
1980	Bachelor	1609 (63)	730 (32)	57957 (90)
	Master	749 (29)	430 (19)	6331 (10)
	Doctor	212 (08)	1154 (50)	277 (00)
1989	Bachelor	1197 (48)	414 (19)	77230 (86)
	Master	935 (37)	383 (17)	11738 (13)
	Doctor	373 (15)	1413 (64)	363 (01)

According to the development of science and technology, master's degree holders became welcomed in the field of engineering. As a result, the number of graduates from master's degree course have been increasing rapidly and this phenomenon means that, in the engineering field, a master's degree has become the minimum requirement for getting a good job in industry.

One of the most important function of universities is degree granting. In Japan, only universities which have graduate schools can grant people master's or doctoral degrees. Even if the environment of research activity is far more better in industry's laboratories than in university's laboratories, a researcher who needs a doctoral degree must submit his or her thesis to a university. The fact that "Ronbun doctors" (doctoral degree recieved by submitting a thesis without residential requirement) are increasing in number in the engineering field (cf. table 6-10) implies that there are not only poorer research facilities in the university sector compared to the industrial sector, but also the unique function held by the university sector[5].

5 It is discussed that the increase of Ronbun doctors means an increase of research capability in industrial sector which conflicts with the function of graduate schools (Amagi, op. cit. pp. 44 - 46).

Table 6-10: The share of "Ronbun Doctors" grantees in the engineering field over time (Source: Monbusho)

1970	50.2 %
1980	55.9 %
1981	56.2 %
1982	60.4 %
1983	62.1 %
1984	67.9 %
1985	65.8 %
1986	66.2 %

6.5 FUTURE PERSPECTIVES

Although the university sector has been losing its share in research performance, providing industry and other sectors with scientists and engineers is still an important function. But this fact again raises a question of the role of the university.

We experienced the separation of research from education and related functions in undergraduate education level, caused by too much popularization of undergraduate education. When combined with the Japenese style employment system adopted by big companies, which is called lifelong employment custom whereby employment is possible for full-time employees only when they are young and just graduated from schools, undergraduate education no longer plays an educational role but means selection through entrance examinations.

Expanding graduate schools has been a trial to separate advanced research from too popular undergraduate education, but graduate education become also popular in some fields. What will graduate school education become? Does popularization work for the better or for the worse of the graduate school system? So far, the graduate school is not very popular among students who want to go into business or into non-research jobs.

Further, the importance of basic research and the role of university research is again taken into account. According to Monbusho, basic research is conducted mainly in the university sector. Monbusho sometimes denotes basic research, scientific research and academic research in the same manner. Basic research is an intellectual creative activity in all the disciplines of natural sciences as well as social sciences and humanities (see also chapter 1). Based on the irresistible human desire to explore the unknown, scientific research aims to find new laws and principles, to establish a methodology of analysis and synthesis, to systematize new knowledge and technology and to open up new fields of science. The fruits of the research gained represent an excellent cultural value in and of itself as a common intellectual property of mankind. At the same time it sustains and enriches our daily life through its utilization and application, thus becoming a driving force for the development of the world. As a centre of science, the university is responsible for maintaining and advancing scientific excellence by developing scientific research systematically in all fields of human, social and natural sciences.

Table 6-11: R&D expenditures at universities by character of work in per cent in international comparison (Source: STA, White Paper on Science and Technology)

		Basic	Applied	Development
Japan	(1987)	54.2	37.4	8.4
USA	(1988)	63.2	24.4	12.4
West Germany	(1983)	76.4	23.6	
France	(1979)	90.0	10.0	---
UK	(1975)	95.2	3.6	1.3

What is needed for Japanese universities is neither shifting from basic research to applied research and development nor being at industry's beck and call. According to the international comparison shown in table 6-11, Japanese universities conduct applied research and development more than European countries and America do. Japan is becoming expected to conduct more basic research corresponding not only to her well economic performance but also to the globalization of technology. The most important task is to establish a new type of identity for the university system, i.e. academic freedom in a "positive sense".

In this sense, from the results of basic research, development of "generic technology" (or common-base technology) which means fundamental technology that realizes a breakthrough and has a broad and long-term effect on other technologlies will be one of the important role that universities should play. University-industry cooperation is also needed but it should be promoted under the initiative of the university side. If not, what will the unique role of the university system become?

REFERENCES

Amagi, I. et al. (1989), 'A Policy-Oriented Study on the Innovation of the Graduate Training in Engineering', *Bulletin of the Institute for Higher Education*, 1.

Kuroha, R. (1984), *Education Reform*.

Monbusho (1980), *Japanese Modern Educational System*, Tokyo.

Monbusho (1988), *The University Research System in Japan*, Tokyo.

Nakayama, Sh. (1982), 'The Development of Universities and the Formulation of a Science Research System', *Bulletin of the Institute for Higher Education*, 3.

Yamamoto, Sh., 'From the Relation between Academia and Others', *IDE (Journal of higher education research)* 314, pp. 44-47.

7 MOLECULAR BEAM EPITAXY - A MESOVIEW OF JAPANESE RESEARCH ORGANIZATION

Lennart Stenberg

7.1 INTRODUCTION

In recent years much attention has been given to Japanese science and technology policy. Japanese programmes involving cooperation between the government and technologically advanced companies have come to be seen as an efficient means of industrial policy and been a source of inspiration for the establishment of similar programmes in other countries.

High regards for the technological strength of Japanese industry and related policies are routinely contrasted with correspondingly low appreciation for the level of basic research in Japan as compared to the United States and Europe. In particular, many Japanese and western commentators alike have expressed highly critical views of the research capabilities of Japanese universities.[1] Although a gradual increase in the Japanese contribution to scientific research is recognized there is still scepticism that Japan will ever seriously rival the United States and Europe in their position in science.[2]

Since around 1980 there has been a shift in policy focus in Japan in the direction of a stronger emphasis on "basic research".[3] As a consequence government policies for

---

1 Okimoto et al. (1987, pp. 416-17), for example, writes about the "second-rate quality of university-based research" in Japan. See also chapter 6.

2 Narin (1988), while providing data showing that Japanese scientists increased their share of papers in the Science Literature Indicators database from approximately 5 percent in 1973 to 7.5 percent in 1984, points to factors that may keep Japan from "ever developing the kind of scientific hegemony that the Europeans had in the early part of this century, and the U.S. had in the middle years".

3 Although basic research has moved to centre stage in Japanese discussions of R&D policy, the *share* of basic research in total R&D expenditure in Japan has hardly shown any noticeable change during the 1980s as all types of R&D have grown at a rapid rate.

support of R&D in firms, government research establishments, and universities have all undergone far-reaching changes during the last decade. New visionary R&D programmes have been launched, new mechanisms for funding of cooperative efforts in basic research have been introduced, research priorities at government institutes have begun to be changed in the direction of more basic research, and initiatives have been taken to strengthen research and graduate education at universities (chapter 6).

In the corporate sector the establishment of new research laboratories for fundamental, or similarly labelled research by a large number of companies has been the most visible sign of a new policy in Japanese industry to take a leading role in the development of new basic technologies. There has also been a mushrooming of new research organizations operated and staffed jointly by several companies. These joint research firms, which in the R&D statistics are classified as "private foundations", represent a new component in the Japanese research system.

In Japan the R&D system is dominated by the business sector to a higher extent than in any other country with the exception of Switzerland, the latter being a special case due to the concentration of research-intensive pharmaceutical firms in that country. An exceptionally rapid expansion of R&D in Japanese industry during the 1980s coinciding with a much slower growth in government R&D expenditure has further shifted the centre of gravity in the Japanese R&D system towards the business sector. As industry has increased basic research at a rate somewhat higher than that for product-oriented R&D the dominance of the business sector is beginning to extend even to basic research leading to concerns in both Japan and abroad about an unbalanced institutional structure for the performance of basic research in Japan.[4]

---

4 Arguments have been put forward, for example, in conncetion with the negotiations of a new science agreement between the United States and Japan in 1988, that Japan is not contributing its fair share to openly accessible basic research. While the main point seems to be that the Japanese government is not funding sufficient research at universities there is an additional fear that expanded basic research in Japanese firms will increase their ability to take advantage of basic research in other countries while not making their own research sufficiently available. For a Japanese discussion of the desirability of a leading role for universites in basic research see Sakakibara (1988a, b).

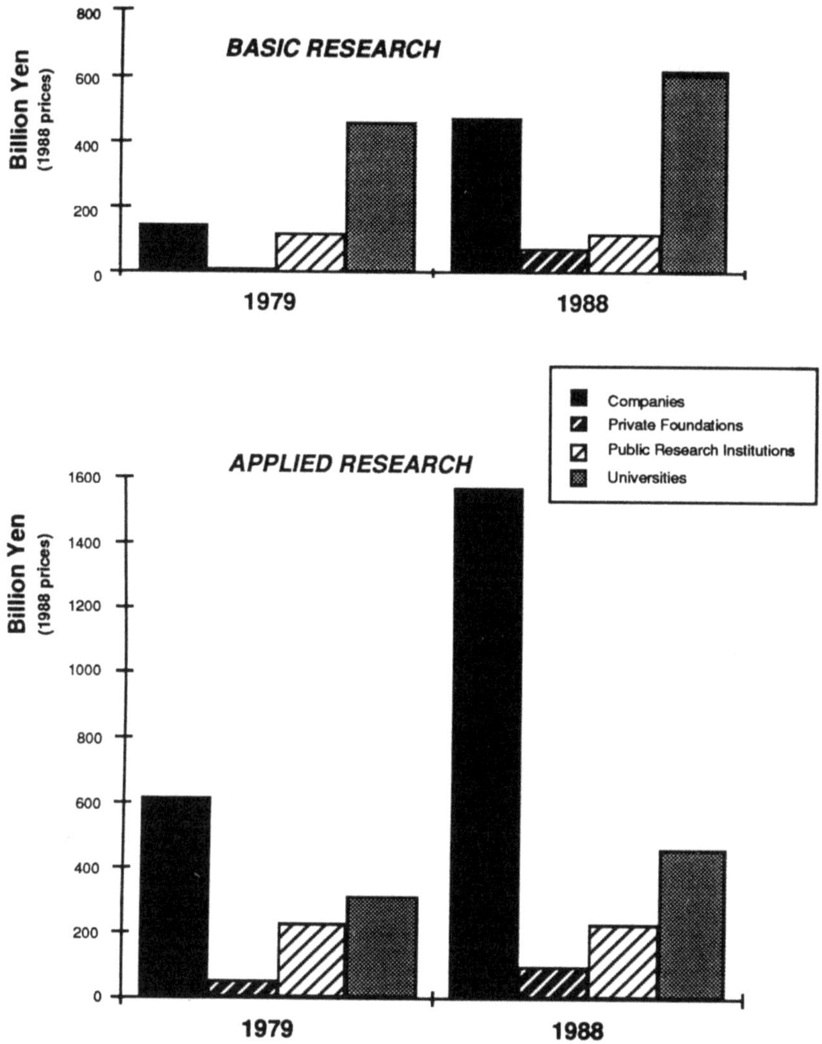

Figure 7-1: Expenditure for basic and applied research (in physical, agricultural and medical sciences and engineering) in Japan 1979 and 1987 by type of research organization in 1,000 million Yen (1988 prices) (Source: Report on the survey on R&D, 1980 and 1989)

Figure 7-1, based on official R&D statistics, gives a quantitative estimate of the expansion and institutional diversification of the Japanese research system that have taken place over the last decade. It needs to be pointed out, however, that the data provided by the R&D statistics is unreliable in several respects. The concepts of basic and applied research have proven to be very ambiguous and therefore less suitable as a basis for surveys of R&D.[5] In the special case of universities the problem may be less one of distinguishing between different categories of R&D than distinguishing research from teaching and other non-R&D activities. This problem is particularly large in Japan where separately budgeted research is a very small part of total research expenditure.[6] Finally, disaggregation of research activity according to scientific or technical field is not possible beyond a very crude level and can even at this level as a rule not be done using the same field classification for different types of research organizations.

The extensive participation of industrial firms in long-term research and the establishment of a large number of joint research firms give the Japanese research system a structure rather different from that in most other countries. Whether this difference will become further pronounced or whether it represents only a temporary aberration from more traditional models according to which universities or, in some countries, other types of publicly funded research institutes are seen as the natural centres for basic research is presently very difficult to judge. The outcome will depend on many factors, most of which defy predictions, such as how deeply Japanese industrial firms will remain committed to investments in basic research, how the joint research firms in the end will be evaluated by their corporate members, the scale of public and private resources that can be made available to universities, and what kind of cooperative relations that can develop between different types of research organizations.

The present paper aims at revealing qualitative aspects of the ongoing changes in the Japanese research system through detailed study of developments in one specific field, Molecular Beam Epitaxy (MBE). The objective is to clarify the role played by companies, universities and other types of research organizations during the course of the evolution of MBE research in Japan, seeking insights that might be generalizable also to other fields. The focus is on the organizational context of the research performed rather than on its scientific and technological content, although the latter aspect will be considered in broad terms.

---

5 Stokes (1982) has for example suggested that basic and applied represent different dimensions and therefore are not mutually exclusive. Work is underway in many places to develop new categories which would be better guides to describing current realities.

6 This problem is discussed in Irvine et al. (1990). The same problem exists for data measuring R&D personnel as there is no data allowing calculation of full-time equivalents.

MBE was chosen because it satisfied several criteria:

- MBE, being one of the most advanced technologies for engineering of materials on an atomic scale, represents a *basic technology* with a wide variety of possible applications.

- MBE has great *relevance both scientifically and from a practical technical point of view:* development of MBE technology requires scientific research which to a large extent is quite basic in nature; MBE is also used as a tool in basic research of physical phenomena. Thus, MBE is a good case in study of the general framework of science-based innovation in this volume.

- The MBE field has attracted the attention of researchers *in universities as well as in industry*.

- The MBE field is a fairly *well delineated field* in both substance and time and is of *suitable size* for an intensive case study.

- The research community in the field has exhibited very fast growth, making the field *suitable for studying questions of recruitment and training of researchers.*

The approach taken, was to study the evolution of the community of MBE researchers in Japan. Using bibliometric data, Japanese researchers active in the MBE field and the organizations for which they worked were identified for different time periods. Similar data were gathered also for other countries to permit comparison of the institutional structure of the research communities in Japan and in other countries.

Information on qualitative aspects of the development of MBE research in the leading Japanese organizations was obtained through interviews. These focused on the broad evolution of MBE research in cognitive and technical terms, the validity of the bibliometric data, the mechanisms for acquiring and diffusing expertise in MBE, the primary research targets and motives for starting up MBE research, and comparisons of MBE research carried out in different organizations and in different groups within a specific organization. After analysis of the interviews, additional bibliometric searches were performed to study subfields of MBE research and also to obtain some data for the evolution of competing technologies.

As a result a fairly coherent picture of the evolution of MBE research in Japan over the last 15 years was obtained. The quantitative data also allows some limited comparisons with other countries.

This chapter first attempts a characterization of the basic features of MBE technology and the general pattern of its development, including some bibliometric data showing the institutional structure and dynamics of MBE research in the United States and world-wide as a reference for analysis of developments in Japan. The bulk of the paper discusses the evolution of MBE research in Japan in some detail, and pays special attention to the objectives and research targets for different organizations, the role of government programmes, sources of research funds for universities, and recruitment, training and careers of MBE researchers. Finally some conclusions are drawn from the MBE case study.

7.2 SKETCH OF MBE TECHNOLOGY AND ITS DEVELOPMENT

Molecular Beam Epitaxy (MBE) is a technology for growing crystals in a highly controlled fashion making possible the growth of very thin layers of high quality crystals, that is crystals of well-defined and very uniform composition.[7] The composition of the layers can be varied during the growth process allowing the fabrication of sandwiched structures, so-called *heterostructures*, which if repeated yield *superlattices*. Such structures give a high degree of freedom for designing semiconductor materials with desirable electronic and optical properties of a great variety by exploiting what has been termed *bandgap engineering*. An essential requirement is, however, that crystals of sufficiently high quality, with few impurities and abrupt interfaces between different materials, can be grown.

Different types of materials may be used, although the technology is currently more developed for some than for others. Initially research concentrated on III-V semiconductor materials, especially structures combining gallium arsenide (GaAs) and aluminum gallium arsenide (AlGaAs). This materials system is still the most widely used and researched, but, increasingly, other III-V materials as well as II-IV semiconductors, for example zinc selenide (ZiSe), and elemental semiconductors such as Silicon (Si) and Germanium (Ge) have attracted attention. Also other materials have

7 For a technical overview of MBE see for example Sakaki (1989). For a more detailed account of current topics see Shiraki et al. (1989). Takahashi et al. (1989) provides a useful summary of recent trends in MBE.

been grown by MBE, for example certain organic materials and, very recently, high temperature superconducting materials.

MBE-grown materials are used both in basic solid-state physics research and for the fabrication of electron devices, for example various types of transistors or lasers. The artificial materials and structures that can be created with the help of MBE produce in many cases physical effects of great scientific interest, effects which may also be exploited in devices for practical use. Some discrete devices, especially high electron mobility transistors (HEMTs) and certain lasers, are already produced commercially using MBE while integrated circuits (ICs) and optoelectronic ICs (OEICs), that is circuits which combine optical and other devices on one substrate, have been available experimentally for several years but still have uncertain commercial prospects.

MBE requires ultrahigh vacuum (UHV) and the technology for achieving this has been gradually perfected since the beginning of MBE research 20 years ago in the United States. The first MBE machines were designed by the researchers themselves and built to custom order by various vacuum companies. In the late 1970s equipment manufacturers began to market the first standard systems. To contain costs some researchers, especially in universities, continued to design their own systems also after that.

Instrumentation to study the samples during the growth process (so called in-situ measurement) is an important part of most MBE machines used in research.[8] The attachment of such instrumentation means that almost all MBE systems to some extent are custom-made as the instrumentation typically will vary from one machine to another. The development of methods for in-situ surface analysis has been and continues to be an important area of MBE research. Needless to say the instrumentation has successively become more varied, plentiful and sophisticated as MBE technology has evolved.

Initially more or less all MBE systems used solid source materials which were heated in crucibles to generate vapour which in turn, when let into the vacuum growth chamber, formed "molecular beams". Since the beginning of the 1980s different types of gas source MBE have been developed which, depending on which element is supplied in gas form, go under the names of gas source MBE (GSMBE), metalorganic MBE

8 The fact that MBE operates under UHV is the reason in-situ monitoring can be used with MBE, something which has not been possible with other crystal growth processes such as MOCVD. This is one key advantage of MBE as a research tool.

(MOMBE) or chemical beam epitaxy (CBE). One of several factors which has prompted the development of gas source MBE is the higher throughput that can be achieved with this method as compared with conventional MBE. MBE has been seen as having a disadvantage in this respect in comparison with other crystal growth techniques, which has been one reason why metalorganic chemical vapour deposition (MOCVD) has tended to be preferred to MBE for volume production. Gas source MBE is also better suited than conventional MBE for growth of certain materials, for example those containing both arsenic (As) and phosphorous (P).

In a sense gas source MBE can be seen as combining certain features of MBE and MOCVD, thereby contributing to the distinction between the two becoming less sharp, something which is also reflected in a growing overlapping of the two research communities with an increasing number of researchers using both MBE and MOCVD and various hybrid or related techniques such as atomic layer epitaxy (ALE).

The MBE technology has been modified also by the addition of auxiliary processes such as laser or electron beam excitation to the basic growth method, thus making possible selective growth, reduction in growth temperature, and increase in chemical adsorption at the sample surface.

The use of photonic beams in combination with a growing repertoire of reactive chemicals introduced through gas source MBE processes gives a range of new options to manipulate the growth process, most of which have so far only marginally been explored. Of particular interest is the possibility that methods could be developed to control the growth process in the lateral direction in order to produce low dimensional structures such as quantum wires and quantum boxes, structures seen as holding the promise of a whole new field of physics as well as of new classes of devices.

Traditionally the growth of crystals and fabrication of microstructures have been distinct processes using different and separate equipment. Recently it has become common to integrate these processes into one piece of equipment combining MBE, electron beam lithography, reactive ion etching, etc., thus making it possible to perform all processing steps without breaking the UHV condition. Just as in the case of gas source MBE this closer integration of different growth and processing technologies also contributes to a blurring of the distinction between MBE researchers and researchers who are involved in the development of microfabrication processes.

The ease with which different materials can be grown epitaxially, that is with the crystalline structure preserved while changing from one material to the other, is dependent upon how well the crystal structure of the two materials are matched to each other. The greater the difference in lattice constants between the materials - the lattice constant being the distance between atoms in the crystal plane perpendicular to the growth direction - the greater are the difficulties in growing high quality heterostructures. As MBE equipment has become more versatile and the understanding of growth mechanisms for different materials systems has increased, ever more challenging combinations of materials have successfully been grown. Since the mid 1980s there has for example been much research concerning the growth of GaAs on Si. So far most MBE research has aimed at growth of semiconductors on semiconductors. There are, however, visions, articulated by, among others, Professor Furukawa at Tokyo Institute of Technology, that it will become possible to grow semiconductors, metals and insulators on each other with a high degree of freedom in how they are combined.

For an outside observer the impression is that there is presently considerable uncertainty as to the future direction(s) that MBE research will take. This is partly related to the, at best, unclear, and according to some rather bleak commercial potentials of MBE for any large volume products, a subject which will be discussed further later. In a situation where nearterm commercial prospects, outside certain niche markets for which the technology has partly already been developed to a mature stage, are not so promising it seems that the interest of many researchers is beginning to converge on the vision of developing MBE for the realization of what loosely may be referred to as "quantum devices".

In a way the very identity of an MBE researcher is also becoming open to question as the different crystal growth and fabrication technologies are becoming increasingly integrated with each other. The identity is also becoming confused for another reason. There is growing number of researchers who use MBE without having any intention of contributing to the development of MBE technology as such. As the MBE technology has become perfected for some materials it has become possible to automate the operation of MBE equipment for these materials so fully that someone with very little experience of MBE can grow materials which may not have the quality required for fabrication of devices but which may still be useful for studies of various physical effects. As the use of MBE in commercial contexts is increasing, there is also a growing use of MBE for routine growth of wafers for device development.

Summarizing the development of MBE technology, one can say that after MBE research was started at Bell Labs in the late 1960s it took around a decade to establish the basic process technology for the most common materials system, GaAs and AlGaAs, to a level which permitted the realization in a laboratory setting of the first practical devices. After further development efforts some devices were put in commercial production during the latter part of the 1980s.[9] Gradually the technology has branched out, as described above, from its initial core, both in terms of the configuration of the equipment and the materials systems to which it has been applied, with a certain increase also in the variety of device applications. The diversity of research themes and approaches have thus multiplied especially in the last 3-5 years. This schedule compares to other science-based technologies; see e.g. chapter 9, table 9-2 in this volume.

Through the counting of publications it is possible to get some quantitative measure of the growth of MBE research and its institutional structure.[10] Of a total of some 6500 publications indexed in the INSPEC data base during the period 1969-1989, 48 and 19 percent came from organizations in the United States and Japan respectively (figure 7-2).[11] In all parts of the world there has been a very rapid growth in MBE publications. In absolute numbers most of that growth has occurred since the early 1980s. The growth was slower in Japan than elsewhere during the indexing period 1981-83 and more rapid during the next three-year period. Publications indexed in a certain year can be assumed to correspond, on average, to research being performed two to three years earlier.

---

9 It should be noted that material for the same device usually can be grown by alternative methods. Lasers are mainly produced by Liquid Phase Epitaxy (LPE), except for sophisticated devices for which MOCVD and MBE compete, with the former so far being the one usually preferred. In the case of HEMTs it appears that MBE is favored over MOCVD by most, but by no means all, firms.

10 For a discussion of the use of bibliometrics to study research communities see for example Granberg (1990), Rappa (1989) and chapter 3.

11 Molecular Beam Epitaxial Growth was introduced as a "controlled term" in the INSPEC data base in 1979. In order to capture MBE publications prior to 1979 the search was performed to include also MBE, and alternative expressions, as uncontrolled terms. The publication set thus obtained was for the period 1979-89 in the case of Japan around 40 percent larger than that retrieved using the controlled term and for all countries together 50 percent larger. An additional reason for choosing a generous search strategy was the fact that the data were to be used to identify as many organizations as possible with any degree of involvement in MBE. The time from publication to indexing in the INSPEC data base varies. Most publications are indexed within one to one and a half year. For papers presented at conferences the time to prepare conference proceedings, which varies widely, must also be taken into account. It appears that on average it takes longer to index Japanese publications than publications from the United States. Publications in the INSPEC data base are only assigned to one address even if the authors belong to different organizations and this is indicated in the original publication.

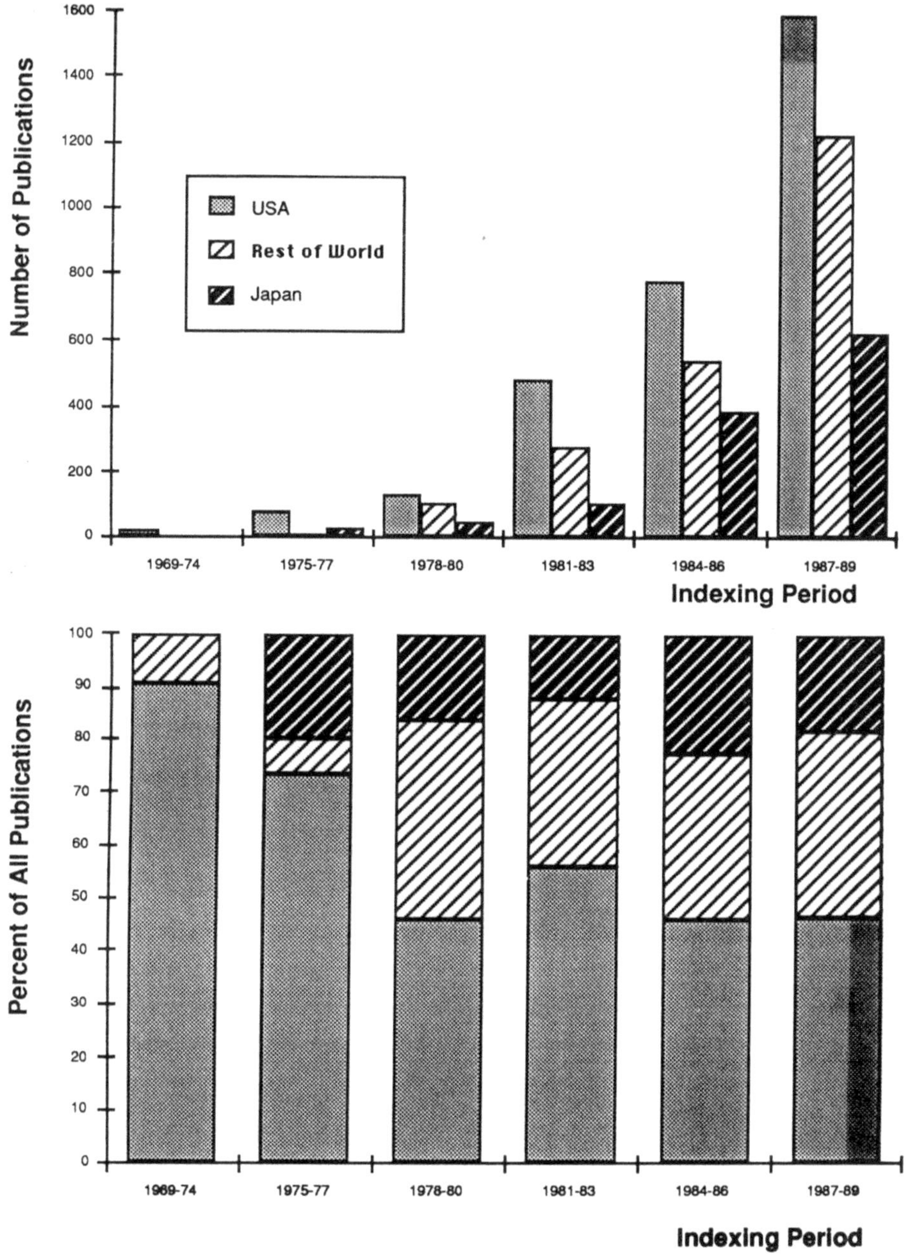

Figure 7-2: Distribution among USA, Japan, and rest of world of publications in the field of MBE according to source of publication

In order to broadly compare the evolution of the MBE research communities in Japan and the United States each research organization was categorized into one of the following six groups:

1. Telecommunications research organizations
2. Three largest industrial firms in terms of total number of MBE publications (excluding those covered in group 1)
3. Other firms
4. Government laboratories or research institutes and cooperative research organizations
5. Three largest universities in terms of total number of MBE publications
6. Other universities (and other higher education institutions)

Figure 7-3 shows how the number of publications developed for each group over three-year periods in the United States and Japan respectively. It should be noted that the scale for the United States is twice that for Japan, roughly corresponding to the ratio between the two countries' populations. Somewhat more detailed data of a similar type for the period 1987-89 is exhibited in figure 7-4. Here the scales are kept the same in order to facilitate direct comparison of publication counts between individual organizations, a kind of comparison which should be interpreted with great caution.

Overall the research communities of the two countries exhibit rather similar development with a strong concentration to only a few organizations during the 1970s and then a rapid broadening of the participation in the 1980s. On a per capita basis the Japanese business sector has a stronger publication record than its American counterpart, while the university sector is weaker. The leading Japanese industrial firms compare well with American firms even in absolute terms while the same cannot be said for the universities. Similar data for European countries, which can not be displayed here for reasons of space, reveal, however, that in the field of MBE, American universities are exceptionally strong. In terms of per capita publication counts Japanese universities fall somewhere between the United Kingdom and West Germany and far ahead of France. In absolute numbers of publications the three largest Japanese universities are about 50 percent larger than their counterparts in the United Kingdom and West Germany. MBE research in the corporate sector is much more limited in the European countries than in either of Japan and the United States.

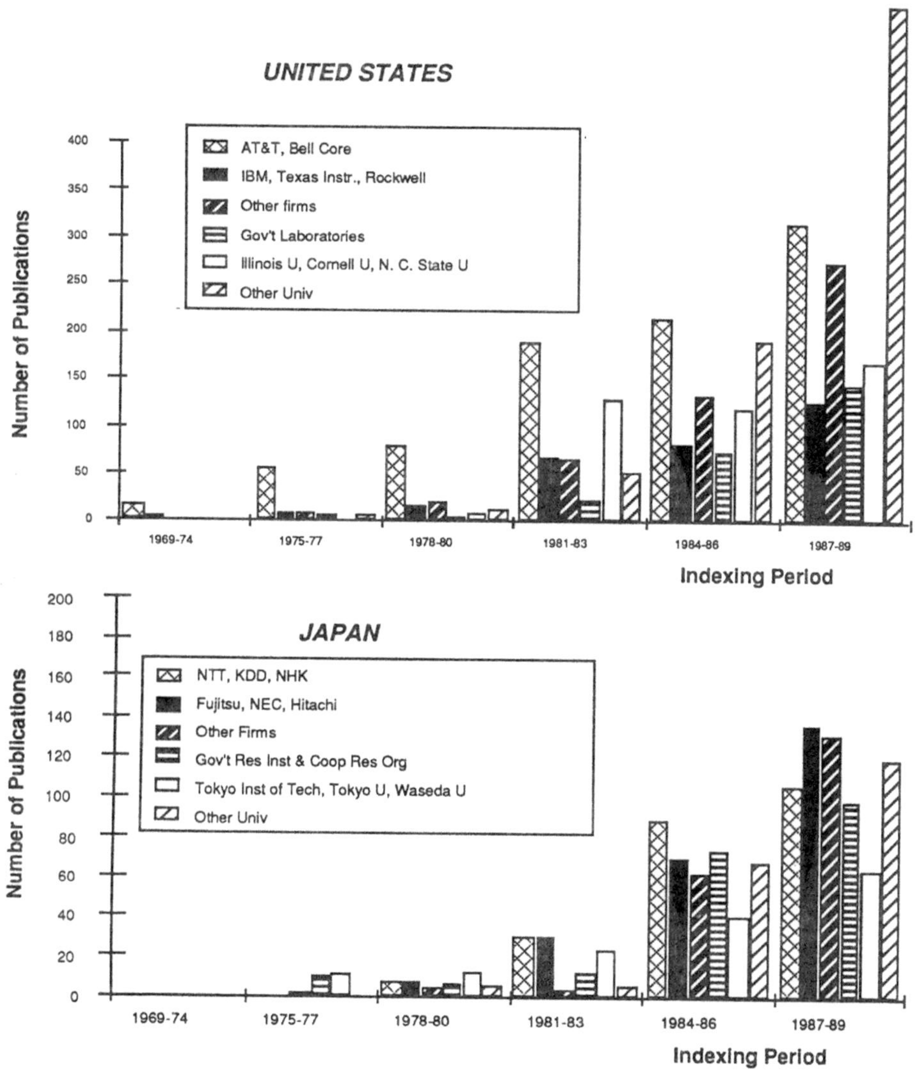

Figure 7-3: Distribution of publications in the field of MBE among organisations in the United States and Japan

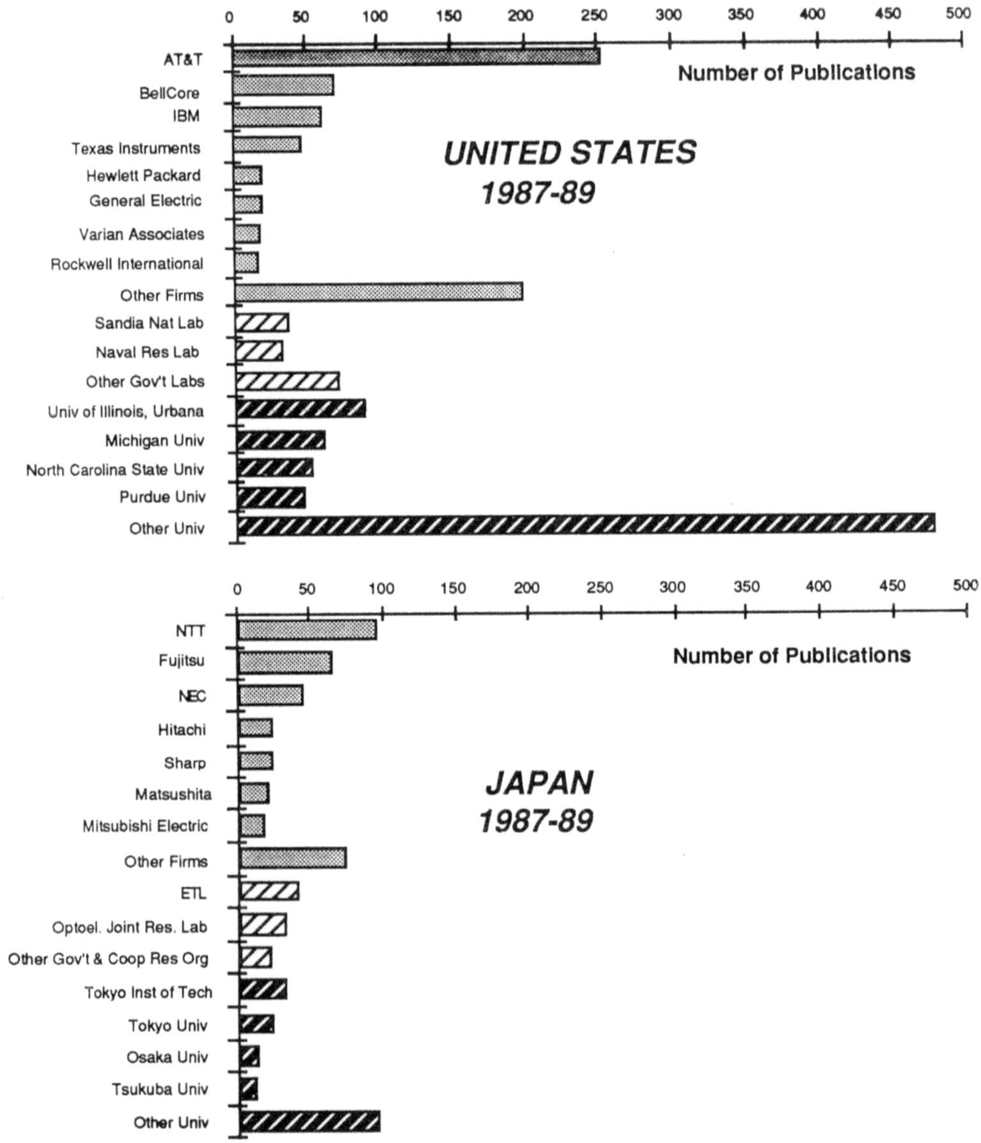

Figure 7-4: Distribution of publications in the field of MBE that were indexed in INSPEC 1987-89 from US and Japanese organisations

7.3 EVOLUTION OF THE INSTITUTIONAL STRUCTURE OF MBE RESEARCH IN JAPAN

THE PIONEERS

1980 represented a turning point for MBE technology. Around this time it became clear that the technology had to be taken seriously while this had previously been very much an open question. In fact around 1978/79 the outlook for the possibility of commercial applications of MBE had, at least in Japan, been at a pessimistic low. Efforts to turn a visionary concept into a useful technology had been underway in the United States for almost a decade, and in Japan a few years less, without yet producing any results that had not already been achieved more easily in other ways.

Prior to 1980 only a handful of companies, three universities and ETL had at all been involved in MBE research in Japan. ETL and Tokyo Institute of Technology (Professor Takahashi) had pioneered the Japanese efforts in the very early 1970s. They were soon joined by Mitsubishi, Matsushita and Fujitsu and around 1977 also Hitachi and NTT started MBE research. Some of Professor Kimata's students at Waseda University had joined the research group at ETL very early for their thesis work and somewhat later Kimata's group at Waseda University started MBE research on very simple equipment that had been built from the meagre funds available. At Tokyo University, Professor Sakaki returned in 1978 after having worked one year with Dr. Esaki at IBM in what was one of the two leading MBE research groups in the United States and in the world. Sakaki was determined to establish MBE research at the Institute for Industrial Science where he had been working since 1974. This seemed to him a good time for a university to get involved in MBE research as Japanese industry was apparently in the process of scaling down its already fairly small MBE research.

Mitsubishi and Matsushita were about to dissolve their MBE groups and management at Fujitsu was also becoming impatient. Hitachi and NTT had started later and therefore might be expected to hold out longer. At Hitachi compound semiconductor research was under general pressure and severely cut. By reorienting MBE research towards growth of silicon crystals instead of the more commonly studied compound semiconductors the researchers at Hitachi were able to obtain management support for continuation of MBE research.

NTT as a public or semipublic organization was expected and able to carry out more long-term research than regular companies and consistent with this does not seem to have been strongly affected by the generally pessimistic views on MBE in the late 1970s. NTT was primarily interested in materials and devices with potential use in telecommunication and therefore set out to explore crystal growth for various kinds of optical devices, more precisely long wavelength lasers, visible light lasers and optical modulators. MBE research on long wavelength lasers and optical modulators was already underway at Bell Laboratories when NTT started up its own efforts. Visible light lasers, a field which Japanese companies dominate today, are mainly used in CD players and laser printers and thus fall outside NTT's primary interest. Research on MBE growth of visible light lasers was therefore later discontinued while the other two themes are still being pursued at NTT. The research just described was carried out in the Optical Device Division. At about the same time MBE research was also started up in the Basic Research Division. This research initially focused on growth of GaSb and GaSbAs and was later to emphasize growth of AlGaAs for multiple quantum well physics and its application to optical devices.

The Electrotechnical Laboratory (ETL), the main electronics laboratory under MITI, played a key role in the launching of the second wave of MBE research in Japan. Its role in the planning and monitoring of three major MITI R&D programmes, each begun around 1980 and supporting MBE research in a different way, may have been more important for MBE in Japan than its own specific research contributions. The two functions are, however, so closely intertwined that they should be viewed together.

ETL's first experience with MBE went back as far as the early 1970s when research on visible light lasers was started as part of the Pattern Recognition Large Scale Project. This project was mainly concerned with software but Dr. Sakurai at ETL who was planning the project had decided to include research on optical elements, including semiconductor lasers, in the project. Semiconductor laser research had more or less stopped in Japan around 1966-67 as a result of a growing conviction that it would be impossible to realize semiconductor lasers that could operate at room temperature.[12] In 1970, Dr. Hayashi, then working at Bell Labs, had instilled some new hope when he demonstrated a double heterostructure (DH) GaAlAs laser at room temperature. As a result semiconductor laser research was resumed in Hitachi, NEC and Toshiba. By including lasers in the pattern recognition project Sakurai was able to provide early encouragement for these research programmes at NEC and Toshiba, while Hitachi

12 This laser, which had been grown using Liquid Phase Epitaxy (LPE) and represented a new concept (GaAlAs heterojunctions had been introduced as late as around 1968) still was far from a practical device. Its lifetime was limited to a few hours which would have to be improved to somewhere in the order of one million hours to allow practical use.

preferred to stay out of the laser part of the project. NEC was given the task of developing single mode lasing and Toshiba to develop long-life lasers.

ETL's task of developing a visible light laser was more challenging and speculative than the tasks allotted to the companies, a division of tasks which was in line with the role defined for ETL. While semiconductor research in the Japanese companies relied on LPE as the growth method, ETL decided to begin experiments also with MBE. As later became clear, the chosen material, GaAsP, due to its content of phosphorous, caused erosion in the growth chamber which seriously reduced the efficiency of the vacuum pumps and made it difficult to grow high quality crystals. Nevertheless GaAsP was successfully grown on GaAs and when the work was discontinued around 1978 a lot of MBE know-how had been accumulated through hard and sometimes frustrating work. New MBE research activities had also been launched, all of them on somewhat unusual materials. The second MBE group to be formed, around 1975, began research on II-VI compound semiconductors, mainly ZnSe and ZnTe. This was followed by the establishment of other groups trying to grow wide band-gap semiconductors (initially AlN and later GaN) and Si respectively.

The fairly hesitant and small-scale MBE research in Japan, and elsewhere, prior to 1980 reflected uncertainty about both whether MBE would be technically successful and if it would be needed considering the availability of alternative crystal growth technologies. Optical devices was not an important target for any of the four industrial companies doing MBE research prior to 1980. The market for optoelectronic devices hardly existed[13] and LPE seemed a more reliable crystal growth technology than MBE. The usefulness of MBE for transistors and other electric transport devices[14] was by no means obvious either. One could argue that the existence of a large market for ICs and discrete electric transport devices gave an economic base for exploratory research on basic technologies for future devices, the size of which was infinitely much larger than that for research on optoelectronic devices. MBE grown materials were, however, expected to be used primarily in high speed devices, which represent but a small portion of the total market for electron transport devices. Furthermore, at the time

---

13 LEDs were put on the market around 1976/77 but were still in 1980 a little too expensive compared to alternative solutions for a real market take-off. Semiconductor lasers were still plagued by too short a life time. Continuously working semiconductor lasers operating at room temperature were therefore not yet commercialized. LPE still showed low reproducibility. Ion implantation had not yet been introduced so the cost of producing lasers was high. The problems of laser life-time were solved around 1981/82 and then the laser market took off.

14 The term transport device is used here to denote electron devices other than optical devices.

military applications (microwave devices) almost totally dominated the market for high speed electronics and Japanese firms did not have a strong position in the US or any of the other major military markets. As a consequence MBE researchers in the Japanese companies found it increasingly difficult to defend their research.

Apart from the need to argue their case with management, the researchers themselves were not so concerned with what devices might ultimately be produced with the help of MBE. They were still struggling to master the basic technology of growing high-quality crystals. The general concept of superlattices, band gap engineering, etc., that had been introduced by Esaki and others around 1970 were so intriguing to most researchers that these visions more than anything else motived efforts to master the technologies that were required to realize the visions and open up a new world of physics and most likely also of device engineering.

The Japanese researchers were by no means alone in having difficulties in growing good crystals, but there seems to be some agreement that until the early 1980s they were behind at least the best research groups in the United States, including Bell Labs and IBM. There was one very good reason for this. The various research organizations in Japan simply had not put the same amount of resources into MBE research as their major contesters in the United States. Another factor which slowed the progress of MBE research in Japan and which only gradually became obvious was that the infrastructure of equipment and materials suppliers in Japan was not up to standards. The vacuum companies had less experience with ultra high vacuum (UHV) equipment than the leading companies in the United States and Europe. The latter typically were more discriminating in the materials they used, thereby eliminating many sources of contamination during growth. When Dr. Cho, the pioneer of MBE from Bell Labs, visited Japan in 1978 it was revealed that different and better crucibles were being used in the United States than in Japan. The Japanese had used "composite BN" crucibles while Union Carbide supplied IBM and AT&T with "Pyrotic BN (PBN)". After the Japanese researchers shifted to the Union Carbide crucibles they were able to obtain much improvement in their crystal quality.

After MBE growth technology became established around 1980 for at least AlGaAs, standard MBE machines became available and traded internationally. Before that, the MBE researchers themselves had usually done the basic design for each MBE machine, and solved the detailed design problems in cooperation with a local vacuum company, the latter subsequently building the machine to order. It is worth emphasizing that the physicists who became involved in MBE, contrary to what would have been their natural inclination, had to pay a lot of attention to technical problems in the design and operation of their machines.

Whatever capabilities were lacking on part of the Japanese suppliers probably was to a large extent a reflection of the as yet less developed market in Japan for the most advanced equipment as compared to, for example, the United States. Today, equipment from the top Japanese companies, Ulvac and Anelva, is considered to be comparable to that from Varian, Vacuum Generator and Riber, the leading foreign suppliers.

In line with the general career patterns in Japan there was basically no mobility of MBE researchers between the different organizations prior to 1980. The only exception we encountered is Dr Naganuma, who graduated from Professor Takahashi's research group at the Tokyo Institute of Technology and then in 1976 moved to NTT's Basic Research Division where he became one of the early MBE researchers.

In 1976 Dr Gonda, leader of the first MBE group at ETL, took the initiative to establishing an informal group of the main MBE researchers in Japan for discussing common problems and exchanging ideas. The group met at least three times a year until around 1981-82.

International exchange appears to have been rather limited during the 1970s. Professor Sakaki had been abroad doing research in the field of MBE, but did not return from IBM until 1978, and Dr Kawai, after having worked at IBM and Cornell University with MBE and related questions of superlattices, was hired by ETL around 1980. During the 1970s, Japanese MBE researchers, of course, read what was published, especially from IBM and Bell Labs, and some researchers made short visits abroad. The occasional visits to Japan of Cho and Esaki were experienced as important events. They were however short encounters. By and large in these early years each research group learnt MBE the hard way through its own trial and error with only rather general guidance from what could be sifted out of the scientific papers published abroad and a few, probably very welcome, opportunities to discuss with other MBE groups in Japan.

THE INVENTION OF HEMT AT FUJITSU

Starting in 1979 the momentum began to build for a second wave of MBE research both internationally and in Japan. The Japanese research community responded particularly intensely to the new research opportunities that were opening up as a consequence of a number of technical breakthroughs. Although they may in some sense only have represented the tip of an iceberg that was ready to come up to the surface, three developments had a particularly strong impact on the perceptions of the potential of MBE technology, two coming out of Bell Labs and a third from Fujitsu.

Dr Tsang, a young researcher at Bell Labs, in 1979 flooded journals with papers showing dramatic improvement in MBE-grown lasers. Inspired by his recent work with LPE, Tsang had used a growth temperature some 100°C higher than what was conventionally considered recommendable. He also used a new technique for cooling the growth chamber which resulted in a better vacuum and lower contamination by oxygen and carbon dioxide during growth. Finally he used Be as p-type dopant. The importance of Tsang's results was that they demonstrated that results of practical interest could be achieved by MBE, and such concrete evidence was much more convincing to management than ever so exciting visions lacking material proof. This therefore raised the spirits not only of the MBE researchers directly involved with optical materials but for those working with transport device materials as well, although there would soon be concrete results also in that field.

In 1978 Dingle and his coworkers at Bell Labs had introduced a new concept, modulation doping, through which it was possible to grow heterojunctions, where electrons that could move long distances without being scattered (high mobility electrons) were confined in a thin layer, forming a so called two-dimensional electron gas (2 DEG). This ability to create high mobility electrons presented opportunities for design of new types of high speed devices and researchers in several groups in both the United States and Europe began to explore these opportunities, but Fujitsu was the first to demonstrate a working device. A patent application for the new device called HEMT (High Electron Mobility Transistor) was filed in early 1980. In May the same year the results were presented at the spring meeting of the Japan Society of Applied Physics and a month later a presentation was made to an international audience at a conference in the United States.

Development of the HEMT had progressed extremely fast at Fujitsu. Dr Mimura[15], who had been doing basic research in device physics at Fujitsu, had read about Dingle's concept of modulation doping in a *newspaper* for the first time in the spring of 1979 and began to give some thought to what kind of new device might be built on this principle. In July his ideas had matured to the point where he had a concrete device concept. He

15 Mimura received his doctoral degree from Osaka University in 1982.

approached Dr Hiyamizu, who had been leading the MBE group since 1975, to discuss the possibilities of realizing the device design. A formal project was started in which Mimura and Hiyamizu worked close together in an iterating process of development of the crystal growth technology and modification of the design of the device. Major problems of reproducibility were encountered until an "air locked system" had been developed early in 1980. The first devices were, however, produced already in October 1979.

A major reason for Fujitsu's success appears to have been the close cooperation between an MBE researcher and a device physicist. Fujitsu was more or less the only company in Japan at the time with an active research programme on MBE growth of GaAs GaAlAs heterostructures. Abroad, most of the groups working on HEMT-like devices seem to have consisted almost entirely of MBE researchers with very little input from device-oriented researchers.

THREE MITI PROJECTS

It took a little while before the dust had settled from the scientific debate about the invention of the HEMT and its genuine novelty was generally acclaimed. This of course eliminated the threat that had earlier been felt that MBE research at Fujitsu might be discontinued. In fact Fujitsu took steps to enlarge its MBE research. A new group was established with the singular task of perfecting crystal growth technique for HEMTs; both discrete devices and ICs. Hiyamizu's group then soon became free to pursue other avenues of MBE research. The momentum to expand MBE research was further aided by Fujitsu taking a leading role in MBE related research in no less than three MITI projects that were started around 1981.

The MITI projects referred to were the Supercomputer Project, the OEIC Project[16], and the Superlattices Devices Project[17]. MITI had already earlier been involved in encouraging MBE research in Japanese companies, mainly through financial support for the acquisition of MBE equipment. At least two companies benefitted from such support, Fujitsu being one.

16 The official names of these two Large Scale projects were "High Speed Computing System for Scientific and Technological Uses" (1981-89) and "Optical Measurement and Control System" (1979-85). The total budgets of the projects were around \$ 120 million and \$ 100 million respectively. (*Throughout this paper \$ 1 = ¥ 150 is used as conversion rate*)

17 The Superlattices Devices Project (1981-90), with a total budget of around \$ 25 million, was part of a new scheme "Research and Development Project on Basic Technologies for Future Industries" launched by MITI in 1981 and initially including 12 projects three of which dealing with new electron devices. Plans for the three electron devices projects are described in Sakamoto (1982).

The Supercomputer Project most directly followed up on the invention of the HEMT. The development of various types of high speed LSIs was an important aspect of this project. Each of the participating six companies received financial support to develop production technology for one or two of three main types of ICs. Fujitsu and Oki were contracted to work on HEMT ICs and reverse HEMT ICs respectively. GaAs MESFETs and ICs based on Josephson Junction devices were the other two types.[18] The target for Fujitsu's and Oki's work was to develop 16 Kbit SRAMs for random number generators and this target has been successfully achieved. In spite of this success and continuing research at both Fujitsu and Oki it remains unlikely, or at least highly uncertain, that HEMT LSIs will ever be used commercially, as they exhibit some operating characteristics (for example, poor drivability) which are very unfavorable compared to alternative ICs.

Discrete HEMTs, that is components consisting of only one transistor, have however found commercial applications. The most important is in tuners for satellite TV broadcasting receivers which make good use of the very low noise characteristic of the HEMT. It is estimated that in 1989 around one million tuners were produced in Japan, each containing five HEMTs. When considering that as many as 10 000 HEMTs may be fabricated on one wafer it becomes clear that the market for discrete HEMTs is still fairly limited in value terms, even if it is expanding at a rapid rate.

The Superlattices Devices Project also focused on electron transport devices but was more speculative in nature than the Supercomputer Project. Its objective was to demonstrate new device concepts. Most of the research, however, actually concerned the development of new basic technologies for crystal growth and device fabrications, with MBE as one of the prime examples of such technologies. Four companies and ETL were involved in the Superlattices Devices Project, and all but Sony would have MBE growth as a crucial part of their research. Sony, which had pioneered MOCVD, a crystal growth technique competing with MBE, remained faithful to its past accomplishments.

The other three companies, Hitachi, Fujitsu and Sumitomo Electric, each worked on different devices and materials systems. Hitachi was supported to direct its already existing Si MBE research to the development a Permeable Base Transistor (PBT) which did not build on the conventional concept of superlattice, but under the label of "lateral superstructure device" was nevertheless accepted. Fujitsu was charged with

---

18 The six companies were: Hitachi, NEC, Toshiba, Mitsubishi Electric, Fujitsu and Oki. Fujitsu worked both on HEMT ICs and Josephson Junction ICs.

developing a Resonant Tunneling Hot Electron Transistor (RHET) in InAlGaAs and succeeded in fabricating not only an individual transistor but an IC with a new function. Sumitomi Electric, which had no previous experience with MBE was given the task of developing a HEMT in $(InAs)_3(GaAs)_1$ with AlInAs as channel layer.

Within the framework of the 3-Dimensional IC Project, running in parallel with the Superlattice Devices Project, there was still another project involving MBE, namely an effort by Oki to grow GaAs on Si. Already in the mid 1980s Oki managed to grow good crystals and is today considered the leader in Japan in the technology of growing GaAs on Si.[19]

It deserves to be emphasized that although all the just mentioned four projects had the demonstration of devices as their targets, these targets were not so important in themselves but served a useful function in directing and setting a standard for the development of the basic technologies of crystal growth and device fabrication.

The Supercomputer and the Superlattices Devices projects both ran over a very long period, nine and ten years respectively. The companies which got engaged in these projects effectively committed themselves to maintain viable research groups for that long a time. For the MBE researchers at Hitachi and Fujitsu, who had very recently worried about the viability of their research, and for researchers in the new groups at Sumitomo and Oki the future must have seemed very bright. Even without the MITI programmes it is likely that the prospects for MBE research would have improved, given the recent breakthroughs, but hardly to the extent achieved by the stability and extra funding of the MITI programmes.

The Superlattices Devices Project also gave a strong boost to MBE research at ETL. As customary in MITI's electronics R&D programmes, ETL had been intimately involved in the planning of the project, with Gonda drawing up the plan, including research at ETL itself. Researchers from three or four different groups at ETL were drawn into the project. A new Riber MBE machine, considered much superior in reliability to the machines already installed at ETL, could be acquired with the new funding made available and was used for growth of AlGaAs superlattices. Another machine, acquired by the VLSI Joint Research Laboratory towards the end of its existence and then transferred to ETL around 1979, was used for Si and Ge MBE. The research at ETL included both the study of the basic mechanisms of crystal growth and exploration of

19 Also MBE research at Sanyo has been performed with support from the R&D project of Basic Technologies for Future Industries (Shiraki et al., 1989, p. 504), but it is unclear under which sub-project.

some new device concepts. An important result was the development of a technique for nearly atomic layer by layer controlled growth, so called phase-locked epitaxial growth, based on intensity oscillations in Reflection High Energy Electron Diffraction (RHEED), the latter a technique usually employed for monitoring of the growth process in real time.

The R&D Project on Basic Technologies for Future Industries, of which the Superlattices Devices Project has been a part, is gradually being renewed as the early projects are phased out and new projects take their place. A Bioelectronic Devices Project was, for example, introduced in 1986 when the Fortified ICs for Extreme Conditions Project was completed.[20] The other two initial electron device projects will both be completed in 1990. There is an interest both at ETL and in some companies in the launching of a new project, which would be a natural continuation of the Superlattices Devices and the Three-Dimensional ICs projects, centering around the concept of quantum devices but it remains unclear whether such a project will be able to compete with other proposals. The fact that similar projects have already been started with support from other ministries is one factor speaking against the project.

The OEIC project aimed at the development of OEICs, that is electric transport and optical devices integrated on one chip, and their application in various systems for control of industrial processes using optical signals. The systems applications came only towards the end of the project while the bulk of the project concerned the development of crystal growth and device fabrication technology for OEICs. The latter was divided into two distinct activities, one being the development of specific OEICs contracted to five companies[21] and the other being research on basic technologies for future OEICs in a laboratory jointly operated and staffed by the five companies receiving special contracts, an additional four companies[22] and ETL. The projects in the individual companies were started already in 1979 while the Optoelectronics Joint Research Laboratory (OJRL) began its operation first in 1981.

The company projects relied for the most part on already established technologies which in terms of epitaxial growth meant LPE. MBE was however included in OJRL's research programme. In the early stages of the planning of the OJRL, which was set up only after considerable resistance from the five leading companies, the idea had been to

---

20 The Superconducting Materials and Devices Project (1988-97), and Non-linear Photonics Materials Project (1989-98) also have relevance for electron devices.

21 Hitachi, Mitsubishi Electric, Toshiba, NEC and Fujitsu.

22 Matsushita, Oki, Furukawa Electric and Sumitomo Electric.

have one of the six research groups work on MBE. As time went on the interest in MBE grew among the companies, not surprisingly considering the breakthroughs already described. There was still considerable hesitancy in most companies to launch any major in-house MBE research programmes especially in the optical materials field. As MBE was beginning to show some promise, the opportunity to explore the technology without having to make any own long term commitments, therefore seemed a very appealing proposition. Several companies expressed a desire to work on MBE in the OJRL.[23]

When OJRL started its operation there were only rather rough plans for the research to be carried out leaving much discretion to the group leaders in their choice of research direction. The group leaders from Fujitsu (Dr Hashimoto), Mitsubishi (Dr Ishii) and Hitachi (Dr Nakashima) all wanted to be involved in MBE research. However MITI/AIST did not want the different research groups to overlap too much. To satisfy MITI/AIST somewhat artificial distinctions had to be made between the groups. As it turned out the three groups doing MBE work naturally focused their work in such a way that their work mixed nicely and became more complimentary than duplicating. Ishii's group had the study of epitaxial growth, mainly MBE but also some MOCVD, including fundamental study of growth mechanisms as its main objective. Hashimoto had a background in ion implantation research and combined MBE and Focused Ion Beam equipment in one machine so that crystal growth and implantation could be done without breaking the UHV condition. Nakashima who, like Ishii, was a laser specialist, decided to develop a multiple quantum well laser, which used a new device structure ("index guided"). This required development of MBE technology and brought Nakashima's group into quite basic studies of surface physics.

The OJRL was a very important step in the evolution of joint research in Japan.[24] Most of the Large Scale Projects have since the beginning of such projects in the 1960s included the establishment of joint laboratories. Their function has, however, almost

---

23 A major cost of participating in the OJRL as seen by the companies was that they had to send some of their own researchers, which could be a very significant sacrifice considering the scarcity of qualified researchers, especially in dynamic fields. At the same time the most effective way of gaining a benefit of research in a joint laboratory was to have own researchers work there and later return to the company. The OJRL seldom received the most ideal researchers from the companies in terms of their immediate research experience. This, however, does not seem to have held back the research to any significant extent. The group leaders, and the specially selected coworkers one of whom each group leader brought from his own company constituted a strong enough backbone, which complemented by a few additional experienced researchers quickly could guide the whole organization to perform very impressive research.

24 Interesting accounts of the operation of the OJRL can be found in Hayashi et al. (1989), Merz (1986), and Shimada (1986).

always been limited to the very end of the projects and consisted in the final integration of the various technologies or subsystems developed during the main part of a project into a demonstration system. The widely discussed VLSI Project (1975-1979) therefore represénted something new in that a joint laboratory was established from the beginning of the project. There were, however, several research sections in the joint laboratory and each was largely staffed with researchers from only one company with the result that the research was joint only to a limited degree. The reason was that the research work carried out was rather close to the market making it difficult for the companies to cooperate in earnest. The OJRL, by contrast, focused on more basic research and therefore could, and did, mix researchers from several companies freely in each section; and there was also a lot of interaction between the groups as they all had their desks in one big room.

The OJRL is generally considered to have been very successful in its mission. It was extremely well equipped and with almost 50 researchers it represented a much larger and more comprehensive research effort than any of the participating companies could muster in the OEIC field at the time. As MBE was one the main topics of research OJRL naturally had a strong impact on the development in Japan of this field. That influence took many forms. An important mechanism for transmitting results from the OJRL to the member companies was the upgrading of the capabilities of the researchers who worked at OJRL and then went back to their companies.[25] Approximately ten researchers, none of whom had any experience with MBE when they came to OJRL, are now actively pursuing MBE research in their companies. Research at OJRL also gave the companies something to lean on when they made decisions about their own MBE research. The research at OJRL set challenging goals and sought innovative solutions, thereby often laying out fruitful directions for future research.

NEW ORGANIZATIONS FOR JOINT RESEARCH

The Optoelectronics Technology Research Laboratory (OTL), which was set up soon after the discontinuation of the OJRL in 1986, is in many respects carrying on OJRL's mission but is also taking the concept of joint research one step further. It is one of the first of a large number of joint research firms that have been established since 1986 with

---

25 While almost all of the researchers went back to their companies, the situation was different for the group leaders. As many as four of them became professors immediately or soon after leaving OJRL. Two went back to their companies but only one to work in the semiconductor field. Iizuka first went back ETL, then joined a company. (One of the group leaders left OJRL early and was replaced which explains why the number of group leaders add up to seven).

financial support from Japan Key Tech Center, a quasi-governmental organization. OTL is planned to operate during 10 years.

The Key Tech Center, operated under the joint supervision of MITI and the Ministry of Post and Telecommunication (MPT), has several tasks but the most significant is to finance joint research between groups of firms. Through equity investment in, or loans to, joint research firms it provides up to 70 percent of the financing necessary to carry out specific research programmes. The Key Tech Center, which has the legal status of a non-profit foundation, relies on a capital fund built up from several sources, the most important of which are dividends from government ownership of NTT shares. The way the Key Tech Center is funded means that it can operate free from the constraints of the government's annual budget. The real strengths and weaknesses of this mechanism for support of joint research, compared to for example the long standing Large Scale Projects, will only become clear when some of the joint research firms have concluded their work and the results of their work as well as the financial arrangements be evaluated by the member firms. There still seems to be only vague notions what will happen financially when the joint research firms have exhausted their initial financing.

The OTL consists of one big group of 20 researchers[26], all looking from different angles at the problem of how to fabricate low dimensional structures, for example quantum wires and quantum boxes. Epitaxial growth, mainly using MBE of one or another kind, is an important research component which is studied in its most fundamental aspects. For this purpose the most advanced analysis equipment has been combined with the crystal growth chambers in ways which, in several cases, have not been tried anywhere before. The research is as basic as that which can be found in any university but carried out under much more favorable conditions. The researchers, widely ranging in age and the research fields and research cultures in which they have worked, have no other obligations than their research while at OTL. The material resources leave little more to be desired. The leadership, represented by the Dr Hayashi, Director, and Dr Katayama, the Technical Director, has a deep knowledge of the topics being researched. With each researcher coming from a different background and many complementary approaches being taken to study the same physical phenomena the OTL seems to be an ideal place for creative research.

One might ask what the member companies expect to get out of OTL. As the utility of quantum wires or the like for practical devices is uncertain at best, the most convincing

---

26 All but one on assignment from one of the 13 member firms: Mitsubishi Electric, Hitachi, Sanyo, Matsushita, Furukawa Electric, Japan Sheet Glass, Oki, Toshiba, Fujikura, Fujitsu, NEC, Sharp, Sumitomo Electric. The Director and the Technical Director are not counted among the 20.

answer is that the companies are seeking access to high quality basic research. The role of the objective to fabricate structures such as quantum wires is mainly to focus the research programme. The expectation on the part of OTL's management is that in the process of reaching for this objective a deep understanding of quantum crystal materials and structures will be gained, an understanding which will be useful also for fabrication of more mundane semiconductor devices in the near term. In developing this philosophy, Dr Hayashi has built on his experience from the OJRL, where he served in a special technical advisory position. The philosophy therefore reflects what he considers to have been the most essential contribution of OJRL.

The Advanced Telecommunications Research Laboratories (ATR), established in 1986 and located in the new Culture and Science City under development near Nara in the Kansai area, is an additional example of a joint research organization which is doing MBE research and which is supported by the Key Tech Center. For several reasons it is likely to have less of an impact in the field of MBE, the main simply being that MBE research is a smaller activity at ATR than at OTL. ATR, formally consisting of three laboratories, has several hundred researchers in total but the great majority of these are studying systems and software oriented matters. In the ATR Optical & Radio Communications Research Laboratory, the Optical & Electronic Devices Group has around ten researchers most of whom are involved in research on semiconductors materials and physics of advanced devices, with MBE as one of the primary technologies. ATR is, like OTL, generously funded and well equipped, but short on researchers. Although ATR probably may have set somewhat less ambitious research targets than the OTL, the differences between them in research organization are probably larger than differences in the nature of the research. The main difference lies in the weaker continuity at ATR than at OTL. While the researchers at OTL are semipermanent and seldom stay less than three years, three years is considered the upper limit at ATR and frequently researchers stay an even shorter time. The fact that also the group leaders at ATR change every three years must cause great problems in setting and adhering to long term research goals.[27]

Japanese electronics companies have almost all their R&D laboratories in the area around Tokyo or in the Kansai area, with a strong and growing dominance by the former. The establishment of ATR is part of a conscious effort to enhance the environment for companies to do R&D in the Kansai area. While OTL draws its

---

27 The head of the whole Optical & Radio Communications Laboratory is appointed for ten years, the originally planned lifetime of ATR.

members from both regions, the Optical & Electronic Devices group at ATR has so far only attracted, and maybe only tried to attract, members from the Kansai area, the exceptions being NTT and NHK which are special cases.[28] Notably, the leading Kansai electronics companies are also members of OTL. Considering the small scale of ATR's semiconductor research, its success will critically depend on how close a cooperation it can develop with research in its member companies and maybe also with research in the universities in the Kansai area. It has a greater need to develop such links than OTL and the development of close cooperation with research in Kansai universities and companies may offer a special opportunity for ATR consistent with the reason for its establishment.[29]

DEVICE DEVELOPMENT AND BASIC RESEARCH IN COMPANIES

Although the MITI-supported projects in all likelihood helped accelerate the second wave of MBE research, this effect became increasingly less important as more and more companies built up and expanded their MBE research.

In the cases of Fujitsu and Hitachi, MITI funding, as already described, contributed to strengthening already existing activities by providing additional funding and guaranteeing stability over a longer term than could typically be taken for granted for research funded purely by internal corporate funds. At Oki and Sumitomo Electric, on the other hand, the MITI projects were launched more or less at the same time as new MBE research groups were being started up. Judging from their publication records MBE research in the latter two companies appears not to have expanded very much beyond the topics covered by the MITI projects.

Other companies which began MBE research did so without any direct financial support from MITI programmes. Mitsubishi Electric and Sanyo were among the earliest to start up MBE research in the second wave, both companies primarily focusing on optical devices. A large number of companies started MBE research around 1982-84. NEC and Matsushita developed rather diverse research programmes with optical devices representing a minor part while Sharp, Rohm and Omron almost totally concentrated on laser development. At Sony and Toshiba MBE research never developed beyond rather small research programmes as both companies early became

28 Among the members of ATR's Optical & Electronics Devices Group are NTT, NHK, Sumitomo Metals & Mining, Sumitomo Electric, Murata, Toyobo, Mitsubishi, Rohm, Sanyo and Sharp.

29 In its major fields of research, ATR is likely to be relatively much stronger and for those the situation would therefore be quite different from that in semiconductor research.

strongly committed to the development of MOCVD, an alternative technology to MBE. In the latter part of the 1980s an additional 5-10 companies, some, for example Toyota, outside the electronics industry, have initiated MBE research for the most part on a limited scale. Japanese suppliers of MBE equipment - Ulvac, Anelva and to a lesser extent Seiko Instruments - have invested resources in the development of MBE technology, usually in cooperation with customers, but do not appear to have had major internal research programmes.

MBE research in the companies cover a broad range of activities which can be characterized in different ways, for example in terms of the materials studied or the device applications at which the research is aimed. Although the ultimate justification for more or less all MBE research in companies is the prospect of new or better practical devices of one or another kind, the link between material growth by MBE and the development and fabrication of devices takes on different forms. On the one extreme crystal growth by MBE is done simply to supply, on a routine basis, wafers for further fabrication of devices aiming at development of devices with improved characteristics. On the other extreme, crystals are grown for the purpose of discovering or exploring some new physical effect which may open up as yet unknown possibilities for new device concepts. Most of the research related to low-dimensional quantum confinement structures can be put in this latter category. HEMTs and certain types of lasers based on MBE-grown crystals have already been commercialized, so MBE research related to such devices has, for the most part, well defined and near term objectives of improving device performance. Most of the MBE research in the Japanese companies falls, as might be expected, between the two extremes.

NTT is generally considered to be the leader in MBE research among Japanese companies with Fujitsu being a good second.[30] From the start NTT has been emphasizing optical devices and Fujitsu been particularly interested in transport devices for ICs, first HEMTs and later resonant bipolar transistors (RBT) and resonant tunneling hot electron transistors (RHET). Already in the early 1980s, Fujitsu, however, began to use MBE also for research on optical devices and was one of the first Japanese companies using this technology for the realization of OEICs. Correspondingly, NTT around 1983, in its LSI Laboratories, began research on various transport devices, the most important of which would become heterobipolar transistors (HBT) which have been developed to the level of ICs.

---

30 NTT was, for example, among the first to develop gas-source MBE in the early 1980s and Horikoshi at NTT has developed a modified version of MBE, labelled Migration Enhanced Epitaxy (MEE), which yields extremely high quality crystals at low substrate temperatures by interrupting the flow of As during the deposition of Al and Ga.

Fujitsu's MBE research has been almost totally limited to III-V compound semiconductors and these have also dominated at NTT. Although NTT has had several groups working on Si MBE, the clear leaders in this field are NEC and Hitachi. Hitachi, which pioneered Si MBE in Japan in the late 1970s, did not begin MBE for III-V semiconductors until around 1983 and then primarily for transport devices. NEC which started its MBE research rather late, around 1984, has since developed a broader and larger research programme than Hitachi with extensive research of Si as well as III-V materials. Although optical devices were initially an important target for III-V MBE at NEC it appears that MOCVD soon became the favored technology for such devices.

Electronic companies in the Kansai area[31] have traditionally been considered as having a more pragmatic view of R&D than the major electronics companies in and around Tokyo[32] and as putting less emphasis on long-term research. The fact that both Matsushita and Mitsubishi discontinued their MBE research in the late 1970s and later developments in these and other Kansai companies seem broadly consistent with this view although the distinction must should not be overdrawn. Matsushita, Mitsubishi and Sharp are well ahead of other Kansai companies in terms of the number of MBE publications. Both Mitsubishi's and Sharp's MBE research have focused on III-V laser materials, while Matsushita has mainly worked on various transport devices. For a while Matsushita had a Si MBE group but that was discontinued. Both Matsushita and Sharp have groups for II-VI MBE.

Three of the top four laser manufacturers[33] in Japan are located in the Kansai area. Rohm Co., which has only published a handful of papers on MBE, all around 1985 and concerning lasers grown by MBE, captured in 1988 the position as the second largest laser producer in Japan after Sharp. Rohm is said to be the only company in Japan using MBE for commercial production of lasers. Sharp has been very successful in producing lasers by LPE, a fact which delayed its start up of MBE research. Research on MBE-grown quantum well lasers which has resulted in practical devices has recently been transferred from the Central Research Laboratory to one of the product divisions. As Sharp has not been very active in MOCVD, judging from the company's publication record in this field, there seems to be a good chance that Sharp will begin commercial

---

31 E. g. Matsushita, Mitsubishi Electric, Omron, Rohm, Sanyo, Sharp and Sumitomo Electric.

32 Fujitsu, Hitachi, NEC, NTT and Toshiba.

33 The market share data, derived from on-line search in *Nikkei Telecom*, refers to the number of lasers produced and ranks the top nine laser manufacturers in 1989 as follows: Sharp, Rohm Co, Sony, Mitsubishi Electric, Toshiba, Matsushita, Hitachi, NEC and Sanyo. Short wavelength lasers which are used in e. g. CD players and laser printers are produced in much larger quantities than long wavelength lasers but exhibit a much lower unit price. The rank ordering of laser manufacturers in terms of value of shipments may therefore be different from the rank ordering referred to here.

manufacturing of sophisticated lasers by MBE. Mitsubishi is unique among the laser manufacturers in that it has had strong laser research programmes for both MBE- and MOCVD-grown crystals, while other companies, as for example Matsushita which has selected MOCVD, have tended to emphasize one or the other of the two crystal growth methods for development of sophisticated lasers. There is however no evidence that Mitsubishi uses MBE for commercial production of lasers.

Many Japanese electronics companies have commercial production of discrete HEMT devices and monolithic microwave ICs (MMICs) and several are using MBE for at least part of this production, with MOCVD and LPE making up the balance. Mitsubishi is known to have developed very effective MBE production technology for these types of devices. Interestingly, a key person in developing this technology has been Dr Sonoda, who was one of the pioneers of MBE at Mitsubishi in the 1970s, but then left the field as the company discontinued its MBE research. Sonoda has later, at Mitsubishi's semiconductor works, revived his interest in MBE, applying MBE to the production of HEMTs. Fujitsu has pushed beyond discrete HEMTs to Large Scale Integrated Circuits (LSI) with several thousand transistors and is the leader in manufacturing technology for HEMT LSIs. There is, however, considerable doubt whether HEMT LSIs will ever become commercially viable products. Hitachi has also made efforts to develop MBE into a production technology and has developed its own production equipment, which although intended for sale on the open market, so far probably is only used by Hitachi itself.

Generally for transport devices that are currently being researched with MBE, HEMTs being just one example, the future commercial prospects of their use in LSI are very uncertain and many observers are sceptical that MBE will have any role in manufacturing of LSI in the next 20 years worth mentioning. Although upsets in this assessment cannot be excluded - devices based on Si-Ge heterostructures is a possible case - the evaluation of the potential of MBE as commercial production technology can be made with greater confidence today than in the early 1980s and as just mentioned it tends to be rather negative, which does not exclude that there may be opportunities for the development of niche markets.

The difficulty in identifying device concepts which offer a really promising commercial potential, may to some extent promote a shift towards more basic and long term MBE research in industry, especially at the present stage when the technology for some of the most obvious candidates for MBE grown devices, such as HEMTs and certain types of lasers, is becoming well established and presenting fewer challenges for the researchers.

The currently strong interest in new concept devices among MBE researchers in Japanese companies could be interpreted in this light.

One of the persons interviewed in this study even suggested that there are too many company researchers in Japan working on what still amounts to very speculative device concepts and that this was producing a risk that research on more realistic GaAs transport devices is being left undone, possibly contributing to GaAs technology declining further in relation to Si technology. This in turn could lead to a new "cold winter" for compound semiconductor research similar to the one experienced in the late 1970s with the result that compound semiconductor technology will effectively have disappeared from the companies before any new concept devices have been realized.

When the expected time horizon for realization of practical devices is long or when there is no conscious link at all with device development it would seem appropriate to classify the MBE research in question as "basic research". With this definition, a rather large part of MBE research in Japanese companies which shows up in scientific publications can be classified as basic research. There has been a general trend among Japanese companies since the early 1980s to expand their basic research. The case of MBE gives some basis for examining the question of why Japanese companies do basic research.

The utility of the specific research results that may be obtained is only one of several factors. The positive image created from pursuing research at the scientific frontier is seen as a powerful means of building confidence in the company among customers as well as among employees and also makes it easier to attract new qualified scientists and engineers. For image creation it is probably more important that a field can be associated with visionary thinking that sets one company apart from the mainstream than that it shows a large commercial potential.

Of course a strong programme of basic research will not be set up to create illusions alone. It will usually be expected to generate basic patents which can be used in different ways, not least in negotiations with other companies, when the latter have patents with the power to block development. Basic research can also be seen as a way of developing competence in a field, a function which may be particularly important in those cases when a field has not yet been well established in the universities and therefore expertise can not easily be introduced in the companies through recruitment of new graduates. Having research competence in a field before it is well established yields a preparedness to sudden break-throughs in the field. It gives the ability to identify and correctly evaluate such breakthroughs both through the competence built up in the company and through the social network relations to expertise in other organizations that are often easier to build around basic research topics than commercially more sensitive R&D. The existence of an inhouse research group of course also means that a company can respond much faster to a new development than if it has to start from scratch, in terms of actually exploiting the new opportunities that are opening up, something which could be of great importance for establishing a prominent position in a new field.

Returning to the specific case of MBE, the technology is continuously undergoing change through an interactive process involving a broadening of the range of materials successfully grown by MBE, experiments with new device applications, changes in MBE equipment and growth methods, and a deepening of the understanding of growth mechanisms. Increasing use of gas sources in MBE, the use of various beams to enhance or localize the MBE growth process, and the combination of MBE, electron beam lithography and reactive ion etching and a growing number of sophisticated analysis instruments in the same UHV equipment are some of the most prominent trends in the most tangible aspects of MBE technology. Together these developments open up a multitude of possible new combinations, the practicality and utility of which are uncertain and quite difficult to evaluate at the present time. In this situation it may be more appropriate to pursue rather basic research than to attempt to organize a research programme around a specific device application. Of course conceivable device applications will to some extent direct the choice of research focus, but considering the motives for corporate basic research, discussed above, the uniqueness of the research target and approach may be as important.

The fact that a company is pursuing a particular research direction more aggressively than other companies, thus cannot automatically be interpreted as a sign that the company evaluates the economic opportunities in this area more optimistically than other companies. It may simply mean that the company has discovered that it has developed a strong research position in this direction and judges that it can receive a high visibility from nurturing that position.

Some of the confusion surrounding definitions of basic research stems from attempts to define categories of research according to the guiding motives of the research. A company which is ultimately interested in producing devices nevertheless will in many cases carry out research which is more or less identical with that pursued by academic researchers who are primarily seeking recognition within the scientific community. This difference in motives is likely to have some impact on what questions are given priority and how the research is carried out. In simplified terms one can argue that the industrial researcher will be content to achieve a practical result, for example to grow a certain material to specified quality, while the academic researcher will want to know how the result is achieved, which in the case mentioned means to understand the growth mechanisms. In a science-intensive field like MBE such a distinction, however, tends to get blurred (cf. also chapters 3, 8 and 9). Many academic researchers, especially in engineering faculties, orient their research with a clear recognition of its potential impact on device development (see also chapter 2). Correspondingly, industrial researchers, while being aware of corporate objectives, will often share the interest of academic researchers in obtaining a scientific understanding of the phenomena they are trying to control, partly because scientific understanding may be necessary for achieving practical results and partly because they value such understanding highly. The intertwining of the academic and the industrial research communities is further aided by the fact that some of the most advanced experimental research is carried out in industrial laboratories. There are examples of industrial laboratories being the first to provide experimental evidence of quite basic theories.

UNIVERSITIES STRENGTHENING THEIR RELATIVE POSITION

MBE research was first established primarily in industrial laboratories and, with only a few exceptions, rather late started up at academic institutions. In this regard the development in the United States and in Japan has been rather similar. In both countries there has been a strong surge of interest in MBE among university researchers, during the latter half of the 1980s. Judging from the bibliometric data discussed earlier, the size of the MBE research community in American universities is however around five times as large as that in Japan, reflecting the much larger number of "research universities" in the United States as compared to in Japan.

Until the mid 1980s only half a dozen universities were doing significant MBE research in Japan. Mentioned in approximately the order of establishment they were: Tokyo Institute of Technology (two groups), Waseda University, Tokyo University, Tsukuba University, Kwansei Gakuin University and Shizuoka University. The different groups grew up with little contacts between them, one exception being contacts between Professor Sano at Kwansei Gakuin University and Professor Sakaki when the former started his MBE research.

During the first half of the 1980s universities also seem to have had rather limited contacts with companies in the MBE field. For one thing the few universities active in the field were for the most part in the process of building up their research so there was still only a very small number of students who had received a doctoral degree for MBE research. Prior to 1985 only a handful new PhDs specializing in MBE moved from universities, largely from Tokyo Institute of Technology and Waseda University, to pursue MBE research in companies. Although there were in addition some MBE researchers who moved to companies after taking a masters degree, and there were some company researchers who spent a year or so at a Japanese university to learn MBE, it seems fair to say that universities until the mid 1980s contributed only marginally to the development of MBE research competence in Japanese companies. The influence from companies to universities does not seem to have been very significant either during this time. This should not be interpreted to mean that academic and industrial researchers were unaware of each others work. There were conferences and other opportunities for researchers to stay abreast of MBE research activities in Japan.

In recent years universities have become a more vital part of the Japanese MBE research community. The number of research groups has expanded considerably, both through the addition of new groups at universities which were already active in the MBE field and through additional universities becoming involved in MBE research. MBE research is today actively pursued at some 25-30 universities in at least 35 different research groups. At Tokyo University alone there are around half a dozen groups. Osaka University, Tokyo Institute of Technology and Tsukuba University each has three or more groups. It is still too early to judge how many of all the new groups will ultimately carry any weight.

While university researchers in Japan often maintain that they try to avoid competing head on with research groups in industry there still is a very considerable overlapping in research topics between these two parts of the Japanese research communities. There are university groups performing significant research in all important materials systems and with most types of equipment configurations. II-VI compound semiconductors and other "exotic" materials are more commonly researched in universities than in industrial laboratories but university groups are well represented also in those areas where industry is putting most of its research efforts. An important exception is deviceoriented research which has seen little participation from academic groups. Device research usually require very high quality crystals, an area in which universities, with only a couple of exceptions, have not come up to the same level as industry, partly due to the latter's superior material resources and partly, and in the final analysis maybe more importantly, because industry has much stronger motives to pursue device research and development. There are, however, examples of some of the leading university groups playing an important role in the early stages of new device development in firms through demonstration of basic features of new device concepts or as discussion partners. As much as device research is not very prominent at universities, there are few pure theoreticians in industry.

A very interesting recent development is that several of the top MBE researchers in industry and at ETL have been recruited as professors to build up new research groups. The most striking case is Osaka University which has hired during a short period no less than four prominent researchers from ETL (Gonda), Fujitsu (Hiyamizu), Hitachi (Nakashima, one of the group leaders at the OJRL) and NTT (Asahi) respectively. Three of them are now working in close cooperation at the Institute of Industrial and Scientific Research. Similarly Tokyo University and Hiroshima University have each hired leading researchers from Hitachi (Shiraki) and ETL (Yao).

This movement of researchers from industry and ETL to universities can be viewed as a result of the strong build up of basic research capabilities in Japanese industry and career choices on the part of industrial researchers. Several companies have dual career systems for their researchers, one leading through the management ranks and the other allowing the researchers to continue with research also at a relatively high age. In practice, however, the latter career seems to be perceived as having less status than a management career and somehow usually not be a natural choice for an able researcher. For industrial researchers who prefer to continue to be directly involved in research rather than proceed to management positions increasingly distant from such involvement, a position as a professor may therefore be attractive when they reach a certain age and have to make career choices. As Japanese industry has expanded its basic research the number of candidates for academic positions among highly qualified industrial researchers has greatly increased.

At the same time there has emerged a growing consensus at Japanese universities that they need to catch up with the research carried out in industry in many new fields. To hire some of the researchers who have been instrumental in building up the strong research capabilities in industry is an effective way of addressing this need. Not only will it directly strengthen the research capabilities of universities but it also significantly improves the conditions for their effective communication with industry. The hiring of industrial researchers to academic positions seems to be on the increase in the electronics field as a whole and not just apply to the MBE field, and could well be an even more general trend.

It cannot be excluded that, in the case of MBE, the movement of researchers from industry to universities also reflects a perception that firms are becoming less committed to MBE as an important basic technology.

Judging from interviews, there seems little risk that a growing number of professors with industrial background will make university research more short-sighted than otherwise would be the case. If anything the new professors coming from industry appear determined to exploit the freedom of academic research to pursue basic research problems, which they may have had to forego in their industrial career, not because research on such problems was not permitted, but because such research had low priority and had to be set aside when more pressing needs made themselves felt.

Just as there has been a movement of researchers from the leading industrial research groups to universities, the early academic MBE groups have increasingly contributed their competence to other groups in universities and companies, through the supply of new graduates and through training of special students from companies in universities. Some of these university trained researchers have come to play a key role in new groups being set up in companies embarking for the first time on MBE research.

Through a combination of bibliometric searches and interviews, the number of doctoral degrees granted for research in the field of MBE was estimated for six early established university research groups.[34] In total, almost 30 doctoral degrees were granted for MBE research in these groups, just over half of which in the two groups at Tokyo Institute of Technology and the rest divided equally between Tokyo University and the remaining three universities taken together. Around two thirds of the degree recipients are continuing MBE research and half of these are working in companies. The degrees referred to are so-called course doctorates (*katei hakase*) which presuppose attendance at a university, usually for three years after a master's degree, including certain course work.

In Japan there exists, however, also another type of doctoral degree, thesis doctorate (*ronbun hakase*; see also section 6-4) which is granted on the basis of a thesis being

---

34 Professor Takahashi's and Professor Furukawa's at Tokyo Institute of Technology, Professor Kimata's at Waseda University, Professor Sakaki's at Tokyo University, Professor Kawabe's at Tsukuba University, and Professor Sano's at Kwansei Gakuin University.

presented to and accepted by a professor at a university. The research work underlying the thesis is usually performed in non-university research laboratories, mostly in companies. Significantly, in engineering, during the decade 1976-86, the number of thesis doctorates approximately doubled while the number of course doctorates remained about level, bringing the thesis doctorates to a level almost twice that of course doctorates. This development suggests that a large and increasing part of the advanced research training associated with the expansion of industrial research in Japan has occurred in the firms. The picture is somewhat modified by the frequent dispatching of company researchers as special research students to universities.

During the latter half of the 1980s, enrollment in doctor course studies has increased notably. In Electrical and Communications Engineering, more than 50 per cent as many doctoral degrees were granted in 1989 compared to 1986. As the number of first year doctoral students was 30 per cent higher in 1989 than in 1986, some further expansion in doctoral degrees should be expected (Monbusho 1987 and 1990). Part, and possibly most, of the increase, however can be ascribed to foreign students. At Tokyo University the number of doctoral students in engineering increased from 579 in 1986 to 653 in 1989. Foreign students made up all but three of the increase bringing the percentage of foreign students as high as 42 per cent (The University of Tokyo 1987 and 1989).

The difficulty of attracting a sufficient number of doctoral students is generally perceived to be a very serious problem for Japanese universities. It is not unusual for a research group to lack doctoral students altogether. In order to improve the situation there have been some suggestions that a new doctoral degree be introduced, representing a compromise between the course and thesis doctorates. According to this idea, the requirements regarding attendance at the university would be relaxed in comparison with the current doctor course, but sufficient to assure that the doctoral student would make a real contribution to the research environment at the university before he was granted his degree.[35] If implemented these ideas could have quite farreaching effects on the research capacity of Japanese universities. Considering the already established system of special research students from companies to universities, and the competition among companies in recruiting new graduates, it is quite possible that Japanese firms would be willing to let their researchers be enrolled in such a new

---

35 A new type of doctorate, a *gakujitsuhakase*, is for example being discussed at the Research Center for Advanced Science and Technology (RCAST) at Tokyo University. Professor Furukawa at Tokyo Institute of Technology has suggested a one year minimum residence requirement for thesis doctorates (Furukawa 1989).

type of doctoral studies while remaining on the firms' payrolls. Some professors have already perceived that management in many companies has begun to show a more generous attitude to their researchers' spending company time to complete doctoral theses.

It is interesting to note, and maybe not surprising, that those three research groups which have been most successful in producing doctorates also have made many of the most original research contributions among the university groups. Professor Takahashi was, for example, among the first internationally to work on MOMBE and has since developed this process further to so called photo-MOMBE. Professor Sakaki first discussed the feasibility of using quantum wires and other low dimensional structures as materials for electron devices and has continued to develop these ideas. Professor Furukawa has used various techniques, including electron beam exposure, to establish growth methods for a number of semiconductor-fluoride-semiconductor structures and used them for realization of new devices.

One major factor deterring universities everywhere from embarking on MBE research has been the high cost of the equipment needed. Operating costs, including costs of wafers, large quantities of low temperature nitrogen and other materials also represent an obstacle. The price of an MBE machine will vary depending on specifications, guarantee clauses, etc., but is in the order of one million dollars. A university department strapped for funds, choosing a simple design with less guarantee of reliable operation, and contracting with a small vacuum company to manufacture the equipment may be able to press the price down to half a million dollars or possibly even less. Even the large suppliers may be willing to offer universities a lower than normal price as a good-will gesture.

Insufficient equipment funds has been experienced as a serious problem at universities in most countries but the situation has often been described as particularly bad in Japan. Generally speaking it appears, however, that the availability to Japanese universities of funds for equipment has improved considerably in the last ten years, to a point where the lack of such funds now appears to represent less of a bottleneck for development of research, particularly in the most prominent national universities, than ten years ago. Although there has been some increase in the total funds for equipment, of more importance may be a seemingly growing willingness on the part of Monbusho and other involved parties to concentrate research funds so as to, for example, facilitate the purchase of expensive equipment.

UNIVERSITY MBE RESEARCH FUNDING

As the mechanisms for research funding constitute a major determinant of the development of university research we will discuss them in rather much detail and attempt to place the conditions for MBE research in the wider context of university research funding in Japan.

The sources of research funds for Japanese universities may be divided into four categories:

- general university funds from the Ministry of Education (*Monbusho*),
- grants-in-aid from the Ministry of Education,
- donations or research contracts from companies or private foundations,
- research grants from other ministries.

Each chair (*koza*) in a *national* university is every year allocated a fixed sum for research expenses, the sum varying between subjects and according to whether there is a doctoral programme or not, but limited to a few ten thousand dollars. These koza funds are used for travel, various office expenses, purchase of literature, materials and small equipment, etc., but are obviously totally inadequate for the acquisition of an MBE machine. They are usually not even sufficient for covering the operating costs associated with MBE research.

Another type of general university funds more important for the acquisition of research equipment is included in the budget for "new requests" which each university annually negotiates with Monbusho. "New requests" may cover everything from funds for a new piece of equipment or a new professorial chair to the establishment of a whole new faculty. As the budgets available for new requests are usually very limited each faculty or attached research institute prepares priority lists of new requests. Facilities and equipment for semiconductor research have increasingly come to be viewed as high priority items in many universities, MBE equipment being but one example.

The establishment of new chairs, individually or as part of new departments, is usually associated with the allocation of special start-up funds, which can be used for the purchase of new equipment. Such start-up funds, amounting to around ¥ 50 million, that is around one third of a million dollars, per chair, enabled Professor Kawabe to start MBE research at Tsukuba University in the early 1980s. When a new professor takes up an already existing chair he may also receive special equipment funds as has, for example, happened in the case of the recent expansion of MBE research at the Institute of Scientific and Industrial Research (ISIR) at Osaka University.

Recognizing the increasing difficulty of furnishing each chair (koza) with adequate facilities and equipment there has been a growing trend in Japan during the last decade of establishing Research Centres representing a major investment in equipment and facilities to be shared by faculty and students from different departments[36] and occasionally even from different faculties. Several new facilities for semiconductor research with large clean-rooms have been established in this way during the latter half of the 1980s. These centres typically contain a variety of semiconductor lithography, crystal growth, fabrication and characterization equipment.

The Research Centre for Ultra-High Speed Electronics at Tokyo Institute of Technology (TIT) which was completed in 1986 at a cost of $ 10 million dollars, equally divided between building and equipment, is an example of how the establishment of a research centre has greatly improved the conditions for MBE research. The centre houses research by six different research groups, most of which are involved in growth of high quality crystals by MBE or MOCVD or their modified versions and microfabrication of nanometer structures for quantum effect devices. It should be pointed out that there are also other MBE machines at TIT, both on the Ookayama campus where the research centre is located and on the Nagatsuda campus of the Graduate School of Science and Engineering.

Individual MBE machines have been installed at research centres also at other universities, without necessarily being the focus of the research. At Hiroshima University a Research Centre for Integrated Systems (RCIS) recently became fully operational. It resembles the CIS at Stanford University which has provided a model, not only in name, but also in that it puts strong emphasis on both design and fabrication of integrated circuits. The RCIS, which focuses on silicon technology, has initially been equipped with one MOMBE for growth of GaAs on Si. This process is needed for growth of the light-emitting diodes and photoconductors included in the "optically coupled three-dimensional common memory", a novel design which is the first major research target of the RCIS. As at TIT, MBE research is also performed outside the centre.

---

36 Each department usually has several koza.

Research centres of the type mentioned became popular during the 1970s. They were planned for a life of ten years, after which period there were no guarantees for their further existence in order to give real leverage for Monbusho, and maybe also for the concerned university, in negotiations regarding new research themes for a centre or redirection of its research funds to other needs. Some of the early centres have been up for review. In the semiconductor field, the Research Centre for Ultrafine Beam Microfabrication at Osaka University, has been formally discontinued and instead a new Research Centre for Extreme Materials been set up in new facilities. The Division for Nanometer Structure Science and Technology in the new centre, however, continues the research from the previous centre. Electron and ion beam technology, fields in which the group has long standing expertise, are developed and used together with other processing and crystal growth technologies, including MBE. Of special interest is the development of a low energy focused ion beam system for doping, etching, and metal deposition to be combined with an MBE to allow uninterrupted ultrahigh vacuum growth and fabrication of quantum effect devices.

The Technological Development Centre at the Technological University of Nagaoka has a more general mission than the centres mentioned so far. Being a small university, opened in 1978 with a charter to emphasize graduate education, many of the research facilities were organized for joint use from the start. The Technological Development Centre, which has an MBE facility for research on blue light emitting devices, has been built to further interdisciplinary research in and out of the university, including joint projects with private companies.

Research centres differ from the so called attached research institutes, which have long been in existence at many Japanese universities, in that the former almost totally rely on faculty and students in the regular teaching departments while the latter have their own permanent organization with professorial and other positions. The new centres, which resemble the research centres and interdepartmental laboratories common at American universities, are seen as a more flexible, and therefore presently favored, alternative to the attached research institutes. The attached research institutes nonetheless remain a very important part of the Japanese universities, not least in the field of MBE research.

There are altogether 37 research institutes attached to national universities working in the field of natural or engineering sciences, most of which were established during the period 1935-1965. They vary in size, The Institute of Industrial Science (IIS) at Tokyo University with a total of 350 permanent staff and another 450 graduate students and visiting researchers being the largest. The IIS and at least two other attached research institutes, The Institute of Scientific and Industrial Research (ISIR) at Osaka University and the Research Institute of Scientific and Industrial Research at Shizuoka University are carrying out significant MBE research.

The ISIR (on a different campus of Osaka University than the above mentioned Research Centre for Extreme Materials) has, as earlier discussed, in recent years built up a very strong capability in the MBE field. Two MBE research groups, working closely together, have no less than five MBE machines, two of which represent used equipment donated from NTT and the OJRL respectively. One of the new machines is a gas source MBE which is being combined with focused ion beam implantation, chlorine gas etching, and a scanning tunneling electron microscope furnishing a very powerful tool for research on quantum devices.

At Tokyo University MBE research started at the IIS in the late 1970s and has more recently been expanded to the Research Centre for Advanced Science and Technology (RCAST), to several groups within the Department of Electronic Engineering and one or two additional groups. RCAST, established in 1987, is a special case in the stream of new research centres. It represents an experiment in university research organization which is monitored attentively by many other universities in Japan.

Like the other centres it aims at encouraging interdisciplinary research and the sharing of costly research facilities. In its staffing policy RCAST falls however somewhere between the old attached research institutes and the typical new research centres by having its own research staff but limiting the time period for appointments. Non-permanent positions are meant to help increase mobility of faculty both inside Tokyo University and to and from other research organizations. Consistent with this philosophy RCAST has also pioneered the practice of endowed chairs which only recently became permitted in Japanese national universities. The endowed chairs at RCAST are established for five years at a time and funded through donations from business firms. As a rule foreign researchers are invited to take up the endowed chairs. Also other universities have begun to establish endowed chairs. In the field of MBE, Hitachi has donated $ 700 000 to Tokyo Institute of Technology to invite a foreign professor to do research on quantum effect devices in cooperation with Professor Takahashi during a three-year period from 1991.

At RCAST there are presently a total of 27 chairs, four of which are endowed. At the start of RCAST it was decided to establish a coordinated research facility, The Process

Centre, for research on electronic or optical materials and devices, fields which represent around a fourth of all the chairs. The Process Centre, mainly paid for by institutional funds and housed in old buildings, is well equipped for crystal growth, device fabrication and characterization. For crystal growth there are three MBE machines and one MOCVD equipment, one of the MBEs being connected with an etching chamber and an other with a focused ion beam implanter. The third MBE is used for growth of silicon crystals. By the fact that Professor Sakaki has appointments at both RCAST and the IIS, good communication is guaranteed between MBE and quantum device research at the two places.

In addition to the research centres and research institutes attached to individual universities, a new system of 14 Inter-University research institutes has been built up under Monbusho, starting with the establishment of the National Laboratory for High Energy Physics (KEK) in Tsukuba in 1971. KEK and the Institute for Molecular Science in Okazaki have become involved in MBE research. These institutes are the only likely candidates for MBE research among the seven Inter-University institutes which are working in areas of science and engineering. The Institute for Molecular Science has recently started research on MBE-growth of phthalocyanine polymer films. During the early part of the 1980s, researchers at KEK cooperated with an equipment manufacturer in trying to develop vacuum chambers for MBE in aluminum, utilizing expertise gained from the use of aluminum in nuclear research. More recently the synchrotron radiation facility at KEK has begun to be used for high-intensity X-ray diffraction studies of MBE-grown materials by researchers from a variety of research organizations. ETL is, for example, installing an MBE machine directly at a port of KEK's Photon Factory for in-situ Extended X-Ray Absorption Fine Structure (EXAFS) measurements, initially for characterization of Ge grown on Si.

Complementing the system of institutional support to universities, research funds are also distributed more selectively on a project by project basis through the grants-in-aid system administered by Monbusho. There are many different types of research grants under this system which has grown very rapidly at the expense of general koza funds which have gradually decreased in real terms. Presently the grants-in-aid system amounts to around $ 350 million which, considering that it serves all fields of science as well as the humanities, looks rather small in international comparison. This is true even if one accounts for the fact that hardly any salaries or overheads are paid out of the grants in contrast to the situation in for example the United States where these cost items may make up as much as 60-70 per cent of a grant.

The regular grants to single investigators may be as high as $ 100 000 per year but are usually much smaller and therefore usually too small to finance expensive equipment. Some simpler MBE machines were, however, financed this way during the 1970s and early 1980s. To meet the needs of today's scientific research Monbusho has devoted an increasing portion of the grants-in-aid budget to larger research grants. In the early 1980s grants for "specially promoted distinguished research" (SPDR) were introduced and more recently a scheme for "priority areas research" replaced and expanded earlier programmes for special project research. Especially the former type of grant has been a very important source of funds for the acquisition of high quality MBE equipment, for example at Hokkaido University (Hasegawa), Tokyo Institute of Technology (Furukawa), Tokyo University (Sakaki), and at Hiroshima University (Yamanishi). SPDR projects are defined by Monbusho as those "requiring an especially large amount of research expenditures, but expected to make significant contributions in research in view of the international recognition of the topic as important for research". (NSF September 1988). Around ten new grants are distributed every year, each grant amounting to some $ 1-2 million distributed over a period of three to five years.

While the SPDR grants are given to single principle investigators and their research groups, "priority areas research" is designed to combine the capabilities of several groups to accelerate the development of research in emerging new fields judged to have a high potential both intrinsically and in terms of their effects on the development of other fields or for the solution of important social problems. (NSF October 1988). Altogether research has been initiated on 62 project themes since the start of this grant category in 1987, some of which have had some content of MBE research.

A particularly interesting case from the point of view of MBE is the project "Basic Studies on Electron Wave Interference Effects in Mesoscopic Structures" begun in 1990 under the leadership of Professor Namba, director of the Research Centre for Extreme Materials at Osaka University (Namba 1990).[37] The project "aims to investigate a new field of physics called mesoscopic physics and establish a basis for applications to electron wave electronics". Mesoscopic physics is a term used to describe the study of quantum effects in one- or zerodimensional nano-structures. A unique feature of the project is the wide range of competences gathered from theorists of quantum physics to

37 "New Functionality Materials - Design, Preparation and Control" started in 1987 and "Metal-Semiconductor Interfaces: A key problem for Ultimate Microelectronics" started in 1989 are examples of earlier priority area projects which also involved MBE researchers. Even earlier, MBE researchers could obtain funds from "Alloy Semiconductor Physics and Electronics", an example of so-called special project research, a grant category which along with part of another category, special long term research, was transformed into priority areas research. The priority area "Fundamental Research on Photo-Induced Process in Semiconductors" started in 1990 under the leadership of Professor Takahashi at Tokyo Institute of Technology also includes crystal growth as an important component, but other growth processes than MBE dominate.

electron device engineers. With a total budget of around $ 4 million over three years, the project is at the high end of the priority areas research themes in terms of size. Given that as many as 35 research groups participate, the scale of funding available to each group is nevertheless fairly limited, even for those groups receiving an above average share of the funds made available through the project. At least a third of the participating research groups employ advanced methods for crystal growth in their subprojects, MBE being the most commonly used. Consequently a majority of the leading university MBE research groups participate in the project.

Specially distinguished research and priority areas research appear to be important vitalizing factors for research at Japanese universities. The former gives a few select researchers exceptionally good working conditions for a limited time period, while the latter gives strong incentives for organization of networks and task sharing within the academic community. With respect to the priority areas research projects one may question how innovative and well integrated the projects really are. It would seem that the usual project period of only three years is too short to allow the establishment of effective collaborative structures, not to mention achievement of the typically extremely ambitious research targets. In this respect some of MITI's research projects extending over a ten year period provide a striking contrast. Researchers taking the initiative to organize and propose such projects naturally have to compromise in trying to achieve the necessary support from potential participants and other researchers who could influence funding decisions. There is therefore a risk that the research plans become *pro forma* aggregations of research already underway and that the selection of themes and the distribution of resources will reproduce already entrenched academic power structures. If these problems can be dealt with, and there is for example some indication that the project periods might become longer, the priority areas research could significantly add to the strength of Japanese academic research.

The proliferation of research centres and the diversification and growth of the grants-in-aid system means that there is more room for initiatives by Japanese university researchers that could influence the material resources available to them for research. The entrepreneurial energies that this more open and hopeful situation is unleashing are probably many times larger than the growth in research funding.

Growing opportunities for research support from the business sector and other private sources work in the same direction. Such support which takes different forms has grown fast during the last decade. Donations are the most important. Money gifts alone amounted to over $ 200 million in 1988, up from less than $ 50 million a decade earlier. There is no statistics over the value of materials, equipment and services given to universities free or sold at high discounts by enterprises. However, judging from acknowledgements in scientific papers such gifts are quite common.

There are several examples of MBE equipment, or funds for the purchase of MBE equipment, being donated from private sources. One of the most significant examples concerns the Kwansei Gakuin University, a private university which has a very small science faculty and no engineering faculty. It became the first university in the Kansai area to seriously get involved in MBE research after receiving a major donation from the Yamada Foundation in 1980. Dr Esaki at IBM, who felt that there was a need to build up a widely accessible resource for MBE research in the Kansai area to promote the diffusion of the MBE technology among Kansai electronics companies, played an important role in bringing about the donation. Professor Sano and his colleagues at Kwansei Gakuin University have built up a well respected MBE research activity and developed important links with many local companies and universities thus to a large extent fulfilling Esaki's vision.

Professor Kimata who started MBE research already during the 1970s at Waseda University had to manage with inadequate equipment funds for a long time. To build the first machine he had only $ 8000 at his disposal which was of course not at all sufficient even for the very simplified design chosen. After some pressure Anelva nevertheless agreed to build the machine as a goodwill gesture. The next two machines were very much better than the first but still had serious drawbacks, for example no preparation chamber, due to shortage of funds. Only around 1987 could really satisfactory equipment be obtained through a donation from the Shimazu company, which has later donated still another machine.

Until the latter half of the 1980s, Waseda University and Kwansei Gakuin University appear to have been the only private universities involved in MBE research. More recently a handful of other private universities have begun to publish in the field.[38] This may reflect the general improvement in the financial situation of private universities which has occurred during the 1980s. Private universities receive some institutional support for research from Monbusho through the "Private University Foundation". This

---

38 For example: Hokkaido Institute of Technology, Sophia University, Nippon Institute of Technology, Osaka Institute of Technology, Okayama University of Science.

however only makes up 10-15 per cent of total general funds for research at private universities, with the remainder covered by income from other sources, primarily tuition fees.[39] Decisions on the use of general funds for purchase of equipment at the private universities are not negotiated with Monbusho, as in the case of the national universities, but left to the discretion of each university.

UNIVERSITY RESEARCH COOPERATION

In a reversal of policy, Monbusho in 1983 began to actively encourage university-industry research cooperation by introducing a "System of Joint Research with Industry".[40] Under this scheme Monbusho finances a part, usually between a fourth and a third, of the cost of research projects carried out at universities with support from industry. In 1988 there were around 600 joint projects underway at a total cost of almost $ 20 million.[41] (Monbusho, No date). One of the most ambitious joint research projects, "Mesoscopic Electronics", is being carried out under the leadership of Professor Ikoma at the Institute of Industrial Science (IIS) at Tokyo University.[42] At a total cost of $2.5 million for the period 1988-90, two thirds of which from ten leading electronics firms and the rest from Monbusho, it is much larger than most other joint projects and is seen as a test case for developing the concept of joint research. The project, which includes MBE crystal growth as an important component has already served to establish

39 Self-financed R&D at private universities in the fields of physical sciences and engineering in 1987 amounted to 174,000 million yen according to official Japanese R&D statistics. As discussed in Irvine et al. (1990) a significant part of these funds, maybe as much as a third or more, in reality cover salaries for non-R&D activities on the part of university researchers. For the same fields, in 1987 private universities received funds for R&D totalling 21,000 million from national and local governments. Most of these funds were provided through the Private University Foundation and a smaller part represent grants-in-aid from Mombusho (for all fields grants-in-aid represent around 5 percent of all government research support to private universities).

40 Monbusho also supports university-industry cooperation in other ways. Special Centres for Cooperative Research have, for example, been established at 18 national universities since 1987. The decision to permit national universities to establish donated chairs has been mentioned earlier. Without support from Monbusho, universities also perform contract research for industrial firms. In 1988 around 2000 contracts brought in almost $ 30 million to national universities. The practice of Japanese companies to send young researchers to universities, in Japan or abroad, as special research students is another very important mechanism for university-industry cooperation. These other avenues of university-industry cooperation may well have been operative in the case of MBE, especially in recent years when the number of universities doing research in the field has multiplied, but the extent remains unclear. As far as the leading university MBE groups are concerned it appears that special research students from companies have mainly been sent to Professor Sakaki at Tokyo University and Professor Sano at Kwansei Gakuin University, but even in these cases the number has been limited to little more than a handful, the two groups taken together.

41 There is some possibility that this figure only includes the contribution from Monbusho, in which case the total amount would be three to four times larger.

42 Cooperation between Toyohashi University of Technology and Toyoko Kagaku Co. regarding laser-assisted metal-organic molecular epitaxy of gamma-sapphire on silicon is another joint project involving MBE research that has been identified.

the concept of mesoscopic electronics in the Japanese research community.

Most of the research is being performed in the Research Group on Semiconductor Heterostructure Materials and Devices, one of nine interdisciplinary groups at IIS. The group received a large (around $ 4 million) institutional equipment grant during the period 1984-87 which was partly used to acquire an MBE equipped with a large number of advanced instruments for in-situ measurements. The joint research project represented a welcome new source of research funds when the institutional grant had run out. Some researchers from Tokyo Institute of Technology and Hiroshima University also have participated in the project. All in all nine professors, including Professor Sakaki, have been involved in the project.

Much effort is spent in the project to achieve effective communication between the university research groups and the participating firms. Each firm has assigned a "cooperating researcher" and several of these worked for some period at IIS. Half-day technical meetings are being held once a month with the involved university professors, the cooperating researchers and other persons selected by the firms. Research results are also being presented to a wider audience through a number of so called forums.

The reversal of Monbusho's policy regarding industry-university cooperation from a rather sceptical view to one of active encouragement is strongly related to the strengthened research potential of Japanese industry. With their increasing involvement in rather basic research it became desirable for Japanese firms to cultivate their relations with academic research groups. Foreign universities, and especially those in the United States, were perceived as more attractive partners than the Japanese universities, in terms of the research level as well as the attitudes and regulations concerning cooperation with industry. Many deals were struck between Japanese firms and research groups at American universities often involving donations of large amounts of money for the construction of new laboratories or the establishment of new chairs. It seemed to many that Japanese universities faced a strategic choice: either they actively tried to become a partner to Japanese industry and thereby derive benefit from its growing resources and research capabilities or they ran the risk of becoming an increasingly marginal factor in the Japanese research system.

There are still sharply conflicting views within the Japanese academic community and seemingly also within Monbusho as to how responsive universities should be to the research needs of industry. Nevertheless, determined efforts are obviously being made to put Japanese university research in touch with and if possible ahead of the development of leading-edge industrial technology.

While Monbusho is presently encouraging industrial presence on university campuses, Monbusho still does not welcome other ministries to engage university researchers in their research projects. While in most other countries universities derive funds from many different government bodies, Monbusho very nearly totally monopolizes government funding of universities in Japan. One would expect that the current policy of the Japanese government to strengthen basic research should favour universities. The unwillingness of Monbusho to give up any control of university research to other ministries, however, forces the latter to spend their funds for basic research elsewhere than at universities. Naturally there are some exceptions, but taken in total they represent an extremely small part of university research funding. In the next section are described some activities of the Science and Technology Agency and ways in which these involve universities. MITI has so far mainly used university professors as advisors in its various research programmes. In one case, solar cell development under the Sunshine Programme, MITI has directly funded university research in a major way. The New Energy and Industrial Technology Development Organization (NEDO), which only a couple of years ago had its mandate broadened to include development of industrial technology more generally, is currently trying to involve Japanese universities in its efforts to organize international research cooperation.

The growing involvement of universities in MBE research is not only a question of improved economic opportunities but also relates to changes in the substance of MBE research. On the one hand MBE is becoming a tool for many physicists who are not interested in MBE per se but employ MBE to produce materials or structures suitable for the exploration of interesting physical phenomena. For the most common materials systems this can often be accomplished without any particularly deep expertise in MBE as the demands on crystal quality may not be so high and the MBE machines since a few years back have become simple to operate. On the other hand the research opportunities for "real MBE research" are expanding at a rapid rate as the MBE process is becoming more versatile through the introduction of gaseous sources, the use of various energetic beams, etc. As a consequence the range of materials and combination of materials that can successfully be grown is steadily increasing to include not only a broader range of semiconductors but also metals, insulators, high temperature superconducting materials, and organic materials. There is furthermore the whole new field of the physics and electronics of low-dimensional structures barely beginning to open up.

These new directions of MBE research are well suited to universities. One reason is that they put a premium on combining expertise from several scientific fields and on integrating theoretical knowledge and experimental research. Understanding of growth mechanisms in photo-assisted MOMBE and analysis of quantum phenomena in low dimensional structures can be cited as examples. Under proper conditions of organization, such as exemplified above, this should give universities a stronger position in MBE research.

Another characteristic of the new areas of MBE research is that no approach has clearly established itself as the dominating one. There is therefore plenty of room for novel ideas, which for the most part can be tried out and demonstrated with the equipment available today at least for the leading universities. Once a new concept is shown to be promising, however, it is quite likely that industrial groups with larger resources will be able to push forward at a higher speed and take a lead.

Until fairly recently there has been a strong tendency among Japanese researchers, in industry as well as in universities, to define research targets by looking at the leading research groups abroad, especially those in the United States. As both the level and the quantity of science and technology in Japan is reaching a par with, and not seldom moving ahead of, the rest of the world Japanese researchers will increasingly themselves have to define new directions of research. This is a particularly suitable role for universities as they have great freedom in generating and exploring new concepts, an activity which is necessary for setting out new research targets. Comparing the situation around 1980 when MBE research began to take off in Japanese industry and the situation during the last couple of years it seems undeniable that Japanese universities are much more prominent in laying out the agenda for future research.

STA LAUNCHES RESEARCH PROGRAMMES TO INCREASE MOBILITY OF RESEARCHERS

Accounting for a fourth of all government R&D expenditure, the Science and Technology Agency (STA) is, after Monbusho, the second largest ministerial level body in Japan in terms of R&D funding and twice as large as the better known MITI. STA is most noted for its promotion of science and technology in prestigious areas such as nuclear power, aerospace and ocean technology. While these areas still consume the bulk of STA's funds, the agency has become increasingly active in more normal fields of basic research, especially those related to life science and materials science, a development which is also reflected in the case of MBE.

STA is currently supporting at least two major research efforts that contain MBE as a key element.[43] One is the Quantum Materials Project which has been part of the Frontier Research Programme (FRP) at the Institute of Physical and Chemical Research (RIKEN) since the start of that programme in 1986. The other is the Sakaki Quantum Wave Project started in 1985 as part of the Exploratory Research for Advanced Technology (ERATO) scheme in operation since 1981 under the auspices of the Research and Development Corporation of Japan (JRDC).

The ERATO programme was designed as a response to concerns that established ways of organizing research in Japan did not sufficiently encourage and support creative research. The programme therefore set out to experiment with new concepts of research organization which challenged ingrained traditions. An individualistic style has been sought by investing a strong leadership role with the Directors of each project and further emphasized by the practice of naming the projects after their Directors. It has also been an ambition to gather under the Directors researchers with heterogeneous backgrounds in terms of disciplines, sector of employment, and country of origin, while at the same time concentrating on young PhD-level researchers in their early thirties. Flexibility in organization has been achieved by renting laboratories, sometimes at several locations in one project, from participating or other organizations, in either case adjusted to where talented people can be found.

On the whole ERATO gives the impression of being a very successful programme. Although the participation of researchers permanently employed at universities and government laboratories, other than the project Directors, seems to have been very limited, important steps have nevertheless been taken to increase the mobility of researchers in the Japanese research system.

Of the 21 projects that had been started before the end of 1989, 14 of which are still running, two thirds are directed by university professors. Around half of the research staff have come from private industry, another third have participated as individuals, and most of the rest have been foreign nationals. Individuals for the most part refer to

---

[43] In addition MBE research concerning ferromagnetic semiconductors is taking place at the National Research Institute for Metals (NRIM), one of STA's institutes.

persons who have just earned their doctorate but are not yet permanently employed. For these researchers ERATO has in effect introduced a system of postdoctoral research positions which has earlier been more or less totally lacking in Japan. As there is no tradition in Japan of postdoctoral positions, it has no doubt been felt risky to accept these positions. The attractiveness of the ERATO projects, in terms of their substance, leadership, resources and stimulating research environment, has nevertheless been sufficient to motivate many researchers to take this risk and thereby contribute to an opening up of the Japanese research system.

Various ERATO projects have been concerned with semiconductor epitaxial growth. Two projects directed by Professor Nishizawa at Tohoku University, Perfect Crystal Project (1981-86) and Terahertz Project (1987-92), have dealt with the development of photostimulated molecular layer epitaxial (PMLE) growth of compound semiconductors, representing an alternative technology to MBE. The Kuroda Solid Surface Project (1985-90) has investigated new techniques for epitaxial growth of Si films.

The Quantum Wave Project (1988-93) directed by Professor Sakaki of Tokyo University, already identified as one of the leading personalities in the Japanese MBE community, is the first ERATO project to include MBE as an important element. The project, having an annual budget of $ 2 million, is performed at three locations, at the Research Centre for Advanced Science and Technology (RCAST) at Tokyo University, at NEC's Tsukuba Research Laboratories and at Matsushita Giken in the Tokyo area. Work at NEC focuses on atomic layer epitaxy (ALE)[44] while work at Matsushita Giken deals with novel quantum material systems including MBE growth of organic materials.[45] The group at RCAST has acquired an MBE combined with etching equipment for fabrication of structures that will permit the study of quantum wave effects. In addition to the two companies mentioned as providing laboratories, another six companies dispatched researchers to the project in 1990, four of which to

44 A technology based on self-limiting features of certain chemical reactions occurring when a solid surface is exposed to vapour during material growth. ALE was pioneered in Finland by T. Suntola to meet the needs of improved ZnS thin films and dielectric thin films for electro-luminescent thin film display devices, ALE has in the 1980s been applied to epitaxial growth of III-V and II-VI compounds, especially layered structures, such as superlattices and superalloys, applications which currently dominate ALE. ALE has also been extended to the growth of single crystals of elemental semiconductors. Japanese research organizations today dominate ALE. According to a publication count done by Suntola, 81 out of 155 papers published during the period 1977-88 came from Japan. The leading Japanese organizations are RIKEN, NEC, ETL, NTT, and the Semiconductor Research Institute in Sendai. The last is a private foundation directed by Professor Nishizawa. His MLE is classified as ALE by Suntola (Suntola 1989).

45 NEC also does research on growth of organic materials using MBE, but this is not part of the ERATO project.

RCAST.[46] Two postdocs from Tokyo University, a master degree holder from Tohoku University and a researcher from the United States add to the research team at RCAST. Additional scientists, especially from foreign countries, are expected to join the three research teams.

The Frontier Research Programme (FRP) employs a somewhat more permanent research organization than ERATO. Firstly, the projects may last as long as 15 years if found to make sufficient progress in the reviews done every five years. This compares with a maximum duration of five years for the ERATO projects.[47] Even if continued, however, the FRP projects are meant to be restaffed after each five-year period. Fulltime researchers are hired on a one-year basis but contracts are renewed for up to five years if performance is satisfactory. Secondly, the FRP projects are all being carried out at RIKEN. While almost all the research in the ten first established laboratories is performed on RIKEN's main campus in Wako near Tokyo, any new FRP laboratories will be set up outside the Tokyo area. The first example of this policy is the establishment in October 1990 of a Photodynamics Research Centre with four laboratories in Sendai under the directorship of Professor Nishizawa. Parts of the new centre build on and extend some of Nishizawa's earlier work in two ERATO projects. The laboratories in Wako benefit from being able to use the resources of RIKEN, a well-equipped research institute with almost 500 research staff and many visiting scientists.

80 researchers worked full-time in FRP as of May 1990, and about as many part-time. Foreign scientists, employees of firms and RIKEN research staff, mentioned in order of their numbers, together represented slightly over one third of the full-time researchers. Of the remaining two thirds around 30 percent were postdocs and the rest new graduates with less than a doctor's degree or researchers permanently resigned from other jobs, mostly in government laboratories. There are hardly any cases of employees of government institutes working full-time in FRP on leave from their institutes. FRP thus even more than ERATO offers temporary employment for researchers who not yet or no longer have a stable job. The laboratory heads, most of whom work only part-time at FRP, are for the most part foreign scientists or Japanese university professors, several of the latter retired from Tokyo University.

Research in the Laboratory for Quantum Materials (LQM) deals with optical properties of quantum well structures of III-V materials, electron transport in quantum

---

46 The six companies are Matsushita Electric, Ulvac, Sumitomo Electric, Toyota, Shimazu, and Asahi Chemicals, the last a producer of Hall effect sensors using MBE.

47 The two projects directed by Professor Nishizawa are largely in the same area and can probably be considered a ten-year project.

wire structures and the development of new evaluation techniques for quantum materials, such as laser spectroscopy of exitons found in quantum well structures. The Laboratory Head, Professor Namba, was until 1988 also Head of RIKEN's Laser Science Group where he around 1980 started a microprocessing project which has functioned as a platform for the establishment of the Quantum Materials Project, which in turn was part of the very early core of FRP. Growth of very high quality crystals is a very important part of the project and several different approaches are used. An MOCVD system acquired in 1984 has been adapted for laser-assisted Atomic Layer Epitaxy (ALE) with good results. No less than three MBE sytems have been acquired during the last two or three years, one of which is using both gas and solid sources and being employed for Chemical Beam Epitaxy (CBE) to study growth kinetics and photochemical reactions in laser assisted growth of GaAlAs and GaInAs. The other two MBEs are designed for the development of growth technique of new quantum materials, such as CuCl/CaF2 quantum wells[48], and for physics research respectively.

The LQM and the Laser Science Group at RIKEN cooperate very closely with Osaka University. Professor Namba has for two decades had a double appointment as Chief Scientist at RIKEN and Professor at the Faculty of Engineering Science of Osaka University. His successor as Head of RIKEN's Laser Science Group, Dr Aoyagi, who has also a key position as Coordinator in the LQM, is furthermore a Visiting Professor at Osaka University, something which gives him an opportunity to work with graduate students. A great number of papers have consequently been co-authered by researchers from Osaka University and RIKEN in the areas of laser assisted crystal growth, microprocessing and nanometer physics. Three postdocs in the LQM have received their doctorates from Osaka University under Professor Namba, while a fourth postdoc obtained his doctorate from Tsukuba University under Professor Kawabe, who initiated MBE research at that university.

Many foreign scientists have worked in the LQM for periods varying from a month to a year which has given the laboratory an extensive network of international contacts. The experiences reported by Gerhard Fasol from Cavendish Laboratory, Cambridge University, gives the impression of the LQM as being well integrated in the Japanese research system. He worked two summers in Japan, partly at the LQM and partly in Professor Sakaki's group at Tokyo University. On the first occasion he also collaborated with Dr Horikoshi at NTT, who supplied him with high quality quantum well samples and during his second visit Toyota Central Research laboratory provided samples.

48 The Laboratory for Photophysics at the new Photodynamics Research Centre in Sendai will research the same materials (RIKEN 1990)

(RIKEN 1988; RIKEN 1989). It should also be noted that formally Sakaki is participating in the LQM on a part-time basis.

In the two MBE-related projects described above, like in other ERATO and FRP projects, STA and its related organizations JRDC and RIKEN obviously collaborate in a significant way with prominent university researchers. Whether these programmes will remain isolated cases or lead the way towards a greater role more generally for STA and other ministries at Japanese national universities is difficult to judge. Universities clearly could benefit from an injection of resources from other ministries than Monbusho, but the latter appears still to resist such a development. University researchers would also benefit from being able to use the often superior facilities and equipment in government and corporate laboratories. Arrangements of this kind are, however, still unusual, especially for professors of national universities, although the cases of informal cooperation appear to outnumber those that are officially recognized and sanctioned.

STA is charged with a special role to coordinate government science and technology policy, a role which has gained in importance partly due to the fiscal austerity of the 1980s and partly due to the perceived needs for structural change in the Japanese research system. While Mombusho and its support of university research falls outside STA's coordination mandate, policy measures designed to strengthen the ties between industry, government and universities are included. STA's coordination role is closely tied to the policies set forth by the Council for Science and Technology (CST), for which STA provides the secretariat, and the economic resources made available through the Special Coordination Funds for Promoting Science and Technology (SCFPST) which was established in 1981 as a means for promoting cooperation among different sectors in the Japanese research system and is managed by STA. Although the amount distributed through the SCFPST, in 1988 around $ 60 million, is fairly small it introduces a new feature into the Japanese research system by offering funding which in principle does not respect ministerial boundaries. The SCFPST was encountered in the MBE community only once when it helped fund the purchase of characterization equipment by a university professor who cooperated with one of STA's institutes as part of a larger materials research project, but there could well be additional cases.

7.4 CONCLUSIONS

The MBE study supports the notion of a growing basic research effort in Japanese industry during the 1980s and demonstrates how this has allowed the Japanese firms to keep abreast of, and actively contribute to, the evolution of a new basic technology which has a close linkage with scientific research. The character of MBE as a basic technology is underlined by the fact that all the major Japanese electronics firms have had some serious programmes of MBE research. The volume and orientation of MBE research varies greatly among firms and is dependent on each firms overall product market profile and its evaluation of, and relative success with, alternative crystal growth technologies. While NTT and Fujitsu have mastered the most impressive MBE research efforts overall, Hitachi and NEC have dominated in Si MBE, Sharp developed high performance short-wavelength lasers and Rohm and Mitsubishi Electric excelled in MBE manufacturing technology, just to mention some examples.

The MBE case also usefully illustrates the supportive role of MITI during the take-off of MBE research in the early 1980s through the timely launching of three new research programmes as well as the importance of ETL in organizing these programmes, a role which was dependent on ETL's own early involvement in MBE research.

The expansion of long-term research in Japanese firms has undoubtedly reduced the leverage of MITI's research support to individual companies. The creation of joint research laboratories appears to offer better leverage for government research funds than subsidies to individual firms as the resources of one such laboratory, if well focused, can yield a research capability that is stronger than that of any single firm. The Optoelectronics Joint Research Laboratory (OJRL) and the Optoelectronics Technology Research Laboratory (OTL) are good examples of how organizations for joint research have evolved during the last decade into a new powerful element in the Japanese research infrastructure, combining a strong commitment to basic research characteristic of universities and very ample material resources usually found only in a few industrial laboratories. The joint research laboratories, although so far seemingly successful, still have to be considered experiments and their permanence will depend on many factors, not least on how research cooperation between universities and industry develop.

It seems unavoidable that the relative importance of ETL will diminish as the number of researchers there are, at best, maintained at a constant level, while the corporate sector, including the joint research firms, is expanding its research rapidly. At the same time the number of research specialties which ETL is trying to cover is increasing. Reduced importance for MITI's research programmes will also tend to undermine the pivotal role ETL has had so far. The official policy is to put increasing emphasis on ever more fundamental research, but it is not clear that this in itself will be sufficient to give ETL a profile which really distinguishes its research from that in industry or in universities. ETL may continue as a very respectable research institution in the electronics field, but it will be constantly challenged by a host of other research organizations in Japan and can probably no longer assume that it has a unique position.

The Japanese university system is actively responding to the opportunities offered by the development of new fields like MBE and to the challenges imposed by the growing strength of company research in Japan. The research funding system has become more diverse and better capable of meeting the demands posed by ever more expensive equipment and the need to achieve critical mass in research efforts. It therefore seems likely that the real research capacity of Japanese universities has grown faster than one would be left to believe from looking at aggregate R&D statistics[49] as, e. g. in chapter 6. Recruitment of highly qualified researchers from companies to become university professors is an interesting mechanism for universities to rapidly catch up with industry in new fields, a side-effect of which is to improve the universities' contact networks with industry. A remaining problem, although there may have been some improvements in very recent years, is the difficulty many research groups have of recruiting doctoral students. Proposals have been put forward to address this problem by introducing a new system for doctoral studies.

While universities played a minor role in MBE research in the early 1980s in Japan, they are ten years later contributing very actively, especially in research related to quantum materials and quantum-effect devices. The pattern is similar in other countries and partly a consequence of changes in MBE technology and related research topics which turned out to expose a cyclic structure as in other cases of science-based innovation (see chapters 1, 3, 8 and 9). The development of MBE technology has, for example, become increasingly dependent on an understanding of the basic mechanisms of the MBE growth process and as the sophistication of MBE technology has grown it has become possible to grow materials and structures which can be used to study scientifically increasingly more interesting physical effects, changes which both have

---

49 A similar conclusion is drawn in Okimoto et al. (1987) for the period 1965-1981 based on study of a sample of catalogues and reports of academic institutions.

served to attract academic scientists to MBE research. Although Japanese universities have responded vigorously to the new challenges opening up, their response has been weaker, in quantitative terms, than that of the American universities but comparable to that of European universities.

As illustrated by the case of MBE, but valid also on a more general level, the Science and Technology Agency (STA) has stepped forward as an important actor on the Japanese scene of basic technology research. It has introduced several innovative schemes aimed at increasing mobility and cooperation among organizations from different sectors in the Japanese research system, and in the process introduced what de facto amounts to postdoc positions, something which earlier hardly existed in Japan.

As the Japanese research system in the MBE field has expanded it has simultaneously become more differentiated and more integrated. Not only has the number of research groups grown but there has also been a creation of new types of research organizations (a general discussion of this issue may be found in chapter 2). As the number of research groups has increased, the potential for organizing research cooperation and mobility of researchers has naturally increased. Although mobility is still low by international standards it has been high enough to create many new linkages between different research groups, linkages which have been further strengthened and multiplied by research cooperation in a number of forms, often promoted by MITI, Monbusho or STA. One important remaining barrier to integration of the Japanese research system is the great difficulties for other ministries than Monbusho to work with universities, a situation which, if it persists, can have a distorting effect on the Japanese government's policies to strengthen basic research.

Some twenty years ago Joseph Ben-David discussed the "scientific gap" between the United States and Western Europe and concluded that this was due to structural inadequacies of the European organization of scientific research (Ben-David 1968). In trying to explain why the United States seemed better able to exploit the fruits of scientific research he proposed that "practical uses of science should be conceived as the result of chance interactions between fundamental discoveries on the one hand and practical interests on the other, which can occur in an infinite variety of ways". From such a perspective of "interpenetration" (chapter 2) the emerging structure of the Japanese research system with a growing multitude of intertwined networks containing a great variety of research organizations, including a particularly good representation of "practical interests", certainly deserves continued attention.

The author wishes to express his gratitude to Professor Ryo Hirasawa, Tokyo University, for providing an excellent working environment for this research and also wishes to thank all the Japanese researchers and research managers who in interviews very generously shared their experience of the development of MBE in Japan. The author would also like to thank the Swedish Board for Technical Development (STU) for supporting this research.

REFERENCES

Ben-David, J. (1968), *Fundamental Research and the Universities. Some Comments on International Differences*, OECD, Paris.

Furukawa, S. (1989), 'Why the Number of Doctor Course Students of the Science and Engineering Fields in Japan has been Saturated,' (in Japanese), *Oyo Buturi*, 58, 4.

Granberg, A. (1990), 'Laser Research in Sweden, West Germany, and Japan', in Sigurdson, J. (ed.), *Measuring the Dynamics of Technological Change*, Pinter Publishers, London.

Hayashi, I., Hirano, M., and Katayama, Y. (1989), 'Collaborative Semiconductor Research in Japan', *Proceedings of the IEEE*, 77, 9.

Irvine, J., Martin, B. R., and Isard, Ph. A. (1990), *Investing in the Future. An International Comparison of Government Funding of Academic and related Research*, Edward Elgar Publ., Aldershot.

Merz, J. L. (1986), 'The Optoelectronics Joint Research Laboratory: Light Shed on Cooperative Research in Japan'; *Scientific Bulletin*, Department of the Navy Office of Naval Research Far East, Department of the Air Force Office of Scientific Research Far East, 11, 4.

Monbusho (1987), *Gakko Kihon Chosa Hokokusho. Koto Kyoiku Kikan Hen.*

Monbusho (1990), *Gakko Kihon Chosa Hokokusho. Koto Kyoiku Kikan Hen.*

Monbusho (Ministry of Education, Science and Culture), 'Daigaku to Sangyokai to no Kenkyu Kyoryoku', Monbusho, Science and International Affairs Bureau, no date. [Has later statistics than the previous document].

Monbusho (Ministry of Education, Science and Culture), 'Research Cooperation between Universities and Industry', Monbusho, Science and International Affairs Bureau, no date.

Namba, S. (1990), 'Digest of Research Project, Basic Studies on Electron Wave Interference Effects in Mesoscopic Structures', Faculty of Engineering Science, Research Centre for Extreme Materials, Osaka University.

Narin, F. and Olivastro, D. (1988), 'Identifying Areas of Leading Edge Japanese Science and Technology. Final Report to International Studies Group NSF', Computer Horizons, Inc., Haddon Heights.

NSF (1988), 'FY 1988 Awardees of Monbusho's Grants for 'Specially Promoted Distinguished Research', 'The Tokyo Office of the US National Science Foundation, Report Memorandum #164.

NSF (1988), 'Monbusho's "Priority Research Area" Grants: 19 New Areas', The Tokyo Office of the US National Science Foundation, Report Memorandum #165.

Okimoto, D. I. and Saxonhouse, G. R. (1987), 'Technology and the Future of the Economy', in Yamamura, K. and Yasuba, Y. (eds.), *The Political Economy of Japan*, Volume 1, Stanford University Press, Stanford.

Rappa, M. A. (1989), 'Advanced Bibliometric Methods Applied to the Study of Emerging Technologies', in van Raan, A. F. J., Nederhof, A. J. and Moed, H. F. (eds.), *Science and Technology Indicators*, DSWO Press, Leiden.

RIKEN, Division of Frontier Research Programme (1988), 'Newsletter, Frontier Research Programme, March 1988, Vol.1, No.4', Wakoshi, Saitama.

RIKEN, Division of Frontier Research Programme (1989), 'Newsletter, Frontier Research Programme, October 1989, Vol.3, No.3', Wakoshi, Saitama.

RIKEN, Division of Frontier Research Programme (1990), 'Newsletter, Frontier Research Programme, October 1990, Vol.4, No.4', Wakoshi, Saitama.

Sakaki, H., 'Molecular Beam Epitaxy', in Malik, R. J., (ed.), *III-V Semiconductor Materials and Devices*, Elsevier, Amsterdam.

Sakakibara, K. (1988a), 'Basic Research Business', *The Japan Times*, May 15.

Sakakibara, K. (1988b), 'University Research Lags', *The Japan Times*, May 16.

Sakamoto, T. (1982), 'Moving Toward Future Electron Devices', reprinted from *JEE July issue*, Dempa Publications, Tokyo.

Shimada, J. (1986), 'Memories on Optoelectronics Project with Love and Sorrow', Electrotechnical Laboratory (unpublished).

Shiraki, Y. and Sakaki, H. (eds.) (1989), *Molecular Beam Epitaxy 1988*, Proceedings of the Fifth International Conference on Molecular Beam Epitaxy, Sapporo, Japan, 28 August - 1 September 1988, reprinted from Journal of Crystal Growth 95 (1989) Nos. 1-4, Elsevier, Amsterdam.

Statistics Bureau, Management and Coordination Agency (1980), *Report on the Survey of Research and Development 1980*, Tokyo.

Statistics Bureau, Management and Coordination Agency (1990), *Report on the Survey of Research and Development 1989*, Tokyo.

Stokes, D. E. (1982), 'Perceptions of the Nature of Basic and Applied Science in the United States', in Gerstenfeld, A. (ed.), *Science Policy Perspectives: USA-Japan*, Academic Press, New York.

Suntola, T. (1989), 'Atomic Layer Epitaxy', in Sinkkonen, J. (ed.), *Treatise on Microelectronics*, Acta Polytechnica Scandinavica, Electrical Engineering Series No. 64, Helsinki.

Takahashi, K. and Callen, E. (1989), 'Fifth International Conference on Molecular Beam Epitaxy (MBE-V)', reprinted from ONRFE Sci. Info. Bul. 14 (1) 89, in Annual Research Report 1988, Takahashi & Konagai Laboratory, Tokyo Institute of Technology.

The University of Tokyo (1987), 'Information on the Faculty of Engineering 1987', Tokyo.

The University of Tokyo (1989), 'Information on the Faculty of Engineering 1989', Tokyo.

PART III

CROSSROADS IN SELECTED TECHNOLOGIES: CASE STUDIES

8 KNOWLEDGE EXPANSION IN APPLIED SCIENCE: A BIBLIOMETRIC STUDY OF LASER MEDICINE AND POLYIMIDE CHEMISTRY

Ben van Vianen and Anthony van Raan

8.1 INTRODUCTION

Technological innovations are to a large extent the result of research and development (R&D) activities. Therefore, the process of interaction between science and technology is an important topic for science and technology policy and R&D management. Systematic investigation of this process is also important for the general understanding of the development of technological innovations.

In this chapter 8, we present the results of the first stage of a project devoted to the study of the science and technology interface. More specifically, we focus on the diffusion of scientific and technological knowledge, as measured by quantitative, 'bibliometric' methods, based on publication and patent data. Diffusion is defined as the use of field-specific knowledge in other fields of science or technology. If this use leads to a further generation of knowledge on the original subject, we may speak about 'knowledge expansion'.

This work is a collaborative project with the Fraunhofer Institute für Systemtechnik und Innovationsforschung (FhG-ISI) at Karlsruhe (see chapter 9). The Karlsruhe group focuses on the technology side by means of patent analysis, whereas in our work the science side is subject of investigation.

Bibliometric analyses are primarily based on data from published material (in particular publications and citations from the serial literature and from patents), that represent different aspects of the scientific and technological endeavour in a quantitative fashion. The development of quantitative methods and techniques (such as citation analysis) for the assessment of scientific performance (Narin 1976; Braun et al. 1985; Schubert et al. 1988, 1989; Martin and Irvine, 1983; Moed et al. 1983, 1985) and technological performance (Narin and Olivastro 1988; Schmoch et al. 1988; Schmoch and Grupp, 1989) and for the monitoring of scientific and technological developments, has been strongly advanced by 'science and technology indicators' research from the 1970s onward. A recent and extensive overview of the field is given in the Handbook of Quantitative Studies of Science and Technology (Van Raan, 1988) and in the proceedings of the first international conference on science and technology indicators (Van Raan et al., 1989).

As indicated before, an important presupposition in quantitative, bibliometric studies of science and technology indicators is that results of scientific work are published in the serial literature (primarily journals), and that the results of technological work are embodied in patents. In many of the basic natural and life sciences, publication in the serial literature indeed is the major way of disseminating research results. In the humanities and social sciences, books and reports ('grey literature') are also important carriers of research results. In the technical sciences again books and reports, but also patents, software, designs, artifacts like prototypes, or even maps. However, recent work in our group (Peters et al., 1988; Nederhof et al., 1988, 1989) shows that for the applied and engineering sciences, as well as for the humanities and the social sciences, international journals do play an important role in the dissemination of knowledge. From these studies we learned that, in good approximation, the applicability of bibliometric indicators for a specific field of science strongly depends upon the extent as to which publication databases and, more in particular, citation databases cover the communication channels used by researchers in that field.

Patent-based indicators are particularly useful at higher levels of aggregation (cf. Pavitt, 1988; Engelsman & Van Raan, 1990) where limitations of patent data are less likely to affect the indicators significantly. At lower levels, patents are at best a partial measure of technological output and change. Moreover, the propensity to patent strongly depends on R&D strategies. It may vary within and between fields of technology as well as in time. Important limitations of the use of citations to scientific literature in patents as an indicator of 'science intensity' are discussed in recent work of our group, Van Vianen et al. (1990), see also chapter 4 in this volume. For 'science-intensive' fields of technology (e.g., electronics, coating, polyimides) the combination of patent and publication analysis may considerably improve the measurement of R&D output (Frumau 1991). For a general overview of bibliometric methods and techniques we refer to Van Raan and Tijssen (1990a, b).

In our current project, we apply bibliometric methods and techniques to investigate the process of knowledge diffusion and expansion in specific fields of applied research. In this chapter we focus on more general knowledge expansion characteristics, in particular the distribution of publication numbers over research (sub)fields, as a function of time. Tracing of diffusion of specific 'pieces of knowledge' by means of citation analysis will be the major topic in the second stage of this project (Noyons and Van Raan, 1991).

In this chapter we study two R&D fields: (1) the application of *lasers in medicine*, and (2) research on *polyimides*. With help of an extensive keyword analysis of scientific publications it is investigated to what extent research in these fields, and thus ('generation' of) knowledge, takes place in other, major fields. As discussed above, the technological counterpart of this study, diffusion of technological knowledge in the same R&D fields, is investigated by the Karlsruhe group by means of patent and, in particular, patent citation analysis (subsequent chapter 9).

The choice of the above fields is based on earlier experiences, especially with patent analysis (Schmoch et al., 1988). The first criterion of choice is a rather practical one: in order to keep the patent analysis for the technological counterpart of this study feasible, the worldwide number of patents (without Japanese domestic applications) in the fields should not exceed 3000 in a period of ten years. Moreover, the number of European patents should be more than 500 in the same period of ten years. A second criterion of choice is that, at least, the basic technology is sufficiently covered by the patent classification system. The third criterion is directly related with the method used: delimitation of the fields must be possible with help of adequately chosen keywords. A fourth criterion is a significant increase of importance of the field in the last fifteen years. The reason for this is that in such cases there will be a reasonable probability of the exploitation of earlier basic research for technological applications. This latter point brings us to the fifth and final criterion: the chosen fields must have a clear linkage with basic research. For an extensive discussion on these criteria in relation to the chosen fields (lasers in medicine and polyimide research), we refer to chapter 9, section 9.3.

8.2 GENERAL METHODOLOGY

In many fields of science important advances in human understanding of the world we live in are attained. Most of the advances are small increments of our knowledge. Some of the advances open new ways for application. More specifically, we focus our attention to technological applications. Obviously, there is some complex selection process filtering from the multitude of scientific advances those which successfully lead to technological applications.

Socio-economic factors undoubtedly play a very important and often dominant role in the further choice of which basic innovations will be pursued, and how. Earlier work shows that after a basic innovation specific delays in the development may occur, leading to 'technological trajectories' (Rosenberg, 1976; Nelson and Winter, 1977; Dosi, 1982). These studies clearly focus more on the technological and economical side. Our project aims at the study of the basic, preceding processes, in particular the knowledge transfer between scientific research and technological activities. A central goal of our research therefore is to find determinants in this early selection process, i.e. factors which play an important role in the selection process. Our hypothesis is that at least some of these factors can be found in the interface between science and technology. In other words, factors which govern to an important extent the interaction between scientific research and technology. More specifically, we hypothezise that characteristics of 'knowledge diffusion', i.e. the use of field-specific knowledge in other fields, and of 'knowledge expansion', i.e. the advancement of specific knowledge by this use in other fields, will be dominating factors in the above mentioned 'early selection process'. We attempt to investigate systematically the phenomenon of knowledge diffusion and expansion in order to identify its characteristics. This paper and the paper of our Karlsruhe co-team (chapter 9) present first steps.

Knowledge diffusion takes place in many ways: from basic research to applied, but also in the reverse direction; from applied research to technological applications, or the reserve; and even more or less 'directly' from basic research to technological applications. But the interaction between science and technology is by no means a 'linear' one, i.e., a scientific discovery A does not lead to applied research B which does not lead on its turn to technological application C. The metaphor of Casimir (1972) describes science and technology as being interlinked together in a spiral-like fashion. De Solla Price (1984) introduced a more poetic metaphor: science and technology are like a pair of dancers, sometimes the one is leading, sometimes the other. Undoubtedly, the interaction of science and technology is dominated by many feed-back processes (often 'induced' by the external environment, i.e. the socio-economic system). Consequently, science and technology interacts in a 'non-linear' way. This is an interesting conclusion because non-linearity implies, as we know from recent developments in mathematics and physics (Gleick, 1987), a chaotic, i.e. unpredictable behaviour. The consequences, however, of this challenging finding are as yet beyond the scope of this paper (for a more general discussion see chapters 1 and 2).

If one is able to find some of the above mentioned determinants, and, in particular, characteristics of knowledge diffusion and expansion, then important tools for research and development (R&D) management may become feasible. We define knowledge diffusion and expansion as the use and further advancement of field-specific knowledge in other fields. In main lines, the operationalization of this definition is as follows. 'Field-specific research' (in our case laser medicine and polymide research) is identified by means of keywords or database classification codes. It will be analyzed to what extend these words or codes are used in publications from other scientific fields, i.e. the number of publications in these 'other' fields characterized by (among others) the chosen keywords or codes.

This collaborative research project consists of five major parts which at the same time define the main methodological principles:

(1) Systematic analysis of the diffusion of technological knowledge by investigating the occurrence of specific concepts (for instance keywords or classification codes) in patents from all technological fields, as a function of time.

(2) The 'scientific counterpart' of the above: systematic analysis of the diffusion of (applied) scientific knowledge by investigating the occurrence of specific concepts (for instance keywords or classification codes) in publications from all scientific fields, as a function of time.

(3) Systematic analysis of numbers, types, and other characteristics of citations in patents to other (earlier) patents. This part of the project should clarify the meaning of patent citations in patents, in particular in terms of the 'technological basis of technological advances'. In particular, the role of highly cited patents will be investigated.

(4) Systematic analysis of numbers, types, and other characteristics of citations in patents to scientific publications. This part of the project should clarify the meaning of citations to scientific publications in patents, in particular in terms of the 'scientific basis of technological advances'. Similar to the patent citations, the role of highly cited publications will be investigated.

(5) The construction of a step-by-step citation history of applied research. This means a 'retrospective' bibliometric study. Starting with basic research, the development of applications is carefully analysed by, at the one side, an analysis of references (the 'citations given') in order to identify the basis of the research concerned. At the other side, an analysis of the citations (the 'citations received') in order to identify the use of the research concerned. Again, the specific role of highly cited publications will be investigated.

The Karlsruhe co-team reports on the results of part 1 and 3 (chapter 9). In this chapter we report on the results of part 2. Work on parts 4 (Karlsruhe and Leiden), and 5 (Leiden) is in progress and not included in this volume.

8.3 DATA COLLECTION

In this section we present an inventory of data (i.e., items in publication records as given by a database) that must be available in order to perform the relevant analyses, and the requirements with respect to search strategies available at the host.

In order to investigate the distribution of publications related to a specific subject area (laser medicine or polyimide chemistry) over other scientific fields, the following items in the database must be available (and searchable): (1) country, (2) title or abstracts or terms (controlled or uncontrolled), (3) classification system, and (4) publication year. With respect to these items, it is of importance that they are properly and reliably unified, and that terminology over the years has not been modified. For instance, the occurrence of all possible variations of the country name during the whole range of years under investigation must be checked.

As mentioned before, it is most convenient if the classification system remains (virtually) constant during the time period covered by the research. In case of drastic changes, a concordance of the classification systems is necessary. If such a concordance is not made available by the database producer, a concordance (only for the higher levels of classification) must be made. In such a case of modifications to the classification system through the years, adjustments are made to the collected data to create a classification system, which is artificially made constant through the years. Starting from the most recent classification system, less recent classification codes, if changed, are translated by means of a match of content descriptors attached to the classification codes. Such a match must be made for each change through the years of the classification system.

Availability of information highly depends on the way the host organization has made the database available. It is not uncommon that a database is made available by different hosts in a different way. At one host other fields might be searchable then in the other. Due to the different possibilities of host database combinations under investigation, different data retrieval strategies and methods of analyses had to be developed.

8.4 EXPLORATION OF THE FIELD OF POLYMER RESEARCH

GENERAL INTRODUCTION TO POLYMER CHEMISTRY

We decided that polymer chemistry was probably an interesting area for this project, because it was regarded as a technologically important growth area. A comprehensive historical overview and a demarcation of polymer chemistry is given in section 9.3. and by Grupp et al. (1990). We here summarize the main lines of this area. Polymers are macromolecules with a molecular weight above 10,000. Industrial application of these macromolecules started in 1868 with the synthesis of celluloid. In 1909 bakelite (a phenol formaldehyde resin) was introduced and in the twenties and thirties well-known polymers such as polyvinyl chloride (PVC), polystyrene, and polyamide (nylon) were developed. In the fifties polyethylene and polycarbonate followed, after the invention of stereospecific polymerization reactions. The sixties show the introduction of high-temperature resistant polymers such as polyimides and polyarylsulfones. In the most recent decades, the seventies and the eighties, the emphasis in polymer research is on the synthesis of polymers with specific properties, such as electrical conductivity.

Economically, polymers are important. About 17% of the German chemical industry production in 1988 concerned polymers. For an extensive discussion of the demarcation of polymer R&D and its subfields as far as the technological counterpart of this study concerns (classification via the International Patent Classification (IPC) system), we again refer to Grupp et al. (1991). Here also a short discussion of the various important subfields is given (fluoropolymers, polyimides, polycarbonates, polyamides, aramides, electro-conductive polymers).

DEFINITION OF THE SEARCH STRATEGY

First of all, the number of polymer publications (i.e., publications containing the keyword polymer+) through the years was examined with help of the database Chemical Abstracts (CA) at the host organization STN. Figure 8-1 shows the (absolute) number of polymer publications for the period 1970 - 1986. A correction procedure converts the yearly figures to match the growth of the number of publications in the database, see figure 8-2. We notice that the number of polymer publications is growing faster than the average number of publications. We may conclude from these findings that in the scientific literature a growing interest in the polymer area as a whole is clearly visible.

For the delimitation of our research to a field of feasible size and interest, a more elaborate analysis was performed to identify possibly interesting (sub)fields within the polymer area. Different keyword data have been gathered independently in order to verify observed trends. The following keyword types can be distinguished: keywords from the abstracts of publications (AB), controlled vocabulary (CV, a fixed set of terms given by the indexer of the database), and title words (i.e., words from the titles of publications, TI). Abstract words (AB) describe the contents of a publication, and thus they provide a relatively complete coverage of the subject area. Major characteristics of CV-keywords are the following: (1) words are given by indexers to describe important subject(s) the article deals with; (2) there is a fixed set of words an indexer can choose from (hence 'controlled vocubulary'); (3) if necessary (the upsurge of new terminology), words are added to the list, usually on a yearly basis. Summarizing, CV-keywords are important, but more restricted to a fixed set of words, which might be lagging behind with upsurging terminology. Title words (TI) not only serve the purpose to briefly describe the subject of the publication, but also, more generally, to attract the attention of possibly interested readers. As there is only limited room in a title, title words give a much less complete description of the subject of the publication than AB- or CV-keywords.

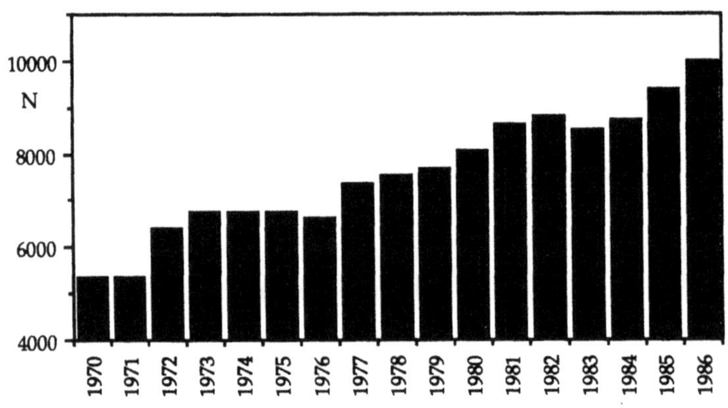

Figure 8-1: Absolute numbers of 'polymer+' publications 1970-1986

In a pilot study the occurrence and possible variations of a large list of AB-, CV-, and TI-keywords corresponding with our subject area have been explored. On the basis of this exploration the most suitable keywords for subsequent analyses could be selected. Further, the most suitable point of truncation of the words to be used in the search can be found. This is an important technical point. The effect of different truncations is shown in figures 8-3 and 8-4. Two cases of similar truncation of CV-keywords are compared. In the case of 'polycar+' and 'polycarbon+' there is a very close correspondence between the two curves. Obviously, 'polycar+' does not leave more room for variation than 'polycarbon+'. Comparison, however, of the 'polysulf+' and 'polysulfon+' curves shows a large discrepancy in the years 1970 through 1977. After that the difference remains approximately stable. The keyword 'polysulfon+' has an offset in the year 1976, while 'polysulf+' seems to have an offset in 1971, thereafter growing steadily. In this case, the upsurge of 'polysulfon+' must be attributed to its introduction into the thesaurus of controlled vocabulary around 1976. The high starting level of 'polysulf+' is due to 'polysulfid+' publications.

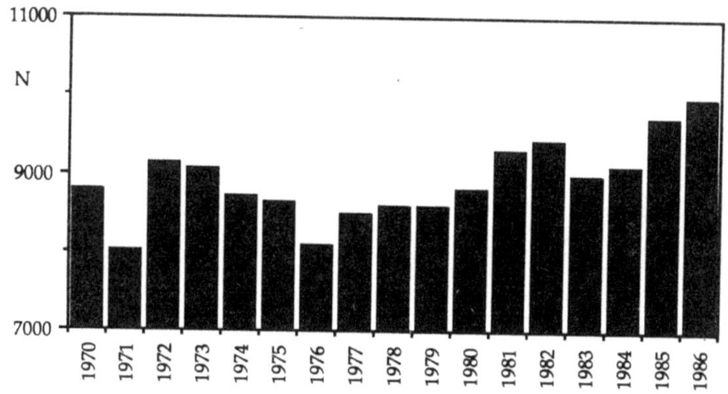

Figure 8-2: Corrected number of 'polymer+' publications 1970-1986 (corrected means N is converted to match publication growth)

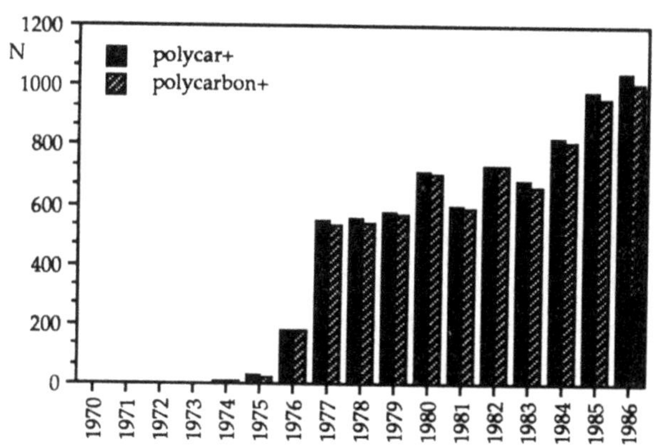

Figure 8-3: Effect of word truncation on article retrieval with 'polycar+' and 'polycarbon+' (CV-keywords)

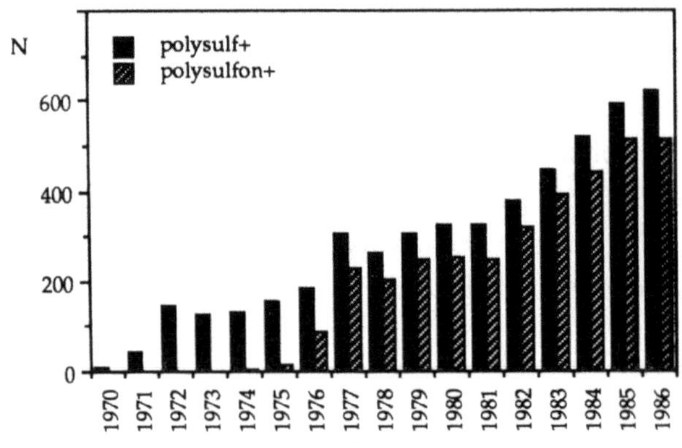

Figure 8-4: Effect of word truncation on article retrieval with 'polysulf+' and 'polysulfon+' (CV-keywords)

Those keywords that could be expected to cover interesting subfields and occur sufficiently frequent, were selected. The trend in number of occurrences of these AB-, CV-, and TI-keywords (i.e., number of publications characterized by these specific keywords) for the period 1970 - 1986 is shown in figure 8-5 for the three most frequent keywords (polyamide+, polyester+, and polymer+), and figure 8-6 for the four subsequent keywords (amid+, polycar+, polyimid+, and polysulf+). The point of truncation is given by the '+' sign, after this point variations may occur. We see that nearly all keywords show a tendency to grow, that 'polymer+' and 'polyester+' are predominant for all types of keywords (AB, CV, and TI). As remarked above, the number of keywords vary strongly between the different types.

Such a careful comparison of AB-, TI-, and CV-keywords gives us clues about possible effects, which would go unnoticed or would be incorrectly interpreted. In particular, we see that the CV-keywords show a relatively strong increase around 1970-1973 as compared to the other types. This is probably an artifact, due to the introduction of the index system. As a consequence, CV-keywords from 1970 until 1973 must be disregarded. Moreover, it can be seen that the introduction of a new word, for example 'polycar+' (about 1976), to the controlled vocabulary results in a rise of this keyword as a CV-word, not paralleled by the same keyword in the other types (AB, TI). Instances of these must also so be excluded from further analyses.

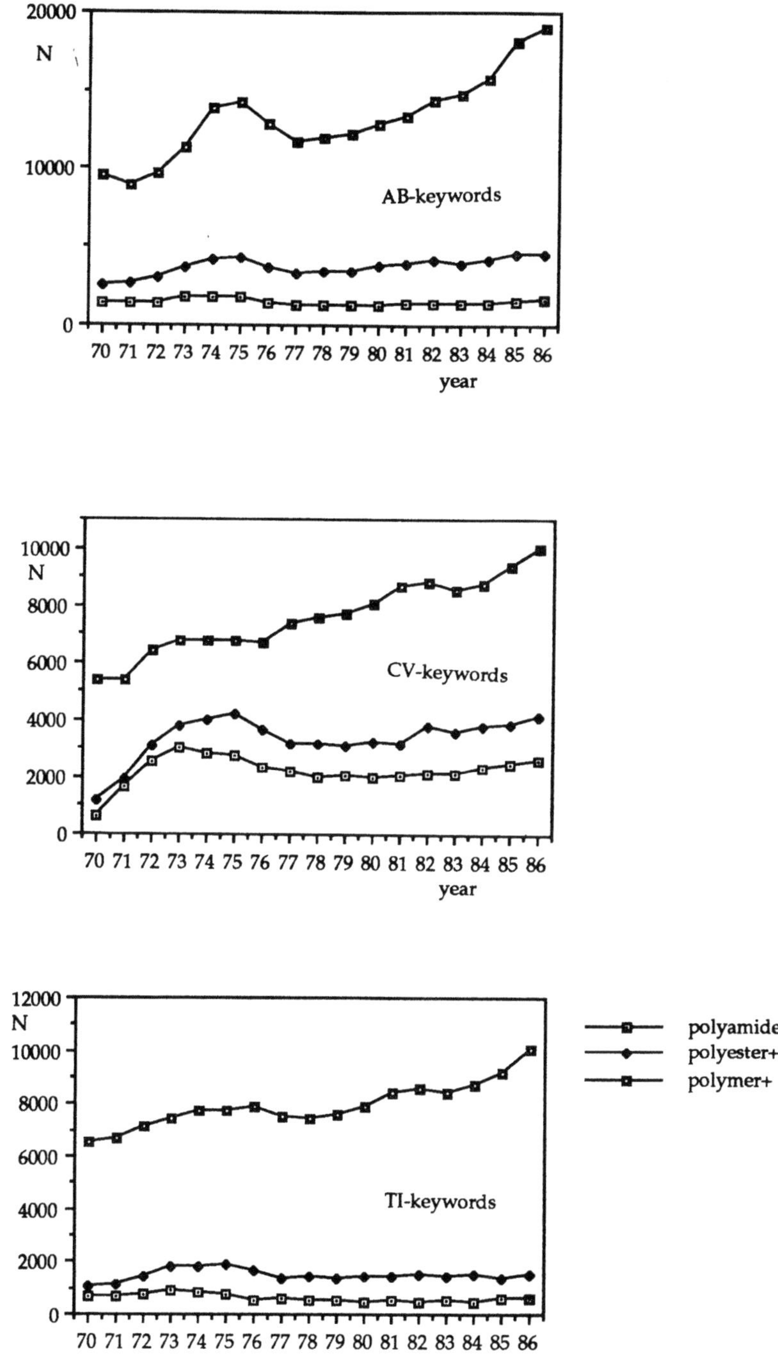

Figure 8-5: Trends of the three most frequent keywords (upper part: AB-keywords, middle: CV-keywords, lower part: TI-keywords)

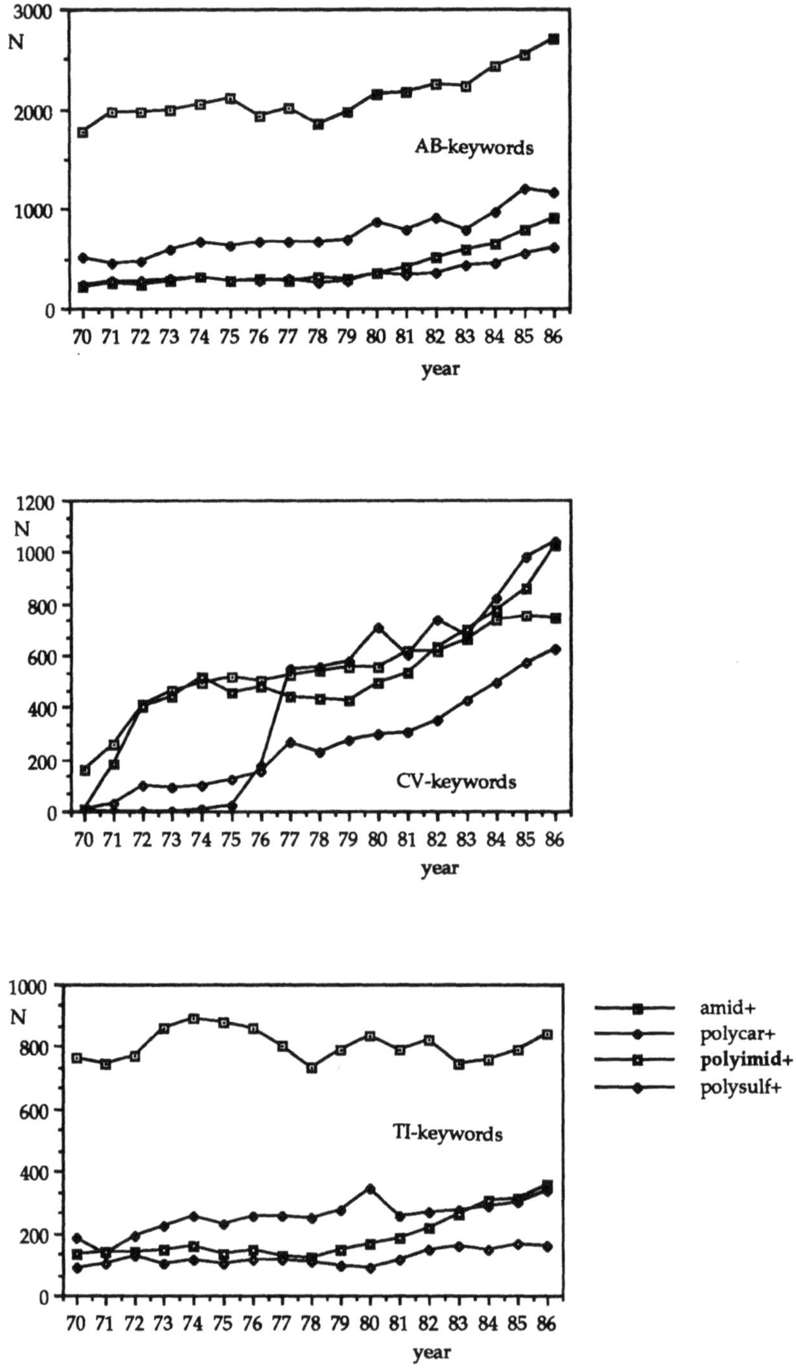

Figure 8-6: Trend of the subsequent most frequent keywords (upper part: AB-keywords, middle: CV-keywords, lower part: TI-keywords)

Generally speaking, our keyword analysis shows that the overall trends for the different keyword-types (AB, TI, CV) are about the same. But some extreme tendencies, like the 1975-76 peak for polymer+ as an AB-keyword, are less prominent as compared to same word in CV or TI. An interesting case of a rather 'subtle' convergence is the trend of the most important keyword for this study, 'polyimide+' (figure 8-6). A slow rise in the seventies shifts to a rapid growth starting in about the year 1980 for all types (AB, CV, TI).

We may conclude from these results that a comparison of keyword trends of different word types through the years provides us with a way to make more accurate statements about the probable origin of the observed phenomena (e.g., the coverage of the keyword by controlled vocabulary). On the basis of findings of the Fraunhofer group and our exploratory results, the field of polyimide research was selected for further analysis.

8.5 TRENDS IN POLYIMIDE RESEARCH

The field of polyimide research is important to study because there are several indications that it is a prosperously growing part of the polymer area. Patent activity in this field is increasing steadily since 1977. Still there is a growing activity in the technological application of polyimides. A long-term trend analysis (1960 - 1987) on international polyimide technology in terms of total patent applications with destination West Germany shows an increase in particular in the sixties, a remarkable decrease in patent activity starting around 1973 up till 1977, and again a (strong) increase from the late seventies (see figure 9-1 in section 9.4). This general trend is confirmed by data from worldwide patent applications between 1975 - 1985 using the databases WPI and WPIL (with exclusion of the Japanese applications at their domestic patent office, because of the 'extreme' Japenese domestic patenting habits).

In this part of the study we focus on the bibliometric counterpart of the polyimide (chemistry) technology field. Publications in chemistry are best covered by the database Chemical Abstracts (CA). Different types of publication carriers are included in this database such as scientific journals, books, conference papers, and patents, all containing chemistry and chemistry-related information. For this study, we restricted the search to publications in scientific journals. Patents were not included because they are delt with in the technological counterpart (chapter 9). Books were also not included, they are much less covered as compared to scientific journals. We assume that important results presented in books usually have been published already to a large extent in scientific journals. Conference papers were not included because, on the one hand, they are less well covered by the database than scientific journals, and on the other hand they usually present preliminary results of which the important ones mostly will be presented later on in a scientific journal. This might, however, not be true for specific fields. In many fields of (applied) research, conference proceedings do play an important role as a means of communication. In fact, for the assessment of research performance in applied fields it appears to be important to investigate how far conference proceedings do cover research results not published in journal articles (Van Raan and Tijssen 1990a, b).

The polyimide technology field is delimited by the IPC patent classes C08G -73/10, /12, /14 , /16 and C08L-079/08 (see Grupp et al., 1991). The bibliometric counterpart of polyimide technology can be identified by a search in the database for publications containing the keywords ('descriptor terms') of these IPC classes. In this way, we collected publication records from the database (CA/STN). These records were further analysed on the mainframe by means of special software in order to recategorize data (if necessary) or to perform additional analysis on the collected data.

Table 8-1: Number of polyimide publications per country, 1979-1988

PY\CO	FRG	JPN	UK	USA	USSR	ALL
1979	6	30	0	97	78	211
1980	2	29	0	87	77	195
1981	1	39	3	98	82	223
1982	5	30	5	93	66	199
1983	6	55	5	114	94	274
1984	6	40	5	118	89	258
1985	11	66	8	137	69	291
1986	14	52	9	160	83	318
1987	20	83	15	229	90	437
1988	23	98	24	295	106	546
ALL	94	522	74	1428	834	2952

We found that more than 85% of the publications in the polyimide field originates from the USA, the USSR, Japan, the Federal Republic of Germany, and the United Kingdom. Further analyses has been restricted to these countries, because the number of publications by other countries are too low.

First, we focus on a more general overview of the field. The growth of the field per country can be observed in the data presented in table 8-1. Between 1979 and 1988 the number of publications increased more than twofold. Large differences in growth can be seen if we compare countries. We shall analyse these differences in more detail later on. The trend through the years (1979 - 1988) of polyimide publications in different chemical (sub)fields (Chemical Abstract sections) is shown in table 8-2. The description of these (sub)fields and their section codes is given in table 8-3. During the period under study a major classification change took place (around 1982). Therefore, a 'backward concordance' has been applied to section codes. This means that section codes in years previous to the classification change, have been modified in such a way that correspondence to the new classification system is as good as possible.

Table 8-2: Distribution of polyimide publications over chemical (sub)fields for the <u>five countries</u> together, as given by CA Section Codes (SC), 1979-1988.

PY\SC	all	35	36	37	38	40	42	71	73	74	76	rest
79	64	0	67	25	14	2	4	1	3	11	20	211
80	62	0	63	24	11	4	4	0	3	8	16	195
81	50	0	61	41	15	9	4	0	9	13	21	223
82	31	21	53	35	7	5	4	1	3	22	17	199
83	42	29	91	31	4	22	4	1	14	22	14	274
84	45	33	90	34	7	6	2	3	8	19	11	258
85	35	30	81	39	7	10	5	12	12	31	29	291
86	57	39	101	24	4	10	7	9	9	36	22	318
87	68	69	131	36	6	8	7	10	15	44	43	437
88	87	82	164	68	7	20	5	14	23	31	45	546
all	541	303	711	548	82	96	46	51	99	237	238	2952

Because of the introduction of section code 36 in 1982, this section code does not occur in the years 1979 through 1981. Actually, the old section 'Synthetic high polymers' (old nr 35) was split into 'Chemistry of synthetic high polymers' (35) and 'Physical properties of high polymers' (36) in 1982. Most publications are categorized in sections 35, 36, 37 ('Plastic manufacture and processing'), or 38 ('Plastic fabrication and uses'). Each of these sections contains more than 10% of the total number of publications.

Table 8-3: Major polymer-research related Chemical Abstract sections (according to the most recent classification)

Section Code	Section name
35	Chemistry of synthetic high polymers
36	Physical properties of synthetic high polymers
37	Plastic manufacture and processing
38	Plastic fabrication and uses
40	Textiles and fibers
42	Coatings, inks, and related products
71	Nuclear technology
73	Optical, electron and, mass spectroscopy
74	Radiation chemistry, photochemistry,
76	Electric phenomena

The most important trends are made visible by using a simplified classification system. Sections 35 and 36 ('Chemistry of synthetic high polymers' and 'Physical properties of synthetic high polymers', respectively) are joined into a new field: 'Polymer research' ('polym.'). Sections 37 and 38 ('Plastic manufacture and processing' and 'Plastic fabrication and uses', respectively) are joined into the new field: 'Industrial application' ('ind. appl.'). The other less frequent sections are combined in the category 'others'. The general trend for the five countries taken together in the period 1979 - 1988 is shown in figure 8-7. 'Industrial application' seems to increase quite similarly as 'polymer research' and 'other'. This increase starts gradually around 1982 and more sharply around 1985 - 1986. We may assess the role of individual countries by comparing the trends in figures 8-8 to 8-12 with each other and with the general trend. A three-year smoothing procedure was used to suppress yearly fluctuations. Large discrepancies between countries can be found. For the USA (figure 8-8), industrial applications are by and large the most important category. Growth in 'polymer research' and 'industrial applications' go quite parallel, while the 'other' category levels off starting around 1986. A very different picture emerges for the USSR (figure 8-9). Starting at a relatively high position with respect to other countries, growth of all areas is lagging behind with that of all other countries. Polymer research is quite predominant, but the number of USSR publications on industrial application is lagging behind dramatically. It even decreases between 1982 and 1986, perhaps starting to increase again in 1986.

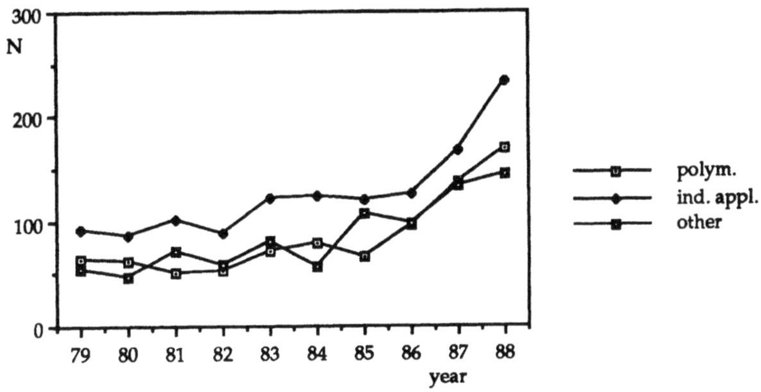

Figure 8-7: Major trends in polyimide publications, all countries together

With respect to Japanese publications (figure 8-10) it can be seen that the 'others category' is relatively large. This is an indication that a relatively large, and further increasing number of polyimide publications is classified outside the 'traditional' polyimide categories. An example is 'Electric phenomena'. Apparently, Japanese polyimide research is moving into specific fields such as research on electrical conductivity of polyimides. Moreover, growth seems to start earlier than for the other countries.

For the Federal Republic of Germany (figure 8-11) 'industrial applications' start relatively early, while 'polymer research' starts relatively late. Looking at the results for the United Kingdom (figure 8-12), we see that 'industrial applications' start rather late. The number of UK publications in the 'other' category is relatively high. Notice that the number of publications of both the UK and Germany in the early period, until 1985, is rather low. The overall results suggest that, by means of the keyword search, predominantly publications are found dealing with industrial application of polyimide chemistry.

In order to compare our results for the different countries with those for the technological counterpart, we use the findings of our co-project team (Grupp et al., 1990; not included in chapter 9). With help of their patent activity indicators (Revealed Technological Performance, RTP indicators), we see that the US are the leading nation, closely followed by Japan. Our results largely confirm this finding, but a stiking difference is that for the 'science side' the gap (in absolute number of publications) between US and Japan is larger. No USSR data are considered in the technological counterpart of this project. The technological performance of both US and Japan is clearly increasing. German polyimide technology performance is rather constant between 1977 and 1986. Research activities are increasing, however, but still on a relatively low absolute level. Compared to West Germany, French patent applications are decreasing since 1979. No French research results are covered by this study, because of the relatively low publication activity of France. The technological and scientific counterpart of this project therefore more or less point in the same direction for France. Britsh technological activities on polyimide applications are fluctuating on a rather low level, throughout the whole period. Like for Germany, we find that the UK polyimide research activities are increasing, although rather late (1986) for 'industrial applications'. The British activity in the more basic 'polymer research' (as far as polyimides concerns) is slightly increasing, but still on a relatively low absolute level.

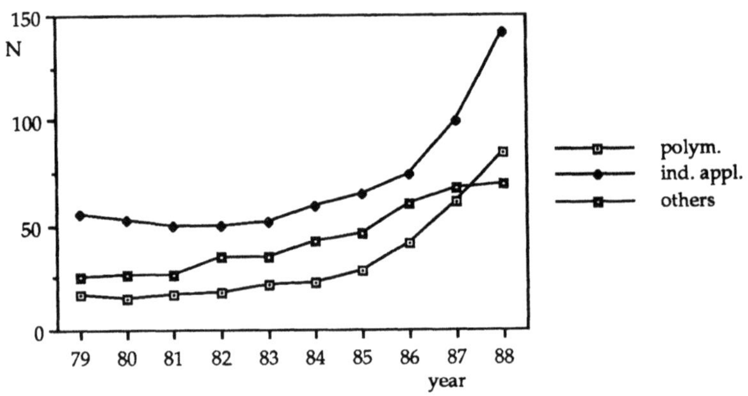

Figure 8-8: Major trends in polyimide publications in the USA

The most important properties of polyimides for electronic (semiconductor) applications are their thermal stability, chemical resistance, physical strength especially at high temperatures, and their dielectric properties. Most of the research on these properties of polyimides will be covered by CA-section 36 ('Physical properties of synthetic high polymers'), which we combined with CA-section 35 ('Chemistry of synthetic high polumers) to the field 'Polymer research', and by CA-section 76 ('Electric phenomena') which constitutes a major part of our 'others' category. Grupp et al. (1990) report on the activities of countries in these semiconductor-related polyimide technology. They again find high US and, especially, high Japanese patenting activities, particularly directed to the European market (i.e., patent applications at the European Patent Office). At a lower absolute level, France, Great Britain and W. Germany show similar activities until about 1981. After 1981 German patenting activity increases, whereas French and British activities decrease. As far as research activities concerns, however, the UK shows an increasing activity, in particular research related to 'industrial applications' (see figure 8-12). It is not clear why this increasing application-oriented research activities are not reflected by patent activities in the same R&D area.

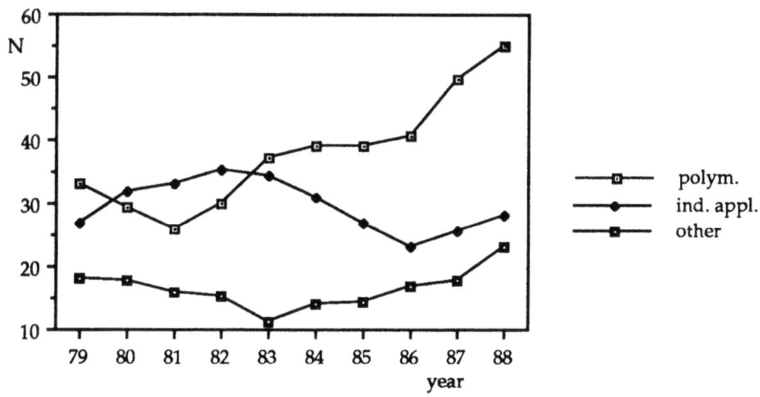

Figure 8-9: Major trends in polyimide publications in the USSR

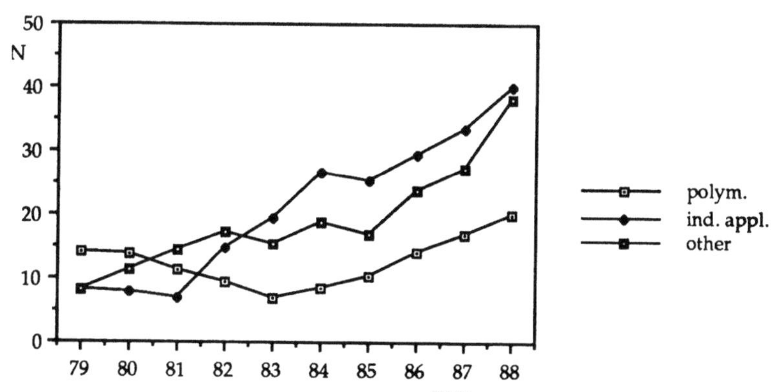

Figure 8-10: Major trends in polyimide publications in Japan

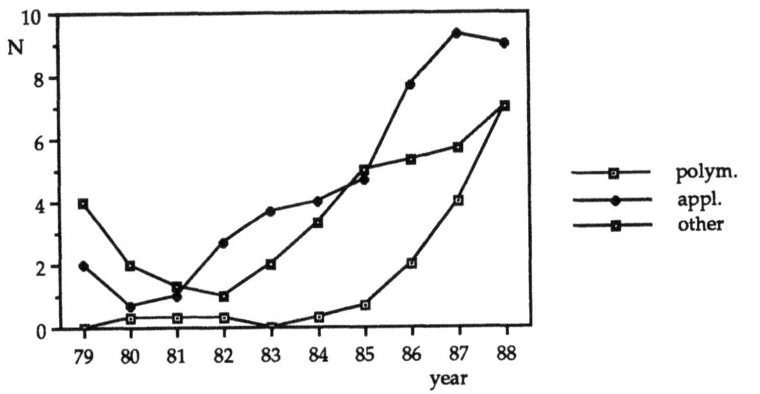

Figure 8-11: Major trend in polyimide publications in the Federal Republic of Germany

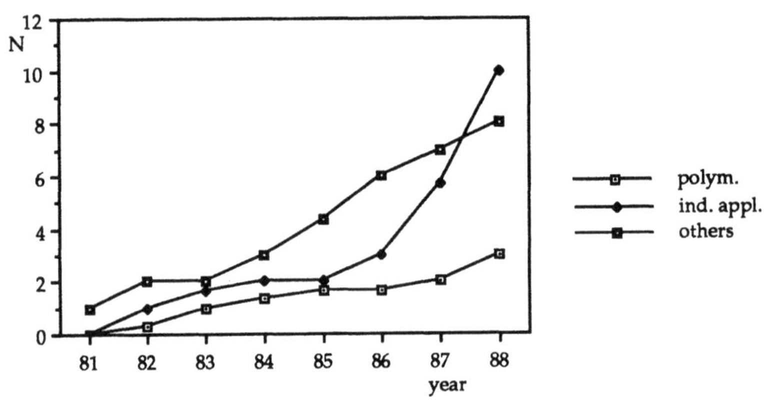

Figure 8-12: Major trends in polyimide publications in the United Kingdom

8.6 EXPLORATION OF THE FIELD OF LASER RESEARCH

GENERAL INTRODUCTION TO LASER RESEARCH

As in the case of polymer chemistry, a comprehensive discussion of the development of laser research is given in the publication of our co-project team (chapter 9). We here briefly anticipate the main lines. Lasers are a source of coherent radiation, i.e. light emitted in one particular small wavelength band (monochromatic, i.e., 'one color') and in a narrow beam. The basic principles of masers (microwave amplification by stimulated emission of radiation) and lasers (light amplification by stimulated emission of radiation) were formulated by Einstein in his famous work (1917) on stimulated emission of radiation by atoms. The first experiments however - in the microwave region of the radiation spectrum - were conducted almost fourty years later: the invention of the maser in 1953 (Gordon, Zeiger, and Townes, 1954, 1955). Seven years later, stimulated light-emission in the visible part of the spectrum was achieved: the invention of the ruby laser in 1960 (Maiman, 1960). A year later, the most well-known laser, the helium-neon gas laser, was constructed (Javan, Bennett, and Harriott, 1961), followed by the semiconductor laser (Burns and Nathan, 1964) and at the end of the sixties the molecular gaslaser (e.g., CO_2), the ion (e.g., argon) gaslaser, and the dye laser. The semiconductor laser now plays a crucial role in, for instance, CD (compact disk) technology. The dye-laser offered the first possibility to tune the wavelength of laser emission over a broad range of the visible spectrum. Together with these developments, new optical phenomena (non-linear optics) were explored, such as frequency-doubling and parametric oscillations. With help of these new optical discoveries, the wavelength region for tuning laser radiation could be extended considerably into the ultra-violet and the infra-red regions. In the last two decades, these new developments in optics and laser physics had a enormous influence on the development of other fields of basic research, in particular in atomic and molecular physics and in chemistry.

Around 1960, in the very beginning, patenting in the field of lasers starts. It is clear that together with the scientific development of this remarkable light source, also applications were foreseen alomost immediately. Initially, the industrial development of lasers was more or less directed to facilitate further scientific research. Broader industrial application started in the early seventies.

On the one hand, we see the application of lasers primarily as a specific light source: coherent light in a very small parallel beam (the use of lasers in geodesy and for other metrological purposes, and, in more recent times, to read compact disks). On the other hand, coherent light concentrated in a very small parallel beam also gives a powerful source of energy, and so we see the application of lasers primarily as a source of concentrated energy: welding and cutting, medical purposes (surgery), and even heating of plasmas at extremely high temperatures to achieve - eventually - nuclear fusion.

As pointed out by Grupp & Hohmeyer (1988), the largest user-market during the second half of the eighties is probably material processing. About 20% of all lasers are used in this field of technology. Almost of the same size is the market for research applications, followed by medical applications and armaments (both about 16% of the market). According to Grupp et al. (1987) the strongest growth in the future laser market will be in the field of optical memories (e.g., compact-disks) and medical applications, followed by material processing.

For this study, we chosed laser applications in medicine as our subject field. Grupp et al. (1990) present a detailed discussion about this choice, together with an extensive overview of related (sub)fields in laser R&D. Laser applications in medicine cover therapeutic as well as diagnostic instruments. It is clear that this field has strong relations with both physics (in particular applied optics) as well as medical fields. A very important and early application of lasers in medicine concerns eye surgery. With laser-ophthalmoscopy retinal detachments (in particular in the case of diabetes patients) are treated ('spot-welding' of the retina). One of the first publications is the work of Smart (1965). A year earlier, Goldman (1964) had published another biomedical application of lasers, the treatment of skin cancer. Shortly after these first medical applications, Thorp (1967) foresees in this review book considerable advances in this field. And indeed, in 1985 the number of patent applications in laser medicine was almost an order of magnitude larger then ten years before. For publications we find an even stronger increase. Therefore, we concluded that laser medicine R&D is a dynamic and relatively young area. For an extensive discussion of the demarcation of laser R&D, and laser medicine R&D in particular, as far as the technological counterpart of this study concerns (classification via the International Patent Classification (IPC) system), we again refer to Grupp et al. (1990) and chapter 9.

DEFINITION OF THE SEARCH STRATEGY

The database PASCAL was used for the study on laser medicine publications. As PASCAL is a multi-disciplinary database, it is supposed to cover all scientific fields. Moreover, a detailed classification system is available. As a point of general information, the total number of publications per year in the PASCAL database for the most frequently publishing countries is given in Appendix 1A. The USA is, as can be expected, the country with the largest number of publications. It should be noted however that the number of French publications is rather large, a clear indication of a bias towards French literature in this database.

Table 8-4: Effect of adding more specific laser medicine related keywords in a search on the number of retrieved publications (database used: PASCAL; country: USA)

year	set#1	set#2	set#3
1979	5	16	33
1980	9	23	44
1981	10	30	58
1982	8	64	71
1983	8	117	147
1984	38	137	163
1985	49	273	292
1986	44	279	307
1987	59	323	355
1988	60	322	351

In PASCAL (via the host ESA) expansion of the keyword 'laser' resulted in a list of many terms connected with the laser medicine field. Therefore, publications have been selected by combining the keyword 'laser' with one or more keywords from a list of medical terms in the above expansion list. An improvement of the search strategy is accomplished by adding stepwise more medical terms to the search strategy, starting with the most frequent ones in the expansion list. This iterative keyword search is performed until addition of more terms fails to improve the search strategy substantially. In table 8-4 the number of retrieved publications for each additional step is shown (period 1979 - 1988). It gives a good impression of the effectiveness of our iterative keyword-search procedure. Set 3 proved to be the most optimal set of keywords for the description of a bibliometric counterpart of a technological area, thus data were collected making use of this set. A comparison of the number of retrieved publications as such (table 8-4) and the number of retrieved classification-codes (i.e., number of publications characterized by one or more of the cc's involved) (table 8-5) shows us that the ratio is about 1:1. Thus, in the PASCAL database most publications are attributed to one classification code exclusively. The number of retrieved publications per country per year are given in Appendix 1B.

Table 8-5: Effect of addition of more specific laser medicine related keywords on the number of CC-codes in retrieved articles (database used: PASCAL; country: USA)

year	set#1	set#2	set#3
1979	10	24	32
1980	19	34	56
1981	17	38	68
1982	9	76	84
1983	11	139	171
1984	38	137	163
1985	50	277	284
1986	44	282	310
1987	64	328	360
1988	62	325	355

The search strategies resulting in the different sets.
set #1:
laser? *and* (biomed? *or* medic?) ?=truncation operator
set #2:
laser? *and* (biomed? *or* medic? *or* surger? *or* surgic? *or* therap? *or* phototherap?)
set #3:
laser? *and* (biomed? *or* medic? *or* surger? *or* surgic? *or* therap? *or* phototherap? *or* tumor? *or* glaucoma? *or* cardiovasc? *or* retina? *or* macula? *or* arter? *or* endoscop? *or* cataract?)

DATA COLLECTION AND DATA ANALYSIS

For each year and country combination, the search strategy described above as 'Set 3' was performed (PASCAL). The occurrence of classificationcodes (cc's) attributed to the selected publications has been analyzed. This analysis resulted in a list of classification codes and occurrence frequencies of these classification codes. An example of a 1989 classification code is 002.B.09.G, which means: 'biological and medical sciences (002); medical sciences (B); ophthalmology (09); lens: diseases (G)'.

The PASCAL classification scheme ('Plan de Classement') is adjusted each year. There was a major modification in 1984. Such modifications are necessary to refine the classification system, to include novel research themes, and to split up fast growing areas. In our example, the 1989 classification code 002.B.09.G was called 346.A.10 in the year 1979, which is clearly different! Thus, in order to follow trends through the years (as far as subject classifications concerns), a concordance of classification codes is needed. Unfortunately, there is not such a concordance list provided by the database PASCAL as there is for example in INSPEC.

We received however, a concordance for general codes on floppy from an internal database at CNRS (Turner 1990). Software was developed to construct a masterfile from this data. The masterfile was carefully checked and corrected manually for some minor inconsistencies. In another computer program, classification code data assembled at the host were matched with the masterfile to link them with unique classification titles that remain the same through the years, if possible. The classification titles are at a somewhat higher level than the classification codes. In the case of our example this means that codes 002.B.09.G (ophthalmology; lens: diseases), and 002.B.09.H (ophthalmology; vitreous body: diseases) are both categorised as 'ophthalmology'.

LASER MEDICINE IN OTHER DATABASES

In BIOSIS the same search strategy for collecting laser medicine publications has been used as the one developed for the PASCAL database. Data were collected by means of the ESA 'zoom' command. Publications in BIOSIS are classified by means of 'concept codes'. Unlike the classification system in PASCAL (and Chemical Abstracts), many concept codes (BIOSIS-cc's) are attributed to one article in BIOSIS. The average amount of cc's attributed to an article varies between about 10 to 14 through the years!

There were, however, no major changes in the classification system of BIOSIS. Therefore, the calculation of the distribution of publications among these cc's was quite straightforward. Unfortunately, it is not possible to differentiate among countries in ESA, so a further subdivision of distribution of cc's per country could not be made yet. The general tables, however, may give us interesting background information about how laser medicine is covered in a database specialised in biological and medical issues. In Appendix 2A an overview is given of the trend of the major concept codes. The description of these concept codes is given in Appendix 2B. As the numbers of publications were widely spread among very many different (sub)fields, a reasonable way to reduce the number of subfields had to be found. The following strategy was devised:

(1) only subfields with more than 1% of the total are included with their full concept code (5 digits); (2) the remaining subfields are truncated to 3 digits (these concept codes have two zero's at the end) and, are included if the new 3 digit concept code contained more than 1% of the total number of publications; and (3) those subfields that do not reach 1% of the total number of publications are placed in the '00000' (rest) concept code.

The trend in the number of laser medicine publications characterized by the major concept codes (with n>4%) for the period 1973 - 1988 is shown in figure 8-13. All (sub)fields grow about simultaneously, no clear field-specific trends are visible (for example growth in one category at the expense of another category). Two take-off's can be distinguished: the first in 1980, and a more pronounced one in 1984. We verified that these phenomena are not an artefact due to variations in the mean number of cc codes per publications through the years. If the data are corrected for these trends, results remain about the same.

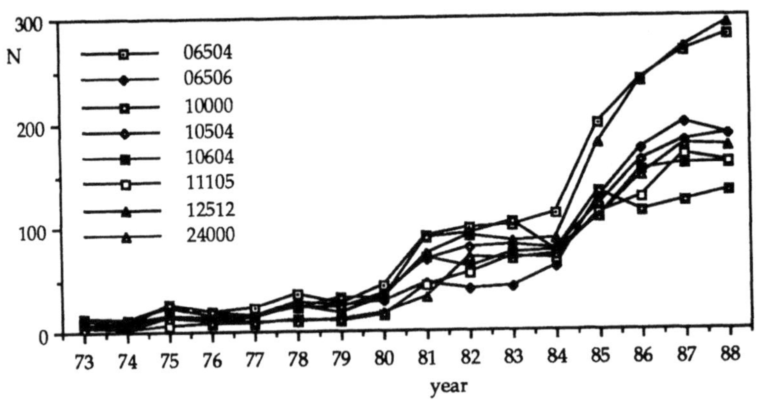

Figure 8-13: Number of laser medicine publications characterized by major BIOSIS subfields

The subfields 'Therapy' (cc= 12512) and 'Radiation and Isotope Techniques' (cc= 06504) occur most frequently. According to the BIOSIS classification, these two subfields should present the two extremes in the continuum from applied to more basic research. Their growth curves through the years seem, however, quite parallel, while a time lag between the more basic and applied subfields would be expected (in conformity with the previous results in PASCAL and Chemical Abstracts). This leads us to the tentative conclusion that the lack of clear separate trends in the data might be caused by the high amount of concept codes attributed to each article.

Table 8-6: Number of laser medicine publications with title words: "$laser$ *and* ($medic$ *or* $surg$ *or* therap$)" in the SCI database (host DIMDI).

country\year	73	74	75	76	77	78	79	80	81	82
USA	0	13	7	8	8	9	19	23	46	77
UK	0	0	0	0	0	1	0	6	4	6
Japan	0	0	0	0	2	0	0	3	2	7
Fed. Rep. Ger.	0	0	4	3	6	4	12	7	13	8
France	0	0	0	0	0	0	9	9	14	11
Canada	0	0	0	0	0	1	1	0	6	7
Italy	0	0	0	0	0	2	0	1	3	9
Australia	0	0	0	0	0	2	0	0	1	1
Netherlands	0	0	1	0	0	1	0	2	0	4
Sweden	0	0	0	0	0	0	0	0	1	4
Switzerland	0	0	0	0	0	2	0	0	1	4

country \year	83	84	85	86	87	88
USA	386	722	662	618	299	374
UK	60	90	125	128	54	94
Japan	30	77	81	66	17	32
Fed. Rep. Ger.	42	125	138	111	47	64
France	36	60	72	52	26	26
Canada	23	42	42	52	18	33
Italy	41	36	31	30	10	16
Australia	10	19	27	10	8	17
Netherlands	12	20	19	31	15	15
Sweden	15	33	31	24	15	11
Switzerland	15	26	33	20	16	10

The years 1973 through 1982 are available in file IS74 at the host DIMDI

The years 1983 through 1988 are available in file IS83 at the host DIMDI

In the Science Citation Index (SCI, Institute for Scientific Information (ISI) Philadelphia, online version SCISEARCH, through host DIMDI, Cologne) an optimal search strategy for laser medicine publications was found by searching for the 'free terms': 'laser- *and* (medic- *or* surg- *or* therap-)'. A drawback of using the SCI database is that its 'free terms' originate *only* from title words. So the chance for retrieving relevant literature is considerably smaller in this database than in databases having available abstract words, keywords, as well as title words. The results are shown in table 8-6, giving the number of retrieved publications in SCI for the period 1973 - 1988. The following can be observed. First, in the years 1973 through 1982 there is very small number of retrieved publications, probably due to restricted search possibilites in that part of the SCI database (IS73). Second, large fluctuations in the number of retrieved publications between years can be found, and this finding is another indication of database 'instability'. And third, a comparison with the PASCAL results show that the number of retrieved publications is indeed relatively low. Conclusion: Unfortunately, the laser medicine field cannot be properly defined by means of a title-word search as available on SCI. Because of this, no further research in the database SCI was performed for this part of the study.

In INSPEC the same search strategy for collecting laser medicine publications was used as the one developed for the PASCAL database. Data were collected by means of the ESA zoom command. Country of origin of the publication is not available for research articles. The numbers of publications retrieved in INSPEC were considered to be too low for further analysis. This is not surprising, as INSPEC is primarily a physics and electronics database, and not a medical one.

8.7 TRENDS IN LASER MEDICINE RESEARCH

We now present our findings with respect to developments in the laser medicine field. These findings are based upon the PASCAL data. Although in the process of establishing a concordance between database years a somewhat higher level of aggregation is reached, laser medicine publications in the PASCAL database are still classified into very many subfields. The proportion of publications assigned to each subfield is shown in Appendix 3A (abbreviation list of cc-titles in PASCAL is given by Appendix 3B). Only subfields with a contribution to laser medicine research of more than 0.50% are included in the table (as mentioned in section 8.6, the ratio cc-codes to publications is about 1:1). For a comparison between countries, the scatter in the above categorisation is still too large. Therefore, a recategorisation was performed. Prominent subfields (n >5%) were retained: 'Gastroenterology', 'Ophthalmology', and 'Biomedical engineering'. In the case of 'Gastroenterology', two subfields covering related subjects ('ENT / Stomatology' and 'Digestive-system surgery') were included. The remaining subfields were reclassified into in the following major fields: Chemistry, Physics, Engineering, Biology, Medicine(-other) (containing the remaining medical categories), and a 'rest'-category of neglectable size.

Table 8-7: Trends in laser medicine publications for 'recategorized' PASCAL subfields.

(a): absolute numbers

category \ year	79	80	81	82	83	84	85	86	87	88	ALL
Biology	2	3	6	6	8	4	10	6	6	10	61
Biom. eng.	9	12	15	9	14	16	37	21	42	36	211
Chemistry	3	0	2	2	3	5	7	5	2	3	32
Engineering	1	3	3	2	1	1	3	3	5	12	34
Gastroenterology	7	11	17	22	23	17	23	19	21	34	194
Medicine (other)	8	11	24	36	56	82	106	102	113	106	644
Ophtalmology	11	10	15	27	26	26	29	23	36	34	237
Physics	9	15	14	10	4	3	2	5	29	37	128
Rest	0	0	0	1	0	0	0	0	2	3	6
ALL	50	65	96	115	135	154	217	184	256	275	1547

(b): relative numbers

category \ year	79	80	81	82	83	84	85	86	87	88	ALL
Biology	3.70	5.48	6.48	6.06	5.26	2.33	4.15	3.30	2.72	4.17	4.06
Biom. eng.	18.52	16.44	13.89	7.58	11.18	10.47	16.60	11.79	15.31	12.82	13.26
Chemistry	5.56	0.00	1.85	1.52	1.97	2.91	2.90	2.36	0.68	1.60	1.94
Engineering	1.85	4.11	2.78	1.52	0.66	0.58	1.24	1.42	1.70	3.85	1.94
Gastroent.	12.96	16.44	18.52	17.42	15.79	11.05	10.79	11.32	7.48	12.18	12.29
Medicine (other)	14.81	17.81	26.85	32.58	41.45	52.33	48.55	53.30	44.22	38.14	41.43
Ophtalmology	24.07	16.44	15.74	25.00	21.05	18.60	14.52	14.15	15.31	13.78	16.69
Physics	18.52	23.29	13.89	7.58	2.63	1.74	1.24	2.36	11.90	12.50	8.06
Rest	0.00	0.00	0.00	0.76	0.00	0.00	0.00	0.00	0.68	0.96	0.34

The result of this recategorization, is shown in tables 8-7 (period 1979 - 1988). With help of table 8-7(a), the relative size of the fields as well as changes in size through the years can be followed. A large number of publications is assigned to one of the medical fields. Clearly, the first large-scale medical applications of lasers were in gastroenterology and ophthalmology. Further applications followed in other medical fields in the period 1981-1985. This is confirmed by our data from BIOSIS, which show an increase of laser medicine applied research activity in the same period. Closer examination of Appendix 3A shows that this increase of laser applications in medical fields is due to a small growth of laser medicine publications in each of the very many different medical fields. The number of publications in more basic research is relatively low. More specifically, the proportion of physics-related publications drops off to zero in the mid 1980s, but comes up again in 1987-1988.

Our results suggest that in the period covered by the database, most publications are pertaining to the application of laser medicine in medical fields. While development of the early new techniques in basic research is predominant for the period 1979-1982, it disappears in the upsurge of applications in clinical medical settings. Interesting in this respect is the 'physics drop' in 1982-1983, and its revival in 1987-1988. This might be an indication that physics developments relevant for the laser medicine field stopped for a while. A closer inspection of Appendix 3A shows us this is largely due to the 'decline' of the field optics. Probably, in the period 1979-1981 some major problems with respect to the optics of lasers, as far as applied in medicine, have been solved. The revival of the field laser medicine in physics, however, is due to the upsurge of biophysics and, to a lesser extent, energy-related topics. This increase of biophysics publications could lead to a second wave in applied medicine areas. New possibilities for the use of lasers in clinical medicine are being developed, possibly resulting in a new increase in use of laser medicine techniques in medical (sub)fields that did not yet use these techniques.

The trend of publication numbers (i.e., publications as classified by the given cc) is shown in figure 8-14 for the USA, France, Germany, and the UK. The overall trends discussed above are roughly found in the data of the individual countries. Interesting though is that the medical application of use of laser medicine in other areas than ophthalmology and gastroenterology starts earliest in Germany (about 1980) and somewhat later in the USA (about 1981). The growth of medical applications of laser medicine research relative to the other (sub)fields (laser medicine publications not classified as medical) is very large especially in the USA. This leads us to the conclusion that laser medicine has now found large scale application outside the realms of the pioneering application fields gastroenterology and ophthalmology.

In order to compare our results for the different countries with those for the technological counterpart of this study, we use the findings of our co-project team (Grupp et al., 1990). With help of their patent activity indicators (Releaved Technological Performance, RTP indicators; full data not given in chapter 9), the following observations have been made. First of all, it appears that the international technological performance in laser medicine is different from that of laser technology in general. Japan is an early starter, whereas the US is a relatively late starter. On the other hand, Japanese activity levels off since about 1981, and the US increase their activity in laser medicine technology from 1982. A similar trend, but at a lower absolute level, is

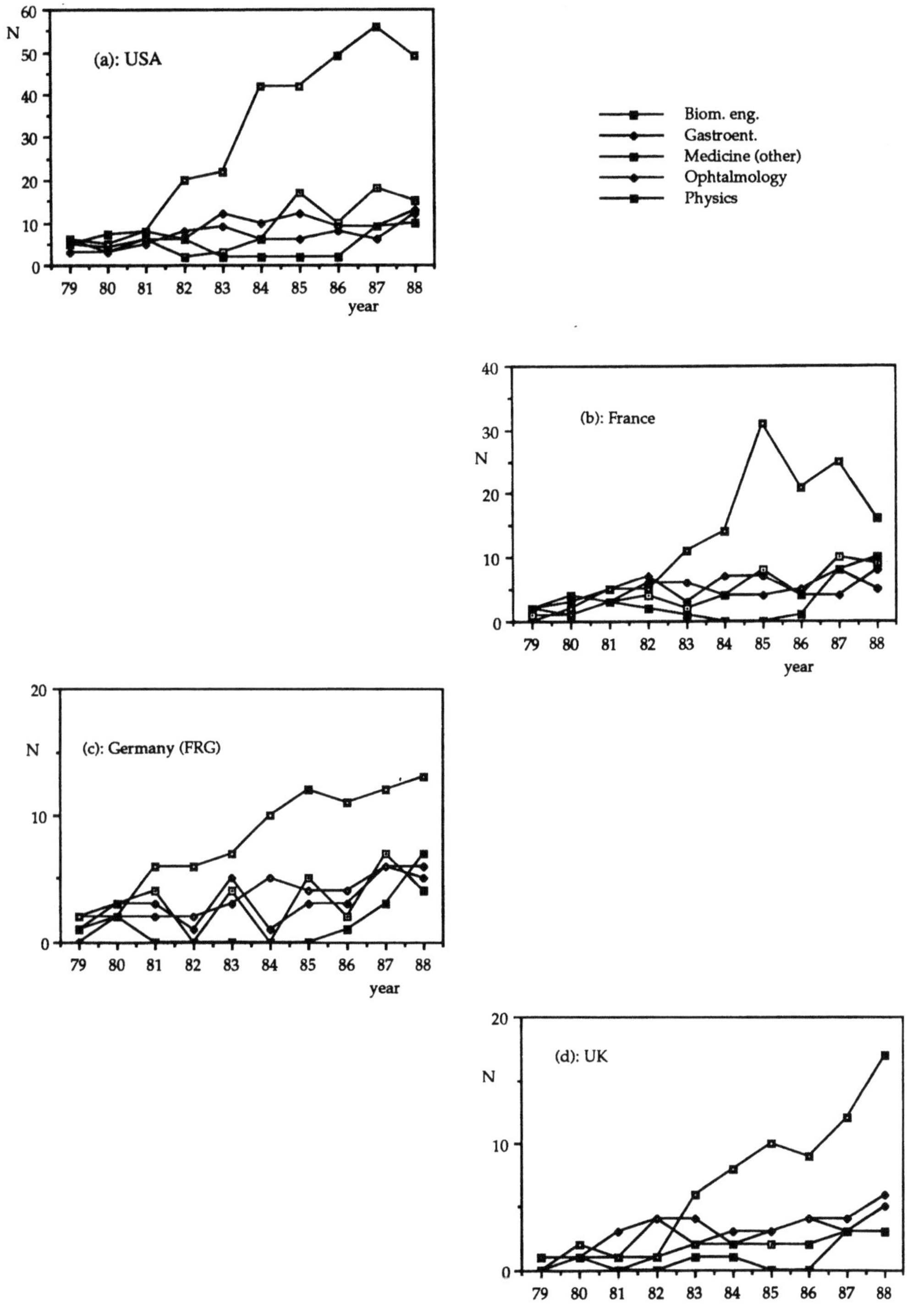

Figure 8-14: Trench in Laser medicine publications, 1979-1988, in four countries

found for W. Germany. Neither Great Britain nor France follow these US and German trends. Their trends are more similar to the Japanese one after 1981. For a correct interpretation of their findings, Grupp et al. (1990) consider the peculiarities of the domestic markets. For the US and Germany, the development of laser medicine is considered to be strongly coupled to the traditional distribution channels in medical equipment. Grupp et al. (op. cit.) assume that the Japanese companies tried to do international business through the channels of laser manufacturers, which probably appeared to be not the most effective way. Consequently, the Japanese lost ground in laser medicine technology.

These patent activities are only partly similar to our findings for the research activities. The US clearly is dominating the field of laser medicine research and almost tripled its research activity since 1981. Germany doubled its research activities in the same period, and the same is true for France, even at a higher level (here again the French bias of PASCAL plays a role). The UK is on a lower (absolute) level in 1981, but reaches in 1988 about the same level of activity as Germany. So, from a research publications point of view, there is no reason to state that France and the UK are on a lower R&D activity level as compared to Germany. Japan indeed shows no large increase in research publication activity, especially since 1985. In this respect, our publication data show a similar picture as found by Grupp et al. (1991) for the technological counterpart. We need, however, more recent data to see how far this lagging behind of Japan in the field of laser medicine R&D continues.

8.8 CONCLUSIONS

Studies of the diffusion and 'expansion' of knowledge by means of publication data require a very extensive exploration of the characteristics of databases and hosts. To bring the raw empirical data in a form suitable for further analysis, many technical problems with respect to classification systems and demarcation of research fields have to be solved. This chapter gives a comprehensive overview and discussion of how to tackle a number of these problems.

In this chapter we present the main lines of our first findings. First the results for the R&D field polyimide research. The general trend of research activity in this field is remarkably similar to the patent activity trends as found by Grupp et al. (1990, and chapter 9). For instance, we find that the general trend in polyimide publications (figure 8-6) shows a relative minimum in the years around 1977 - 1978, and after that (1979) a revival. According to chapter 9 (figure 9-1), the revival in patenting activity starts a few years earlier, around 1977. This would mean that a revival of industrial application (or at least the 'output' of this revival) precedes an increase of research publications. Research activity in the polyimide field seems to have a further revival around 1982 (figure 8-7). 'Industrial applications'-related research represents the largest part. This further revival (1982) is not visible in the patenting activities. On a somewhat lower aggregation level we find the same difference: The UK clearly shows an increasing polyimide research activity (figure 8-12), but this is not reflected by its patent activity. For the other countries a similar situation is found. The above findings may be a striking illustration of De Solla Price's model of the interaction between science and technology: in this case technology is leading, and, in a further stage, science. It is clear, however, that a more precise analysis is necessary before empirical findings can be related to any theoretical model. This more refined 'diffusion' analysis is the major part of the ongoing next (bibliometric) stage of this project not included in this chapter.

Our findings confirm the results of our co-team that the US are the leading nation in polyimide research, followed by Japan. A striking difference, however, is that our results indicate that the 'gap' between the US and Japan (in absolute numbers of patents and publications) is larger for the 'science side' than for the 'technological side'.

In laser medicine research, the general trends of publication and patenting activities are quite similar. In fact, this field shows no 'dips' in activity, there is a steady increase from the early seventies, and both in publication activity (figure 8-13) and in patenting activity (figure 9-3) the onset around 1980 of a stronger increase is visible.

We find a very strong diffusion of laser medicine research in the medical disciplines. This is clearly illustrated by table 8-7. If we compare the ratio of laser medicine publications in a specific medical field, ophthalmology, and laser medicine publications in the general category 'medicine-other' around 1980 with the same ratio a few years later up till now, we see a large increase for 'medicine-other'. This phenomenon is well illustrated by our finding that before 1980 laser medicine publications were for the largest part published in ophthalmology journals, whereas around 1982 the general journal 'Lasers in Surgery and Medicine' rapidly establishes its position as the major journal of the field.

Our results on research activities in the field of laser medicine are only partly similar to the findings of our co-team for the 'technological side'. As far as research activities concerns, there is no reason to believe that France and the UK are on a lower R&D

activity level as compared to Germany. Japan indeed shows no large increase in research activity, as far as reflected by the number of publications.

From these first results of our study it is clear that a more detailed bibliometric tracing of the use of scientific knowledge is necessary to gain further insight into the process of knowledge diffusion. We are currently exploring the use of citation analysis for this purpose.

*

This project was supported by the Stiftung Volkswagenwerk Hannover. Grant number II/63999.

REFERENCES

Braun, T., Glänzel, W. and Schubert, A. (1985), *Scientometric Indicators. A 32 - Country Comparative Evaluation of Publishing Performance and Citation Impact*, World Scientific, Singapore.

Burns, G. and Nathan, M.I. (1964), *Proc. I.E.E.E.*, 52, p. 770.

Casimir, H.G.M. (1972), *Proceedings of the European Physical Society* (EPS), Wiesbaden.

Dosi, G. (1982), 'Technological Paradigms and Technological Trajectories', *Research Policy,* 11, p. 159.

Einstein, A. (1917), 'Zur Quantentheorie der Strahlung', *Phys. Z.,* 18, p. 121.

Engelsman, E.C. and van Raan, A.F.J. (1990), *The Netherlands in Modern Technology: A Patent-Based Assessment*. Report 5A in the Policy Studies on Technology and Economy (BTE) Series (ISSN 0923-3164), Ministery of Economic Affairs, Den Haag (The Hague).

Frumau, C.C.F. (1991), 'Choices in R&D Business Portfolio in the Electronic Industry', to be published.

Gleick, J. (1987), *Chaos: Making a New Science*, Penguin Books, New York.

Goldman, L. (1964), *New Scientist,* 21, p. 284.

Gordon, J.P., Zeiger, H.J. and Townes, C.H. (1954), *Phys. Rev.,* 95, p. 284 and (1955), *Phys. Rev.,* 99, p. 1264.

Grupp, H. and Hohmeyer, O. (1988), 'Technological Standards for Research-Intensive Product Groups and International Competitiveness', in A.F.J. van Raan (ed.), op.cit., pp. 611-673.

Grupp, H., Hohmeyer, O., Kollert R., and Legler, H. (1987), *Technometrie: Die Bemessung des technisch-wirtschaftlichen Leistungsstandes*, Verlag TÜV Rheinland, Köln (Cologne).

Grupp, H., Reiss, T., and Schmoch, U. (1990), 'Knowledge Interface of Technology and of Science - Developing Tools for Strategic R&D Management', FhG ISI Report to Volkswagen foundation, Karlsruhe.

Javan, A., Bennett, W.R. and Herriott, D.R. (1961), *Phys. Rev. Lett.,* 6, p. 106.

Maiman, T.H. (1960), *Nature, 187,* p. 493.

Martin, B.R. and Irvine, J. (1983), 'Assessing Basic Research: Some Partial Indicators of Scientific Progress in Radio Astronumy,' *Research Policy,* 12, pp. 61-90.

Moed, H.F., Burger, W.J.M., Frankfort, J.G. and van Raan, A.F.J. (1983), *On the Measurement of Research Performance: the Use of Bibliometric Indicators*, Centre for Science and Technology Studies, University of Leiden, Leiden.

Moed, H.F., Burger, W.J.M., Frankfort, J.G. and van Raan, A.F.J. (1985), 'The Use of Bibliometric Data for the Measurement of University Research Performance, *Research Policy*, 14, pp. 131-149.

Narin, F. (1976), *Evaluative Bibliometrics: The Use of Publication and Citation Analysis in the Evaluation of Scientific Activity,* National Science Foundation, Washington D.C.

Narin, F. and Olivastro, D. (1988), 'Technology Indicators Based on Patents and Patent Citations', in A.F.J. van Raan (ed.), op cit., pp. 465-507.

Nederhof, A.J. (1988), 'The Validity and Reliability of Evaluation of Scholarly Performance', in: A.F.J. van Raan (ed.), op cit., pp. 193-228.

Nederhof, A.J., Zwaan, R.A., de Bruin, R.E. and Dekker, P.J. (1989), 'Assessing the Usefulness of Bibliometric Indicators in the Humanities and the Social Sciences: A Comparative Study', *Scientometrics*, 15, pp. 423-437.

Nelson, R. and Winter, S. (1977), 'In Search of A Useful Theory of Innovation', *Research Policy,* 6, pp. 36-76.

Noyons, E. and van Raan, A.F.J., to be published.

Pavitt, K. (1988), 'Uses and Abuses of patent Statistics', in A.F.J. van Raan (ed.), op cit., pp. 509-536.

Peters, H.P.F., Hartmann, D. and van Raan, A.F.J. (1988), 'Monitoring Advances in Chemical Engineering,' in *Informetrics 87/88*, edited by L. Egghe and R. Rousseau, pp. 175-195, Elsevier Science Publishers, Amsterdam.

De Solla Price, D. (1984), 'The Science-Technology relationship, the craft of experimental science and policy for the improvement of high technology innovation', *Research Policy,* 13, pp. 3-20.

Rosenberg, N. (1976), *Perspectives on Technology*, Cambridge University Press, Cambridge.

Schmoch, U. and Grupp, H. (1989), 'Patents between Corporate Strategy and Technology Output: An Approach to the Synoptic Evaluation of European, and West German Patent Data', in A.F.J. van Raan, A.J. Nederhof, and H.F. Moed (eds.), op cit., pp. 49-67.

Schmoch, U., Grupp, H., Mannsbart, W., and Schwitalla, B. (1988), *Technikprognosen mit Patentindikatoren*, Verlag TÜV Rheinland, Köln (Cologne).

Schubert, A., Glänzel, W. and Braun, T. (1988), 'Against Absolute Methods: Relative Scientometric Indicators and Relational Charts as Evaluation Tools', in A.F.J. van Raan (ed.), op cit., pp. 137 -176.

Schubert, A., Glänzel, W. and Braun, T. (1989), 'Scientometric Datafiles: A Comprehensive Set of Indicators on 2649 Journals and 96 Countries in All Major Science Fields and Subfields 1981-1985', *Scientometrics*, 16, pp. 3-478.

Smart, D. (1965), *New Scientist,* 26, p. 570.

Thorp, J.S. (1967), *Maser and Lasers: Physics and Design*, Macmillan, London.

Turner, W.A. (1990), 'Internal CNRS report', personal communication.

Van Raan, A.F.J., Nederhof, A.J. and Moed, H.F. (eds.) (1989), *Science and Technology Indicators: Their Use in Science Policy and Their Role in Science Studies*, Proceedings of the Leiden Workshop on Science and Technology Indicators, DSWO-Press Leiden.

Van Raan, A.F.J. and Tijssen, R.J.W. (1990a), *An Overview of Quantitative Science and Technology Indicators Based on Bibliometric Methods*, Technology/ Economy Programme (TEP), OECD, report nr. 27769, Paris.

Van Raan, A.F.J. and Tijssen, R.J.W. (1990b), 'Numerical Methods for Information on Aspects of Science: Scientometric Analysis and Mapping, in Ch. Oppenheim, J.- M. Griffiths, and Ch.L. Citroen (eds.), *Perspectives in Information Management, Vol. 2*, Butterworths London.

Van Raan, A.F.J. (ed.) (1988), *Handbook of Quantitative Studies of Science and Technology*, Elsevier Science Publishers, Amsterdam.

Van Vianen, B.G., Moed, H.F. and van Raan, A.F.J. (1990), 'An Exploration of the Science Base of Recent Technology', *Research Policy,* 19, pp. 61-81.

Appendix 1: Country-coverage of the PASCAL database

A: Total number of publications per year for major countries

	usa	gbr	jpn	deu	fra	can	ita	aus	nld	swe	che
1979	87513	21115	17288	22032	22344	11362	9748	6134	3984	4238	4433
1980	124418	29457	22874	28566	32315	15076	13973	7508	5704	5450	5782
1981	149276	34819	27156	30459	35478	17263	13007	8873	6877	6275	5925
1982	104274	24829	21556	22026	25265	11735	9491	6120	5096	4509	4201
1983	156385	35296	29917	31952	33713	19390	11544	9185	7167	6858	6231
1984	119167	25792	22895	22137	26003	15069	8729	7111	5661	4737	4088
1985	133451	27916	26308	23650	29795	16607	9615	7959	6465	5910	4404
1986	133458	29217	28858	23019	29446	16281	9378	7714	6652	6155	4631
1987	156387	34743	34934	28616	34639	19734	11495	9170	8014	6933	5462
1988	138875	32179	31885	25571	33875	17473	10755	8177	7623	6468	4995

B: Number of publications with the keywords of set #3 per year for major countries in PASCAL.

	usa	gbr	jpn	deu	fra	can	ita	aus	nld	swe	che
1979	33	5	2	11	12	1	1	0	0	3	0
1980	44	7	6	12	13	2	4	2	0	0	3
1981	58	6	10	26	27	2	4	0	1	2	1
1982	71	14	23	18	44	5	7	1	7	0	6
1983	147	15	12	23	41	9	16	2	7	1	1
1984	163	24	11	23	51	2	6	0	5	6	7
1985	292	31	24	34	26	10	12	2	7	5	14
1986	307	32	9	40	70	9	10	1	12	6	10
1987	355	48	25	63	87	25	18	7	9	11	12
1988	351	53	25	88	78	13	19	4	10	4	12

Country abbreviations in Pascal since 1979 :

usa:	USA	gbr:	United Kingdom	jpn:	Japan
deu:	Fed. Rep.Germany	fra:	France	can:	Canada
ita:	Italy	aus:	Australia	nld:	Netherlands
swe:	Sweden	che:	Switzerland		

Appendix 2A: Number of laser-medicine publications characterized by BIOSIS concept codes with n>1% of the total.

cc\year	73	74	75	76	77	78	79	80	81	82	83	84	85	86	87	88	89	all
00000	19	16	64	49	45	61	69	54	129	210	282	213	283	281	305	303	246	2629
01000	2	6	10	6	12	27	15	20	36	49	49	39	74	70	72	71	54	612
02500	5	3	5	7	3	12	8	11	15	18	27	19	36	43	42	36	25	315
06504	14	12	25	18	23	36	27	44	90	98	100	111	199	239	266	282	199	1783
06506	11	11	23	15	15	29	24	28	45	40	41	60	127	173	198	186	113	1139
10000	12	6	26	19	15	23	32	32	89	92	103	76	133	114	123	132	104	1131
10500	6	2	10	13	6	9	9	11	18	29	43	43	40	69	54	42	42	446
10504	7	3	16	14	16	22	22	35	67	79	82	78	119	163	181	186	116	1206
10600	2	4	11	10	8	8	7	10	39	24	27	15	26	30	42	42	27	332
10604	5	10	16	10	14	25	17	34	70	61	74	76	108	152	158	158	113	1101
11100	9	4	14	5	10	22	14	26	42	51	65	59	42	59	82	97	63	664
11105	4	2	5	8	7	11	9	16	43	55	70	68	111	127	168	160	101	965
12500	5	5	10	7	6	19	6	11	38	36	41	31	52	61	60	58	43	489
12504	2	0	2	5	6	12	5	10	31	33	25	19	39	67	60	94	62	472
12512	6	7	14	12	15	24	17	33	74	91	85	87	180	237	270	293	180	1625
13000	0	3	11	6	6	11	4	7	23	28	38	21	41	47	50	58	40	394
14000	2	0	1	4	3	9	0	10	29	28	24	14	54	62	71	104	51	466
14500	5	3	11	6	13	8	8	25	52	44	52	39	63	99	77	95	80	680
14508	4	2	7	4	9	18	8	20	46	35	40	26	61	78	82	110	66	616
15000	2	5	12	4	8	19	9	28	62	48	65	42	59	90	88	92	72	705
16000	2	0	1	0	4	3	3	9	18	16	11	16	30	51	76	49	32	321
16500	2	0	0	1	1	1	5	10	14	17	45	48	44	34	34	26	29	311
18500	4	2	3	5	2	4	3	8	28	22	17	20	26	25	52	45	22	288
20001	6	6	15	6	14	19	13	23	40	41	49	35	91	107	85	102	74	726
20006	8	10	17	10	12	24	16	24	43	44	49	37	103	128	117	132	83	857
20500	5	1	7	5	3	9	5	3	26	23	64	28	45	40	45	54	58	421
22000	2	1	4	11	2	6	6	17	45	37	45	50	55	80	98	121	43	623
23000	0	8	10	11	5	3	4	14	39	36	25	11	23	38	47	51	18	343
24000	4	1	13	10	10	9	11	17	32	70	68	71	107	148	178	175	111	1035
ALL	155	133	363	281	293	483	376	590	1323	1455	1706	1452	2371	2912	3181	3354	2267	22695

Appendix 2B: Titles of major BIOSIS concept codes, and brief descriptions.

00000	Rest category.
01000	Methods, Materials and Apparatus, General.
02500	Cytology and Cytochemistry.
06500	Radiation Biology. Retrieves biological and medical studies on: electromagnetic radiation (usually ionizing radiation wavelengths below the ultraviolet range) particle emissions and other radiation.
06504	Radiation and Isotope Techniques. Scope: Studies on: the methods, apparatus and instruments used in radiation biology research.
06506	Radiation Effects and Protective Measures. Scope: Studies on: the effects of external radiation -- light, ultraviolet, X-rays, high frequency radiowaves, microwaves -- on any organism of part of an organism. It can also be used to find studies on: the radioprotective effects of drugs, chemicals, biochemical compounds of other materials.
10000	Biochemistry.
10500	Biophysics. The codes in this group retrieve studies on: the application of the principles of physics to the study of biological systems and their components in any organism.
10504	General Biophysical Techniques. Scope: Studies of biophysical apparatus, methods and techniques.
10600	External Effects. Studies on: beneficial or adverse effects of external light, pressure, sound waves, electric, magnetic, gravitational, physical and mechanical phenomena, temperature or humidity on any organism or part of an organism
10604	Light and Darkness. Scope: Studies on: the effect of the presence or absence of light.
11100	Anatomy and Histology, General and Comparative: The codes in this group retrieve studies on: the structure of animals and humans -- including gross, comparative, radiologic, microscopic and ultramicroscopic anatomy studies.
11105	Surgery. Scope: Studies mainly on humans relating to organ part removal for therapeutic or prophylactic purposes. Also clinical surgical methods and procedures.
12500	Pathology, General and Miscellaneous: The codes in this group access studies on: animal and human diseases affecting multiple tissues and body systems.
12504	Diagnostic Pathology. Scope: Studies and techniques used to determine the cause of disease or methods used to distinguish one disèase from another.
12512	Therapy. Scope: Studies on: all chemical, biochemical and physical techniques and methods of treatment for the diseases or disorders of animals and humans.
13000	Metabolism.
14000	Digestive System.
14500	Cardiovascular System: The codes in this group retrieve studies on animal and human cardiovascular systems. They access information on cardiovascular methods, structures, functions and diseases.
14508	Blood Vessel Pathology. Scope: Studies on blood vessel diseases and any related structural and functional changes in blood vessels caused by the disease.
15000	Blood, Blood-Forming Organs and Body Fluids.
16000	Respiratory System.
16500	Reproductive System.
18500	Integumentary System.
20001	General and Methods. Scope: This code retrieves methods, apparatus, instruments and techniques used to study sense organs associated structures and functions.
20006	Pathology. Scope: Studies on disease of sense organs and associated structures and any related structural and functional changes.
20500	Nervous System.
22000	Pharmacology.
23000	Temperatare: its Measurement, Effects and Regulation.
24000	Neoplasms and Neoplastic Agents.

* this table is derived from the BIOSIS previews guide: concept codes; scope notes

Appendix 3A: Trend (relative) of laser-medicine publications characterized by specific PASCAL cc-titles (only cc-titles with all > .50%).

cc title\year	79	80	81	82	83	84	85	86	87	88	all
An. chem.	0.00	0.00	0.93	0.76	1.32	0.00	0.41	0.00	0.34	0.96	0.51
Anesthesia	1.85	2.74	3.70	5.30	3.29	2.33	2.49	1.42	2.37	2.56	2.68
Biochemistry	1.85	0.00	0.00	0.00	0.00	2.33	1.66	1.42	0.00	0.00	0.68
Biomed. eng.	18.52	15.07	13.89	7.58	11.18	10.47	16.18	11.79	14.58	12.14	12.90
Biophysics	1.85	0.00	3.70	2.27	1.97	0.58	2.90	1.42	1.02	2.88	1.94
Cardiology	0.00	0.00	0.00	1.52	2.63	4.65	3.73	6.13	5.08	5.11	3.82
Dermatology	3.70	1.37	6.48	6.82	9.21	4.65	4.56	5.66	4.07	2.88	4.85
Dig. surgery	0.00	0.00	0.00	0.00	0.00	2.33	2.90	1.89	1.36	2.24	1.48
El. micr. biol.	1.85	4.11	1.85	2.27	1.32	0.00	0.00	0.00	0.00	0.00	0.63
Electronics	1.85	1.37	0.93	0.76	0.66	0.00	0.83	0.47	0.00	0.32	0.51
Endocrine	0.00	0.00	0.00	0.00	0.00	2.91	1.24	2.36	1.69	1.92	1.37
Energy	0.00	0.00	0.00	0.00	0.00	0.00	0.00	0.00	1.36	1.92	0.57
Energy/ Fuels	0.00	0.00	0.00	0.00	0.00	0.00	0.00	0.00	2.37	2.56	0.86
ENT/ Stoma.	5.56	6.85	8.33	9.85	10.53	2.91	2.90	2.83	1.69	3.19	4.51
Fluid mech.	1.85	0.00	0.93	0.76	0.00	0.00	0.00	0.00	2.03	1.92	0.86
Gastroent.	7.41	9.59	10.19	7.58	5.26	5.81	4.98	6.60	4.41	6.71	6.28
Genit. surgery	0.00	0.00	0.00	0.00	0.00	0.58	1.24	0.47	1.36	0.96	0.68
Gynecology	0.00	0.00	0.00	0.00	0.00	4.65	4.56	3.30	3.05	3.19	2.57
Head surgery	0.00	0.00	0.00	0.00	0.00	5.81	3.73	3.77	2.71	2.24	2.40
Hematology	0.00	0.00	0.00	0.76	0.66	1.16	0.00	1.42	0.34	0.96	0.63
Mech. eng.	0.00	0.00	0.00	0.76	0.00	0.58	0.41	0.00	1.36	1.28	0.63
Metals	0.00	0.00	0.93	0.00	0.00	0.58	0.41	0.47	1.36	1.60	0.74
Metrology	0.00	1.37	1.85	0.76	0.66	0.00	0.00	0.94	1.69	3.19	1.26
Nephrology	5.56	1.37	2.78	3.03	3.29	2.91	1.24	1.42	0.68	2.24	2.05
Neurology	0.00	2.74	3.70	0.76	5.26	1.74	3.32	2.36	1.69	1.28	2.28
Neurosurgery	0.00	0.00	0.00	0.00	0.00	2.33	4.56	3.77	1.02	1.28	1.71
Ophthalmol.	24.07	16.44	15.74	25.00	21.05	18.60	14.52	14.15	15.25	13.74	16.67
Optics	12.96	17.81	7.41	4.55	1.32	0.58	0.83	0.94	1.69	0.64	2.74
Orth. surgery	0.00	0.00	0.00	0.00	0.00	2.33	0.41	0.94	0.34	0.64	0.57
Pharmacology	1.85	1.37	0.93	4.55	1.97	1.16	0.00	0.94	2.71	0.64	1.48
Photochem.	3.70	0.00	0.00	0.76	0.66	0.58	0.83	0.47	0.34	0.32	0.57
Plastic surgery	0.00	0.00	0.00	0.00	0.00	2.91	2.07	3.30	1.36	1.60	1.48
Pneumology	0.00	2.74	2.78	3.03	3.29	1.16	2.90	2.83	2.37	1.60	2.34
Publ. health	0.00	0.00	0.00	0.00	0.00	0.58	1.24	0.47	1.36	0.00	0.51
Reproduction	0.00	2.74	3.70	5.30	8.55	0.00	0.00	0.00	0.00	0.00	1.48
Resp. surg.	0.00	0.00	0.00	0.00	0.00	1.74	0.83	1.42	0.68	0.64	0.68
Surgery	0.00	0.00	0.00	0.00	0.00	1.74	0.83	0.94	1.36	0.32	0.68
Tumors	0.00	0.00	0.00	0.00	0.00	2.33	2.49	2.83	2.03	1.28	1.48
Vascular surg.	0.00	0.00	0.00	0.00	0.00	0.58	0.83	0.94	1.69	0.96	0.74
Vert. neurop.	0.00	1.37	0.93	1.52	0.00	1.16	0.83	0.94	0.68	0.00	0.68
Viral diseases	0.00	0.00	0.00	0.00	0.00	0.00	0.83	1.42	0.68	0.96	0.57

For full cc-titles: see abbreviation list cc-titles (Appendix 3B).

Appendix 3B: Abbreviation list PASCAL cc-titles.

short cc-title		full cc-title
An. chem.	=	Analytical chemistry
Biom. comp.	=	Biomedical computer applications
Biomed. eng.	=	Biomedical engineering
Cardiology	=	Cardiology/ Cardiovascular disease
Dig. surgery	=	Digestive system surgery
El. micr. biol.	=	Electron microscopy on biological structures
Endocrine	=	Endocrine pancreas
ENT/ Stoma.	=	ENT/ Stomatology
Fluid mech.	=	Fluid mechanics
Gastroent.	=	Gastroenterology/ Digestive diseases
Genit. surgery	=	Genital tract and mammal surgery
Gynecology	=	Gynecology , Obstetrics , Andrology
Head surgery	=	Head / neck surgery
Mech. eng.	=	Mechanical engineering/ Mechanical construction
Metals	=	Metals/ Metallurgy
Nephrology	=	Urinary system disease/ Nephrology
Ophthalmol.	=	Ophthalmology
Orth. surg.	=	Orthopedic surgery
Photochem.	=	Photochemistry
Plastic surg.	=	Skin plastic surgery
Polym. chem.	=	Physicochemistry of polymers/Macromolecular chemistry
Publ. health	=	Public health hygiene
Resp. surg.	=	Respiratory system surgery
Tumors	=	Tumors treatment
Vert. biology	=	Vertebrates biology
Vert. neurop.	=	Vertebrates neurophysiology

9 AT THE CROSSROADS IN LASER MEDICINE AND POLYIMIDE CHEMISTRY: PATENT ASSESSMENT OF THE EXPANSION OF KNOWLEDGE

Hariolf Grupp and Ulrich Schmoch

9.1 INTRODUCTION

Investments in R&D are always precarious, because both the technological and even more the commercial success are not certain. On the other hand, not investing in R&D involves the risk of a complete failure of an enterprise. There is a great need to develop tools for strategic R&D management. One of the difficult problems of R&D planning is the question, under which circumstances the results of basic research are needed and, if so, how are they successfully transferred into a broader technical application. In order to work on these particular problems, both the previous and this chapter describe the knowledge interface between science and technology by bibliometric and patent indicators, wherein the bibliometric studies were presented in chapter 8 and the patent analysis in this chapter (for a more detailed treatment of methods and results see Grupp, Reiss & Schmoch (1990); a methodological stand-alone version not referring to the other chapters of this volume has been published by Schmoch (1991)).

The basic assumptions of cyclicity in the evolution of new technologies have been discussed in chapter 1. The importance of a phase of technical realization which often anticipates industrial development for commercial markets has been stressed there. As both polyimides and lasers are assumed to be science-dependent (see section 4.6, figure 4-12) these technical fields are further investigated here (for selection criteria see section 9.3). In order to classify their potential development phases, section 9.3 highlights very briefly important historic events related to their evolution.

Scientometric methods may help to better understand the economics of innovation (Callon et al. 1991, p. 155). One obvious method for tracing the knowledge transfer from science to technology is the analysis of citations in patent search reports, but in the respective literature the role of citations is not sufficiently clarified. In section 4.2 the differences between citations to other patents and those to the non-patent literature were illuminated, the motives of establishing the latter type are explained there in detail. As figure 1-1 in chapter 1 sketches, literature citations in patents point to possible links to theory or model development or to earlier phases of technical realization outside of industrial R&D laboratories. Patent citations, however, were assumed to represent knowledge transfer within the industrial development, innovation and imitation phases and thus may be used to illuminate the knowledge expansion going on within the socio-cognitive subsystem of technology (chapter 2). The meaning of patent-to-patent citation is treated in more detail in section 9.2.

The results of this investigation are outlined in sections 9.4 (cyclicity), 9.5 (science-technology relation), and 9.6 (technological knowledge expansion). In these sections the case study on lasers and the one on polyimides are treated in parallel although the two cases have been selected precisely because they seem to be quite different examples of science-based technologies. The juxtaposition facilitates comparisons of what might be common to both cases. Some conclusions from chapter 9 are drawn in section 9.7.

Before starting the discussion of historic time trends (section 9.3) and related patenting (section 9.4), in the next section some methodological aspects of patent citation have to be considered.

9.2 MEANING OF CITATIONS IN PATENT SEARCH REPORTS

This section is largely an excursion into patent citation research methods, as little has been published on the functions of patent citations. While patent time trends in lasers and polyimides are the subject of section 9.4, patent-to-patent citations will be used to study the expansion of knowledge within the technological communities in section 9.6. The following remarks are based on interviews with patent examiners at the Japanese, German, European, and American Patent Offices and with patent attorneys in Germany and Japan.

For the comprehension of patent citations first of all it is necessary to realize the structure of patent claims. A patent claim describes the area of protection for the patent and is composed of a certain number of technical features. As the basic criterion for the approval of a patent is novelty, some of these features have to be new or at least the special combination of known features must be novel. Besides the first claim, which is called the main claim, the applicants generally list some or even many subclaims, wherein the special arrangements of the invention are described in more detail (Schmoch 1990).

In the examination procedure, the patent office examiner has to look for prior documents which describe the same or nearly the same features as the application concerned. Only if he or she finds no relevant documents the application is accepted for award of a patent. Consequently, there are two important types of citations (Grupp, Reiss & Schmoch 1990):

- Documents of particular relevance, which already show essential features of the invention in the main claim in another invention or at least question the innovation or inventive step of these features, if taken alone. This type of citation is marked with the letter X in the search reports of the European Patent Office (EPO).
- Combined citations which are particularily relevant if the features of one cited patent are combined with the features of another cited document. This type is marked with the letter Y in EPO search reports. For this reason there are always at least two such references in a search report.

The citations can refer to the main claim as well as to a subclaim. It is obvious that a citation referring to a subclaim which includes only one special feature can be distant from the central object of the invention. Furthermore, there exists a third important type of citation:

- References concerning the technological background which show the general state of the art prior to the current application. This type is marked by A in an EPO search report.

The A references provide general information for the applicant and are important first of all for the assessment of the so-called inventive step which forms the second basic criterion for the approval of a patent. In many cases only a rough similarity between an A reference and the central features of the invention exists. A fortiori, the examiner judges the inventive step not only by A documents, but by the similarity of X or Y documents to the subject of the invention, too.

The above considerations clearly show that not all references have a strong linkage to the examined patents, but that there are different degrees of linkage. There may be a low similarity in the case of Y documents, which are only important in combination with other documents. X references to subclaims can be linked to the central object of the invention in a weak manner, and also the proximity of most of the A type citations is rather low. Therefore, only a few references trace back specifically to the direct technology (or even science) sources of a patent. But in a broader sense, nearly all citations originate from relevant technical fields and are therefore interesting for the analysis of the expansion of technological knowledge.

Considering the examination procedures, it is obvious that highly cited patents must not in each case be technically and economically important. Many documents are highly cited because they comprise a good description of the prior art from a didactic point of view and are therefore used as technological background in the examinations. Therefore, further investigations are necessary to verify the extent to which highly cited patents are used as X, Y or A references. Without paying attention to the distinctions between X, Y, and A types of citations, it has been shown statistically that in *samples of patents* "the number of times a patent document is cited may be a measure of its technological significance" (USPTO 1976, p. 167). Ellis, Hepburn & Oppenheim (1978) traced the patent citations from sets of patents to identify key discoveries and technologies. They showed that in areas where technology developed over a relatively long period of time - as is the case both for lasers and polyimides - a key cluster of basic technology is difficult to determine as the pattern of technology sources is less focused than in rapidly developing fields. Trajtenberg (1990), in the case of computed Tomography scanners tested by empirical analyis that "patent counts and citations may play an important role in studying the very emergence of new products which seems to be the period when most of the important innovations occur" (op. cit., p. 185). Further he found that patent citations may be indicative of the economic value of innovations. Very recently Schmoch et al. (1991) compared patent citations from space and robot technology in order to describe spin-off effects. These authors conclude that "the method of analyzing patent citations makes it possible to trace the route taken by technology transfer (...) and to reveal more clearly the structures on which this is founded" (op. cit., p. 362). Albert et al. (1991) validated the meaning of patent citations directly in the field of silver halide technology (Kodak photography patents granted in the year 1982). They thereby supported the utility of patent citation indicators in assessing the technological importance of a company's patent stock in this case.

All the above-mentioned investigations were based on the US patent system with the notable exception of Schmoch et al. (1991). However, the EPO is the only office which is marking the categories of references explicitly. Baré (1981) published some data on the early experience of the EPO with citation categories. According to this investigation, the X documents amount to 11 % of the total citations, the background category A is important in 38 % of all cases. But in this early stage there were a large number of documents which were not classified at all, because the Y type was not introduced yet. New investigations on patent applications in the fields of laser medicine and of polyimides in electronics lead to a quite different picture: according to table 9-1, 13.2 % of the citations fall into the category X which is in concordance with Baré's findings. But 71.1 % are classified as type A which is a surprisingly high share. If X and Y citations are considered as references of high or medium proximity to the examined document, only 28.9 % of all citations represent linkages to direct sources. These findings are in agreement with the results of Dunlop and Oppenheim (1980), according to which two thirds of all official citations have no close linkages to the examined applications. Their results are based on one-by-one inspections of patents at the United States Patent and Trademark Office (USPTO) related to X-ray tomography. From this evidence it may be concluded that patent citation analysis is a valid tool for the analysis of links within the industrial development phase among all innovation-oriented functions if statistical sets of patents represent homogeneous sub-fields of technology but not necessarily for a particular single invention.

Table 9-1: Categories of patent citations in patent search reports of the European Patent Office for the fields of medical lasers and polyimides in electronics, 1980 to 1986

Category of citation	Lasers		Polyimides		Total	
	abs.	%	abs.	%	abs.	%
X	113	12.5	95	14.1	208	13.2
Y	161	17.9	87	12.9	248	15.7
A	628	69.6	493	73.0	1121	71.1
Total	902	100.0	675	100.0	1577	100.0
therein: D	72	8.0	60	8.9	132	8.4

All these considerations exclusively concern citations of the patent office examiners, not of the inventors, because only the examiners' citations are available in data banks. Only a few of the inventors' references are taken over by the official search report; in the random sample of table 9-1, 8.4 % of all citations originate from inventors (category D). Therefore, further research on the relevance of citations given by inventors is necessary.

The different categories of citations are applicable for patent references and for non-patent references as well. The methodological considerations presented so far solely relate to patent-to-patent citations. The motives for citing non-patent scientific documents are discussed in section 4.2. How do the patent-to-patent citations relate to patent-to-literature citations in number?

Obviously, there is a certain influence of the size and the structure of the internal documentation of a Patent Office on the size and the structure of the citations in the search reports. For example at the EPO, there is an average share of non-patent citations to all citations of only 6 to 7 % depending on the respective priority year. At the end of 1986, the internal documentation of the EPO included 8.1 % non-patent documents, and at the end of 1988 8.2 %. These - of course quite general - figures indicate that the examiners reproduce more or less the structure of the search documentation in their search reports with a sligth preference for patent citations. They have access to more non-patent sources than they actually need and use.

9.3 SELECTION OF CASE STUDIES AND ABSTRACT OF HISTORIC EVENTS IN LASER AND POLYIMIDE EVOLUTION

The selection of examples was made on the basis of several partly methodological, partly pragmatic criteria (e. g., sufficient number of patents). First of all, two case studies should be performed for the science-oriented fields of innovation which have the highest possible number of differences in their technological character. This is done in order to distinguish general findings from technology-specific results. As the interest of the project was a methodological one, the case studies should also be historic so that an adequate interpretation of the actual state of knowledge would be possible. On this background, the general fields of laser technology (from physics) and polyimide technology (from chemistry) were chosen.

It is obvious that these two fields are too large for detailed studies so that suitable sub-fields had to be defined. The number of patent documents in these sub-fields should not be too small to achieve a valuable random sample and not too large to avoid problems of data handling. Furthermore, the existence of appropriate symbols of the International Patent Classification (IPC) and of appropriate keywords was necessary for a good description of sub-fields. For the general problems of matching patents to R&D subjects see Grupp, Schmoch & Kuntze (1991, pp. 424-426) and Grupp (1990, pp. 220-221) who suggested a combined search strategy. In section 9.4 it is explained how to arrive at a distinction of the core technology as opposed to the applied technology (i. e. products and systems). In sections 3.5 and 3.6 the selection criteria for the sub-fields are discussed which led to a detailed study of medical lasers and polyimide applications in electronics. In this section, however, an abstract of historic events in the overall fields of lasers and polyimides is given.

Optics is a field of scientific investigation which has been followed since ancient times. The law of reflexion dates back to Euclid (300 BC); Newton and others tried to understand optics from physical principles in the 17th century. Hertz contributed the experimental evidence that light is part of the electromagnetic spectrum, and Einstein transposed the quantum theory by Planck 1916 into the quantum theory of light. Only one year later again Einstein postulated the possibility of stimulated light emission. It took more than a decade for this to be experimentally proved by two German physisists in 1928; the first instruments were constructed in 1954 with (invisible) microwave radiation. The Nobel prize was awarded in 1964 to an American and two Soviet professors for their parallel development of an ammonia "laser". The first visible laser beam was emitted from a ruby laser in 1960; the first gas laser, a helium-neon laser, became reality a year later (for bibliographic details see section 8.6). In 1962 the semiconductor laser followed. The lead in the development of instruments has been held ever since the Second World War by American physicists, though Soviet physicists had a major share in describing the concept and making suggestions for the realization of appropriate devices. Following the CO_2 laser in 1964 in the Bell laboratories, the dye laser (in 1966) was simultaneously invented in the United States and the Federal Republic of Germany. In recent years Japanese companies took the lead in some types of semiconductor lasers (Grupp & Hohmeyer 1988).

Whereas the phase from theory to first technical realization took 43 years (37 in the case of microwave radiation), in the sixties nearly every year a new lasing medium was discovered. Already in 1962 the first laser company (Spectra Physics) introduced a laser product being the forerunner of several other newly founded firms. The first lasers were overwhelmingly used for military or R&D purposes. Between 1972 and 1976, an expansion in wavelength (ultra-violet light), an increase in efficiency and in power was achieved.

In the mid sixties the laser market was worth about 50 million US $. The market grew particularly strong towards the end of the seventies. It is estimated that production increased then by a factor of 8 in 10 years (Beyer et al. 1985). In the second half of the eighties the largest user market is probably in material processing: 20 per cent of all laser beam sources (with a total turnover of about 750 million US $ per year) are used in this sector. About the same size must still be ascribed to the market segment in R&D applications followed by medical applications (about one sixth of the total market) and armaments (about the same; see Grupp & Hohmeyer 1988, p. 637).

The evolution of polyimides has an even longer history than coherent light. While theory preceded the making of coherent light, in polyimides, conversely, theoretical explanations followed synthesis. Natural resins were used long before the first industrial polymer, celluloid, has been synthesized in 1868. In the year 1908, Bogart & Renskaw in the USA happened to isolate polyimides by polycondensation without an appropriate theoretical explanation (Riediger 1984, p. 92). The scientific basis to understand polymers was not clarified until 1922 when Staudinger proposed the macromolecule and a linear structure of polymers. He first began to turn "the study of polymers from a black art to a reasonably comprehensive branch of chemistry" (Reuben & Burstall 1973). His breakthrough represented a true "Scientific Revolution" (Walsh 1984, p. 223). However, Staudinger was not primarily working for the benefit of theory but he wanted to meet the industrial requirements to substitute natural polymers by industrial ones. His work at Freiburg university was sponsored by IG Farben. For industry, the traditional molecular theoretical models could not explain much about polymers so that synthesis of new polymers was severely hampered (Holland 1988). This is a clear case of transfer of "knowledge about problems" from industrial technology to science.

Another milestone in polymer science was the stereospecific polymerization using inorganic catalysts by Ziegler & Natta in 1953. Based on new theoretical models systematic research to synthesize more specific polymers followed. In the Du Pont laboratories (USA) Edwards & Gresham isolated highly molecular polyimides 1955; commercial production was started in 1961. Between 1964 and 1968 the variety of special polyimide materials was broadened, and in the early seventies further product diversification took place (Riediger 1984, p. 98): an improvement in thermal stability was desired. In the seventies also the search for "super polyimides" being stable up to 600°C began. Polyimides became more and more important to the electronics industry which began the substitution of circuitry by semiconducting chips. This substitution required minimization, precision, heat stability and strength on the side of the materials used. Polyimides with their extreme thermal stability combined with physical strength, chemical resistance and low dielectric losses ideally fulfill these requirements. They are also used for making very large-scale integrated semiconducting parts as they may also serve as photoresists.

In terms of production figures polyimides are the most important class of highly temperature resistant polymers. In the seventies their production increased by a factor of 10. Today they are used in all industries in which the exclusive combination of thermal and electrical properties are important (electronics, motors, aerospace, aviation and so on).

Table 9-2 compares the most important events in the history of both laser and polyimide technology. Some features of other science-based developments, e. g. molecular beam epitaxy (chapter 7), seem to compare to this general time perspectives quite well. A discussion of the historic time trends in related patenting follows in the next section.

Table 9-2: Important historic events in laser and polyimide technology in perspective

	Lasers	Polyimides
Period between theory development and experimental proof	11 years	-14 years (isolation preceded theory)
Period between theory development and first technical realization	43 years	33 years
First technical realization	1960	1955
First commercial production	1962	1961
Broadening in scope of application	1972-76	early 70's
Growth of markets ...	in the late 70's and 80's	in the 70's
... within a decade by a factor of	~8	~10

9.4 CYCLICITY OF LASER AND POLYIMIDE DEVELOPMENT

In concordance with the coupled model of innovative functions (chapter 1, figure 1-1) it is worth-while to separate application-oriented functions (innovation, imitation, diffusion) from the core of technology. Therefore, the distinction between core technologies and respective applications is introduced. In this dual model, it is assumed that the core technologies are strongly linked to science, whereas the related product applications draw on applied research and mainly industrial development.

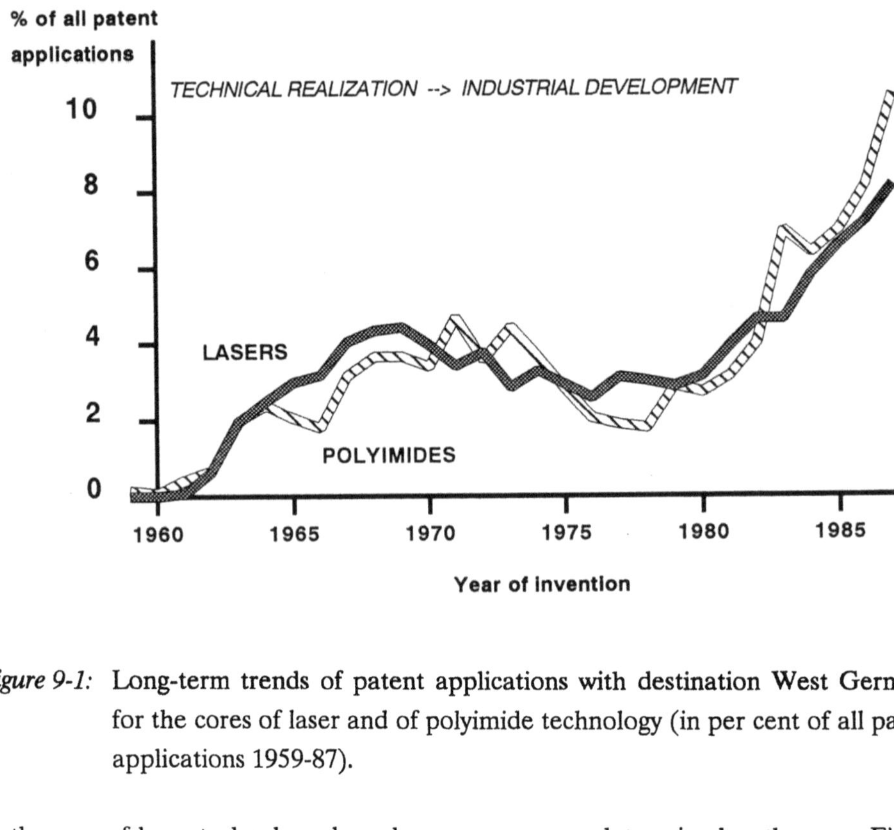

Figure 9-1: Long-term trends of patent applications with destination West Germany for the cores of laser and of polyimide technology (in per cent of all patent applications 1959-87).

In the case of laser technology, laser beam sources are determined as the core. Figure 9-1 shows the evolution of laser beam sources since their first technical realization in 1960 (see table 9-2). The investigation was performed for patents with destination West Germany, as this market is the largest and most representative one in Europe. In order to avoid double counts, a definite legal range (territory) must be specified in patent analysis (Schmoch et al. 1988). Further, the market segments of lasers and polyimides are different in size. Therefore it should not be expected that the patent propensities for lasers and polyimides will be similar. For this reason all patent applications between

1959 and 1987 (n = 3716 for lasers and n = 1176 for polyimides) were normalized to 100 per cent in figure 9-1.

The first invention was registered with priority in 1961, the year which followed the first practical realization of a ruby laser. Until the end of the sixties a steady increase can be observed which is followed by a decline until the middle of the seventies. Then, there is a sharp recovery of the invention rate for laser beam sources.

Unfortunately, it is not possible to describe the same period completely for products or systems, i. e. applications of lasers. This is because for the demarcation of the various laser systems, keywords are necessary which are not available in patent data banks as far back as 1960. Nevertheless, a respective combined search is feasible from 1973 onwards. Figure 9-2 shows that a substantial number of inventions related to laser systems can be observed starting from the middle of the seventies with steady increases. The relationship between systems and core inventions is progressively growing until the beginning of the eighties and is very slightly falling thereafter.

Since 1980 the number of system inventions nearly tripled the number of core inventions, whereas in the early seventies they exceeded them by only 30 % (figure 9-2 is indexed to this starting level). Thus the first wave of patents in the laser core can be related to the phase of technical realization and basic appropriation of this new technology. It has to be noted that even in this early science-oriented phase there were an important number of product developments. Already in the phase of technical realization it was clear that laser technology would be an area of commercial interest so that even the first inventive results were protected by industrial property rights. As was indicated in section 9.2, in the period of growing technical realization (1960-1968) nearly every year a seemingly powerful new laser medium was invented and protected.

After 1968 the innovation-oriented activities slowed down because many newly realized laser concepts did not meet the cost, easy handling and utility requirements for commercial application. Semiconductor research, e. g., had more or less stopped in Japan around 1966-67 as a result of growing conviction that it would be impossible to realize products that could operate at room temperature (see section 7-3). Most laser systems in the first wave were used either for R&D itself or for military purposes, i. e. not under standard industrial conditions (in terms of safety, reproducability, size, weight, environmental conditions in production facilities). The people to operate lasers then were overwhelmingly scientists and engineers (Grupp 1986, p. 9), but not operators, craftsmen, physicians, and labourers. Some first pioneer laser manufactures started commercialization too early in the seventies and went bankrupt between 1973 and 1975 (e. g. Optics Technology, Spaarays, Carson Labs, Laser Sciences, Laser Energy, Osram). Thus further progress from the side of science was required.

Between 1972 and 1976 from the scientific side an expansion in laser wavelength, in quantum efficiency and in light power was achieved, i. e., *after* various laser sources had been successfully realized and *while* new scientific discoveries broadened our understanding of coherent light, the actual rate of inventions decreased (figure 9-1). To give an example, in 1970, at Bell labs, a Japanese researcher demonstrated a new semiconductor type operating at room temperature. As a result, semiconductor laser research was resumed by Hitachi, NEC and Toshiba (see section 7-3). The number of system-oriented developments remained at the low level as before. Both core and product developments stagnated (see figure 9-2) until about 1976. Right thereafter both the number of system inventions increased (figure 9-2) as did also the number of core inventions (figure 9-1 does not include systems technology).

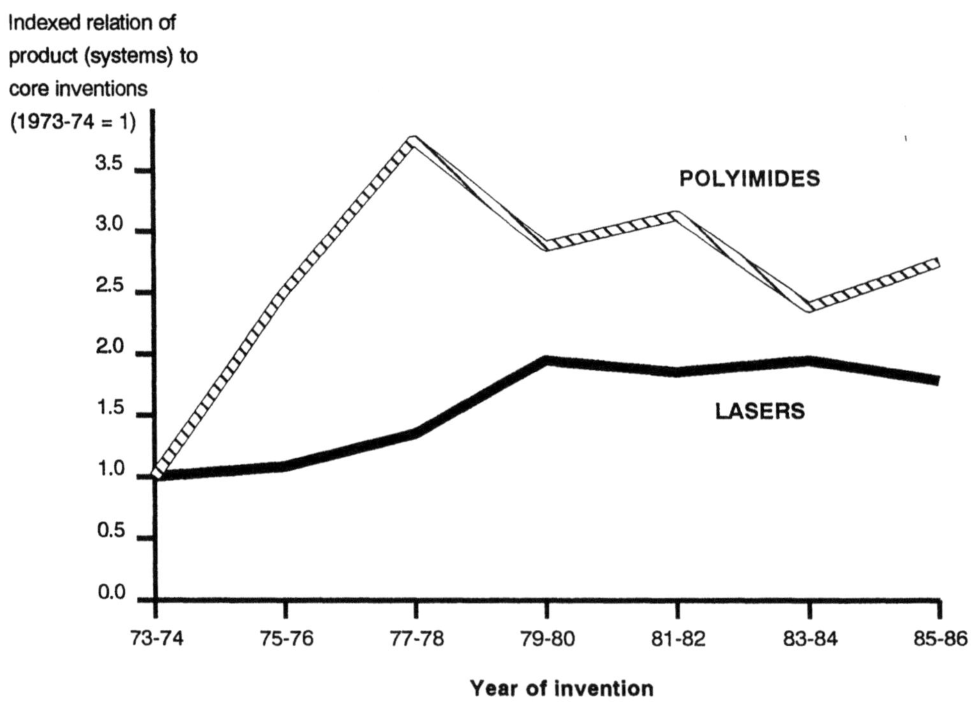

Figure 9-2: Indexed relation of product (systems) to core technology in terms of patent filings with destination West Germany for lasers and polyimides (index for 1973-74 is set equal to 1. Actual relation: 0.08 (polyimides), 1.3 (lasers)).

The second take-off is induced by a growing number of realistic, marketable products which required special modifications of the beam sources. New basic properties were made exploitable by scientific efforts. Thus, the recovery of patents of the core since the middle of the seventies is simultaneously due to a "technological pull" from product requirements as well as to a "science push".

In the case of polyimides it is more difficult to define a clear core area when compared to laser technology. Polyimides are mainly classified in the chemical section of the IPC, first of all, in the class for organic macro-molecular compounds. If an industrial R&D facility explores new varieties of polyimides, it tries to achieve certain characteristics which are mostly oriented towards concrete product innovations. These strategic new products are often not explicitly mentioned in the respective patent documents, but only the improved features of the materials. Nevertheless, a remarkable number of polyimide-related patents exist which are classified outside the central chemical IPC classes. Thus, all patents within the chemical classes were defined as core technology (mainly the synthesis of polymers) bearing in mind that an important number of product-oriented patents could not be separated from this sample. All polyimide patents having at least one classification out of the chemical class were labelled as product technology.

The long time trends (figure 9-1) show quite equal starting points for the cores of the laser and of the polyimide technologies around 1960. The sequence of patent cycles for polyimide chemistry is also similar to laser beam sources with the first wave of about 15 years length and a second take-off in the eighties. The relation between product-oriented polyimide inventions and polyimide core technology (synthesis) has the same characteristic as in the laser case, too (figure 9-2). The only remarkable difference is the lower absolute level of pure product technology (around 10 % in 1973-74 up to 30 % in 1977-78) which may be due to the high number of hidden application-oriented patents in the core. Therefore, the relations in figure 9-2 were indexed (1973-74 set equal to 1) to make them directly comparable with lasers. Nevertheless, even bearing in mind the difficulties with the polyimide classification, the differentiation between core and application seems to be a suitable tool for the description of the science-technology interface in this field as well. The interpretation of the trends in figures 9-1 and 9-2 has much in common with the laser case. Callon et al. (1991) also found, by using a different scientometric technique, that in polymer chemistry between two peaks of overall development the "cluster (...) undergoes significant internal development" (p. 186). Common aspects are discussed at the end of this section.

Polyimides were isolated under modern conditions in 1955. The first patent priority claims were filed by 1959. After Du Pont started production in 1961 a variety of special polyimide materials were invented between 1964 and 1968 (see section 9.2). This phase of growing technical realization clearly coincides with the first patent wave in figure 9-1. In the beginning of the seventies some mass product markets tended to be saturated and the opportunities for further product innovation dropped (Holland 1988, pp. 79-80). Several polymer chemists claimed then that the time of ever new polymer discoveries has ended (Riediger 1984, pp. 38-39). Further scientific progress was expected for the eighties, e. g. active polymers for pharmacy, a new "materials science" merging from metallurgy, ceramics and polymers and a new "life science" including biomaterials (op. cit., p. 40). But for the seventies the rumour of a crisis in polymer science was going around. At the same time new demand dictated new expectations on technology (see section 9.2). Mass polymers were no longer suitable to meet the demand of the aviation, aerospace, defence, electronics, and pharmaceutical industries. The industrial chemical R&D facilities reacted to the new demand profile by introducing special polymers and high-efficiency fibres in small batches. Because of their excellent electrical and thermal properties, polyimides, in particular, carried the hopes for a recovery of polymers. The development of "super polyimides" began in the seventies. This recovery is in concordance with the patent data in figure 9-1. An end to the second dynamic phase is not in sight. In the eighties the rate of invention apparently accelerated further under the new paradigm of "material science" - as predicted.

The laser and the polyimide cases were selected because they were assumed to represent quite different science-based technologies. Why are the above findings so similar?

The striking correspondance between the laser and the polyimide cases can be interpreted as a strong indication that typical patterns of the development of science-intensive technologies have been found. It is an important aspect that in the two case studies there is a time lag of about 15 years between the first technical realization and its broader application. Furthermore, there is no clear distinction between a science-intensive and a technology-oriented phase as suggested by simple linear, sequential phase models (see chapter 1, section 1.2). In both case studies the stagnation phase in the seventies is characterized by *bipartisan* pressure: new scientific achievements enlarge the potential of solutions to be exploited, and altered demand from industrial user branches requires improved technological specifications. Seemingly, judging from only two cases, if a broader introduction of a new science-based technology on the market takes place then new activities in basic research are also stimulated. Another case exposing a cyclic development is fuel cell development (see chapter 3, in particular figure 3-1). Further investigations on the long-term trends of patent applications for even older technologies like polyamides should be made in order to increase our knowledge of such generic aspects in science-based innovation.

Another observation is the obvious representation of the early science-linked R&D activities *in patents*, i. e., in industrial property rights. This is common to both cases explaining the highly significant correlations which often can be found between patent and bibliometric analyses in science-intensive areas (Schmoch et al. 1988, pp. 253-278, and also Grupp et al. 1990a, pp. 127-128). In this context, the work of Coward and Franklin (1989) has to be mentioned who tried to describe the interface between science and technology by matching patent inventors and paper authors (see also chapter 3 of this volume). They define the well-known concept of "strategic science" as located at the interface between "applied" and "not-yet-applied". The notion of "strategic science" with a slightly different meaning traces back to Irvine & Martin's (1984, pp. 2-5) "strategic research".

The above findings are thus in concordance with the approach of general cyclicity in science, technology and innovation (see chapter 1, section 1.1, and also Holland et al. 1984). If there is something like a discrete, stepwise overall progress in science and innovation then the period before 1980 or so is characterized by "Fordist" mass production, i. e., a golden age of growth and full employment leading inevitably to a crisis of structural adjustment (Freeman 1987, p. 70). For the special case of polyimides, not so much for lasers, a period of mass production was the case, indeed. After the turning point, i. e. in the 80s and 90s, Freeman (op. cit., p. 71) and others depict an information and communication period with cheap chips replacing the cheap mineral oil from the Fordist wave, and diseconomies of scale. For the key factor "chip" under the new paradigm both lasers and polyimides will be required as was discussed above. "Third generation" biotechnology (with biomaterials) and fine chemicals come into the picture of the new wave. "The new paradigm affects the chemical industry profoundly in many ways. (...) Special plastics, such as polymer alloys and *polyimides*, and composite materials will increase their share of production at the expense of the bulk commodity materials" (Freeman 1990, p. 85; emphasis added). In this interpretation the two cases analysed may be seen as representatives of the new science and innovation paradigm.

Compared to the similarities the differences between the two patent trends seem to be unimportant. Yet, the increase in inventive activity in the second wave is steeper for polyimides than for lasers (compare figure 9-1). Without embarking upon correlation analysis between patent and economic trends (for regression studies of other cases at full length see Schmoch et al. 1988, pp. 297-346, and later Grupp et al. 1990a) it seems to be clear that this is related to the different economic upturn at that time. As is noted in table 9-2, the polyimide and the laser markets grew by a factor of 10 and 8, respectively, in 10 years. It remains to be seem how long and at which growth rates both upturns will last in the nineties.

Another minor difference lies in the fact that - contrary to the somewhat earlier and steeper upturn in polyimide patents - the peak of polyimide patent activity in the first wave was flatter and later. This is most likely an effect of industry structures. Polyimide inventions are brought about mainly by R&D departments of large chemical concerns as a result of strategic R&D planning. Laser industry, on the opposite, is characterized by many newly founded technology-oriented small firms some of which took out a good deal of early patents but were never able to convert their science-oriented technology into profits. A number of laser bankrupts are known (see above), whereas nothing is published on deficits of the chemical trusts in the polyimide business. In conclusion, the large chemical companies seemingly have imposed a momentum of inertia into the early innovation phase in the polyimide field, but have earned earlier economic income as compared to laser firms with a more furious early invention phase and more reduced economic success up to now.

9.5 SCIENCE-TECHNOLOGY RELATION IN LASER MEDICINE AND POLYIMIDES IN ELECTRONICS

For investigations on the knowledge interface between technology and science, special sub-fields within the two main areas lasers and polyimides were determined. In the case of laser technology, the partial fields of laser metrology, laser material processing, and laser medicine were examined. The systems containing laser beam sources differ considerably when applications in metrology, medicine, or material treatment are aimed at. *Laser metrology* includes mainly instruments for geometric and geodetic surveying as well as speed measurement. By running over some hundred titles of related patents it was found that commercial application is mostly seen in the industry sectors of mining, civil engineering, vessel construction, and general machinery. These sectors of industry presumably do not offer good prospects for insights into the science-technology interface (see, e. g. chapter 4, figure 4-2). Further, this field of laser application does not exhibit very strong growth: the number of inventions world-wide doubled between 1975 and 1985. *Materials processing* by help of lasers is mostly centred around welding, cutting and drilling. Sectors of application are the metal as well as the semiconductor industry. The number of patent documents internationally in 1985 was about four times as large as in 1975. Pretests did not show strong ties between technology and science as most of the work is dedicated to sophisticated engineering like the suppression of unwanted surface reflections. *Medical laser applications* cover the full range of therapeutic and diagnostic instruments (see section 8.6). Presumably there are strong ties to physics as well as to medicine. In 1985 the number of patent applications world-wide was eightfold compared to 1975.

In the seventies, there were only a few patent applications for medical lasers on the German market which were of no major consequence (see figure 9-3). Starting from about the invention year 1979, there is a sharp increase up to today which in the first years of the eighties is due to Japanese inventions and later on to US and German activities. I. e., in the sub-field of laser medicine Japan is the headstarter whereas in the overall laser field it is the US. It must be concluded that this field of application is quite dynamic, quite young, and quite close to science. Although certain moments of licence can not be eliminated, it seems to be a good starting point to select the medical lasers as the sub-field of choice for a detailed analysis.

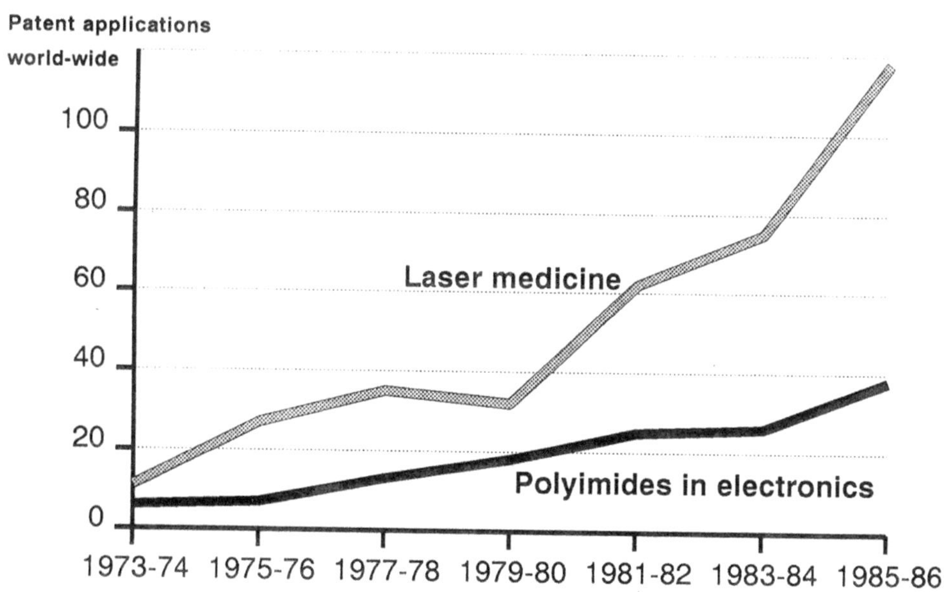

Figure 9-3: Patent applications world-wide with destination Western Germany in laser medicine and polyimides applied to electronics 1973 to 1986

For polyimide technology, the two main application fields in optics and electronics have a reasonable size and a relatively high growth rate. Both are science-intensive. Since with lasers a technology related to optics has already been analysed, semiconductor electronics was selected for further detailed investigations (excluding non-semiconducting electronic appliances). For this sub-field, a steady increase of patent applications beginning after 1976 can be observed (see figure 9-3). The United States and Japan are the most important countries followed by Western Germany. France and Great Britain nearly completely stopped the respective patent activities.

At first sight, the determination of citations to non-patent literature seems to be promising in order to describe the knowledge transfer from science to technology. However, the considerations on motives of citations to non-patent literature, as discussed in chapter 4, section 4.2, show that not all these citations are really linked to science. A second practical problem is the limited availability of non-patent citations in data banks (Schmoch 1991). A further methodological problem lies in the fact that the number of non-patent citations compared to patent citations is low (see section 9.2) so that there is the danger that random samples are too small. Finally, the distribution of the number of references to other literature is extremely skewed which often leads to quite erratic results (van Vianen et al. 1990). In this situation, the share of patent applications with at least one citation to scientific literature has proved to be more meaningful than the mean quota of citations per patent application.

According to the outdated linear or sequential model (chapter 1, section 1.2), one would expect that in the early phase of an emerging technology, the science intensity would be high and would diminish with its ongoing establishment in the market. This hypothesis is not supported by the results for the two selected sub-fields for which very stable trends are found (table 9-3). If citations to other literature really indicate a link to science, a constant relation between non-patent citations and the number of respective patent search reports points to a growing science activity in the field of laser medicine and polyimides in electronics in absolute terms, as the number of citing patent documents is sharply increasing (see figure 9-3). These findings would fit in with the overall considerations on cyclic trends of the core and product technologies. However, the examined period of invention years in table 9-3 is not long enough due to the fact that the EPO began operation only in 1978 (the non-patent citation analysis is based on EPO search reports). In particular, it was not possible to cover the restructuring decade between the two development waves of laser and polyimide technology, respectively. But because laser medicine and electronic polyimide usage are young sub-fields of scientific and technological activities, the data for 1978 and before are low in absolute numbers anyway (see figure 9-3; for laser medicine also the data for 1979-80 are insignificant).

Table 9-3: Non-patent literature citation in laser medicine and polyimides in electronics (patent applications at the EPO only); given is the percentage share of search report with one non-patent citation at least

Period	Laser medicine			Polyimides in electronics		
1979-80	*)			57		
1981-82	37			63		
1983-84	38			57		
1985-86	38			57		
For comparison	All technology	All lasers	Medicine	Polymer synthesis	Polymer processing	All electronics
1985-86	25	63	14	23	22	50
*) Data statistically insignificant						

Comparing the different science-intensities as measured by non-patent citations in patents in table 9-3, laser medicine with about 38 % and polyimides in electronics with about 57 % of all patent search reports referring to scientific results seem to have a distinct linkage to science, as these percentages are clearly above the actual average value of 25 % in all technologies. Laser medicine is intermediate to the science-intensity of medicine (low; including hospital beds and other mechanical inventions) and to the one of lasers (high; including R&D lasers). The average science linkage in the greater laser and the biomedical sectors, respectively, is not uniform but exposes a silhouette structure. This is even more true for polyimides applied to electronics. This science-related sub-field ranks clearly above polymer synthesis and polymer processing (about average compared to all technologies) but also tops the very science-intensive area of electronics. To whichever greater sector of technology it is attributed to, it towers above the sector. This again contradicts any idea of uniformity of science-dependence of whole sectors of technology and underpins the notion of a silhouette structure of the science-to-technology interface (compare also figure 4-12 in section 4.6).

These results are at least a plausible indication for the validity of non-patent citations as indicators for linkages to science (see also the whole of chapter 4), as reportedly about

10 % of new products and processes depend on academic research somehow (Pool 1991, p. 661, citing unpublished work by E. Mansfield who asked the executives of a random sample of 76 major US companies). Out of these, some less R&D-intensive imitative innovations presumably are not based on inventions at all. The academic linkage of the invention-based new products and processes, i. e., that one expressed in patent documents, should therefore be expected on a higher level than 10 % (in table 9-3 it is 25 % for average *patented* technology). The cited source (op. cit.) further reports that the dependence on academic research varies from industry to industry and triples for drugs at the high end. Given the baseline for average *patented* technology at 25 % according to table 9-3, the high end should be around 75 %, a figure which looks like an envelope figure to non-patent literature citation percentages compiled in the table.

Nevertheless, it is problematic to compare directly different technologies in a quantitative form. In the presented case studies, it is not known whether the publication behaviour in the field of laser physics and medicine is comparable to the situation in polymer chemistry and electronics, for there are quite different scientific communities who are active in these areas. However, if the notion of a silhouette structure of the science-technology interface is true then there are differences *within* scientific communities sticking to the same publication behaviour. Further investigations with citations to non-patent literature on the micro-level seem to be necessary. Related work is in progress and will be published elsewhere (Noyons et al., 1992).

9.6 TECHNOLOGICAL KNOWLEDGE EXPANSION IN LASER MEDICINE AND POLYIMIDES IN ELECTRONICS

The approach to describe the knowledge expansion (for a definition see section 8.2) with the evolution of technology is based on the analysis of the technological distribution of patent citations to patents. According to the proposed sequential phase model (which is not supported here from theoretical deliberations, see section 1.2), one would expect a growing independance of the respective applied, i. e. product- or systems-oriented technology in the course of time with a decreasing influence from core technology (laser beam sources, i. e. coherent light generation, or polyimide synthesis, respectively).

The findings on the overall relations between core and applied technologies presented in section 9.4, see figure 9-2, however, seem to suggest a lasting importance of the core of technology, contrary to model expectation. Yet, the data samples used in section 9.4 are quite inhomogeneous in terms of applied technology. In this section the focus is narrowed to laser medicine and polyimide materials used in electronic components only as in the previous section 9.5.

The expansion of knowledge within the socio-cognitive subsystem of technology (as distinct from the one of science; see chapter 2) is examined by patent-to-patent citations. The meaning of these links within technology in general terms is being considered in the methodological section 9.2. There it has been made clear that most of these references point to relevant technological background fields of the examined patent application but only in directly to technology sources in a narrow definition.

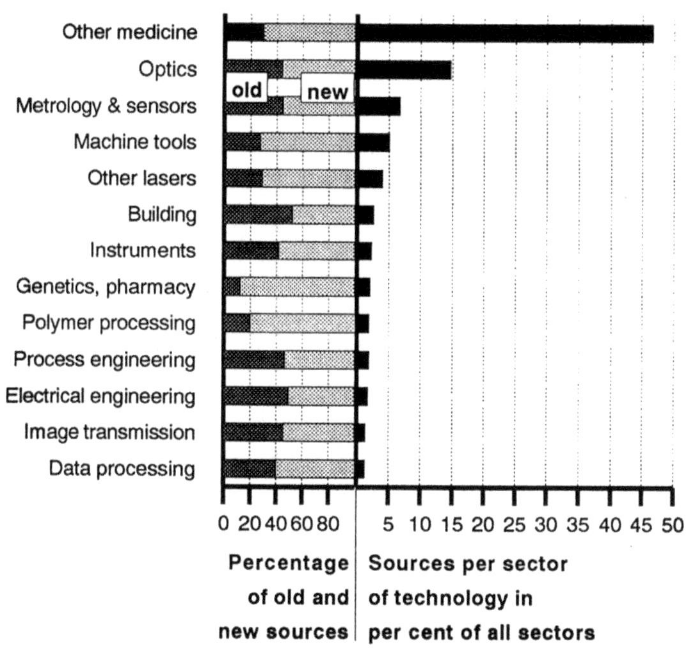

Figure 9-4: Source sectors and age of technological knowledge of the 13 most important sectors in support of laser medicine technology as measured by patent-to-patent references in the search reports from the invention period 1985-86 (old sources: 1980 and before; new sources: 1981 and thereafter).

Figure 9-4 sketches the technology source patterns of 1985-86 medical lasers whereby the 13 most important source sectors are mentioned (summing up to 93 % of all sources). To classify the sources of technological knowledge the same sectorial model was used as in chapter 4 (see table 4-1 and explanation given in section 4.2).

Corresponding to the figure, the assumption of a continous influence of the source technology "lasers" is generally supported. The proportion of citations to laser sources did not diminish compared to earlier invention periods. However, the absolute number of references related to beam sources is quite low compared to the fields of metrology and especially optics. Thus, the expansion process within technology contains a lot of supplementary information compared to the dual model of core and applied technologies. Obviously, the progress in medical lasers heavily depends on progresses in optics and metrology; for surgical lasers, the field of material processing ("machine tools") is important, too.

The most important knowledge fountain-head is biomedicine itself. Nearly half of the prior art in laser medicine is related to medical equipment. As coherent light in many cases substitutes other light sources but does not change the purpose of the apparatus (e. g., in ophthalmology), the utility of the inventive step is proven in terms of better performance characteristics of the laser product. But otherwise, the advances of the traditional non-lasing equipment know-how are exploited for the laser-related innovation. In 1985 the rate of penetration of laser equipment in the US was already 65 % in ophthalmology (amongst nearly 15,000 physicians questioned), 30 % in otorhinolaryngology, 25 % in gastroenterology, and 20 % in obstetrics and gynecology (Halter 1985, p. 55). The fine-tuning of laser inventions to given leading-edge medical technology is thus reflected also in the broad coupling to medical know-how sources.

An interesting question is the age of cited technology sources in particular with respect to the two waves of overall laser development. The source patents related to 1985-86 sample patents are thus divided into old and new citations so that the dynamics of the respective areas become obvious. In figure 9-4, the knowledge sources are separated into older documents before the publication year 1981, i. e. patents from the restructuring period in between the two main waves of activity (see figure 9-1), and newer ones after and including this year, i. e. patents from the second upturn of overall laser activity. The time coverage of older patents is limited by the data bank coverage, in this case to the invention year 1973. In reality, there exist much older citations (Claus & Higham 1982).

According to figure 9-4, sources concerning medicine, machine tools (material processing), laser sources, genetics & pharmacy, and polymer processing are relatively new whereas optics, metrology, building (light guides), and instruments are based on relatively older documents. The basic hypothesis of a shift of the technology base from laser beam sources to medicine appears once again in a quite different light. Laser beam sources never played a dominant role, but nevertheless the incorporation of the *most recent laser beam generation* knowledge does play a role. The age of beam generation citations is as young as that of medicine. This is a very important result as the sequential phase model suggests that there is "supply" of basic science-based technology which is "pushed" into innovations by talented entrepreneurs later on. However, the data from medical lasers support the notion of cyclic or coupled innovation-oriented R&D processes whereby a synchronized exploition of the latest basic core technology takes place *together* with a stimulation of actual progress also in the core of technology caused by the demand from the users' side for better, more appropriate technical specifications. To carry the argument to extremes: the basics of technology from the heroic first wave of technical realization is largely disregarded in the present phase of a broad expansion of technological knowledge and commercial success.

Is there already technology diffusion *from* laser medicine to other fields? To which extent are laser medicine patents co-classified with patent codes for other technical sectors? First, one must state clearly that the new and dynamic field of laser medicine is not very diffusive compared to the total field of laser technology (Grupp, Reiss & Schmoch 1990, p. 56). Presently, the technical sectors medicine, optics, metrology, and machine tools alongside with lasers and genetics & pharmacy profit from laser medicine knowledge. It is interesting that the diffusion sectors are nearly the same as the technology source sectors. This applies for optics and metrology as well as machine tools. Before looking more deeply at the technological diffusion from laser medicine, the question arises to what extent the findings reflect the real-world source and diffusion structure of the referring sample patents. There may be distorting factors like particular citation behaviours of patent examiners and the structure of the search documentation of the patent offices. For example, in the case of a few citations to a certain field the question is left open, whether the sample really has few technology links, or whether there is a low citation propensity by the examiners in this field, or - above all - whether there are only a few documents available which may be cited.

These methodological problems are discussed elsewhere (Schmoch 1991). They can largely be solved by a combination of the technology diffusion analysis with the technology source analysis of a given set of sample patents in a two-dimensional representation. Under the assumption that the influence of the distorting factors is similar for both dimensions they cancel out and one may differentiate between supporting, developing, and mainstream sectors of the related sub-field of technology, in this special case of medical lasers. The *supporting sectors* introduce new elements into laser medicine more frequently than they benefit from knowledge of medical laser technology. *Developing sectors* are less important as suppliers of technology, but laser medicine has become a driving force for them.

Figure 9-5 shows the results for the most important sectors in the case of laser medicine. Two cross-sections for laser medicine inventions from 1979-82 and from 1983-86 are combined. Medicine itself is the largest developing sector (see small insert figure), whereas the second largest sector, optics, is a supporting one. Measuring technology changed its role from a developing to a supporting sector, whereas the opposite is true for material processing (machine tools). The field of laser sources is mainstream, but not prominent in absolute figures. Other interesting sectors are electrical engineering (no more supporting), genetics & pharmacy (clearly developing), and instruments.

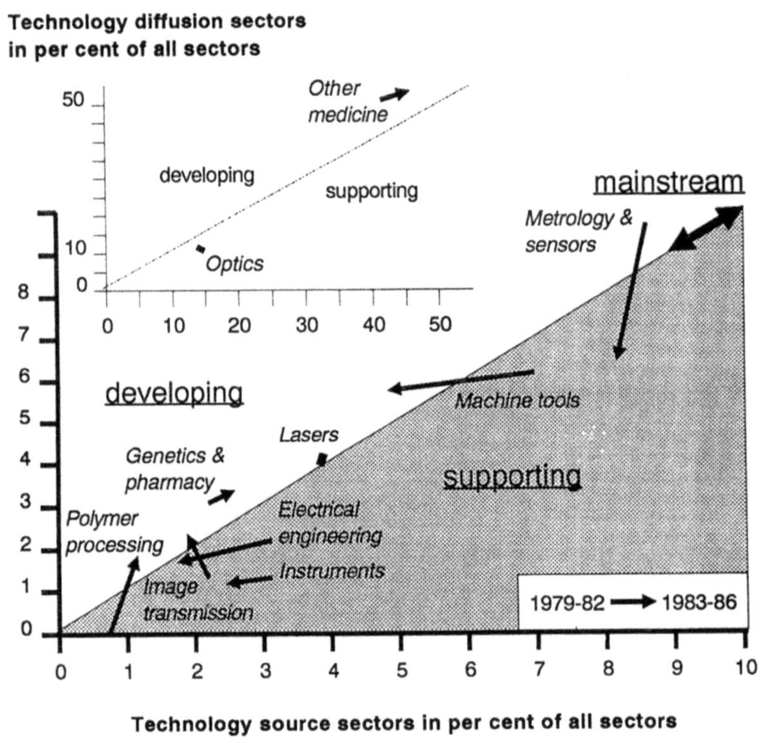

Figure 9-5: Supporting and developing sectors in laser medicine as structural parts of the expansion of technological knowledge. Cross-section comparison of the invention periods 1979-82 and 1983-86 of sample laser medicine patent applications

From the juxtaposition of developing and supporting technologies in figure 9-5 the sequential model of a one-way transfer of science and technology into entrepreneurial practice must be clearly rejected. The expansion of technological knowledge around laser medicine is cyclic, indeed. The following examples are based on an interpretation of selected patent abstracts. Laser medicine not just benefited from, but also influences the development of laser beam source for instance in producing better pulse lasers so as to avoid too much heat transferred to the body. Another impetus is beam optimization for eye surgery where the precision requirements from the side of medicine are different as from the side of other applications. Whereas many laser systems require high power lasers, medical applications typically emit low power radiation, for instance for bio-stimulation of tissue. Medical applications thus influence the development of laser beam sources in the low power direction. A special problem with beam propagation is the very tiny optical fibre required for instance for blood vessel treatment. Two beams must be transported with high quality. Measurement problems introduced from the side of laser medicine touch for instance into the observation of relative movement between the upper and lower jaw in dentistry or in automated visual deficiency screening systems and range up to topography measuring apparatuses in ophthalmologic surgery.

Laser medicine seems to become a key technology of its own which enables other developments, although it is a small speciality. The borderlines between the perception of laser medicine as a subject of diffusion from laser physics and as the driving force of diffusion into other fields are vanishing. The speciality of laser medicine does unfold such a variety of interesting and complex structures as part of the science-technology interface that it must be regarded as a new scientific and technological area of its own.

Compared to medical lasers, polyimides used in electronics draw on a completely different technology source pattern which is not just different by sectors but also different in structure (figure 9-6). There is a clear top sector for knowledge (namely "other electronics" as compared to "medicine" in the case of medical lasers) but then a rather broadly distributed array of technology sources with polymer chemistry leading. In the past decade electronic know-how became more important whereas polymer synthesis (polymer chemistry), the presumed core of polyimide applications, ranked always second - other than in the case of laser medicine. Further important sources of technology are coating (surface treatment), optics, polymer processing, and paper & printing (of special interest; see below).

It has been stressed in section 9.3 to which extent polyimides have become useful to the electronics industry after the substitution of circuitry by semiconducting chips. Until today the technological knowledge directly from electronics is the most important constituent of polyimide development. Electrical engineering (8th rank), information storage (11th rank) and data processing (12th rank) are important, too. Together with metrology (sensor technology) the broader electrical sector adds up to nearly 50 % of all technology sources. The 13 sectors indicated in figure 9-6 cover 95 % of all technology sources.

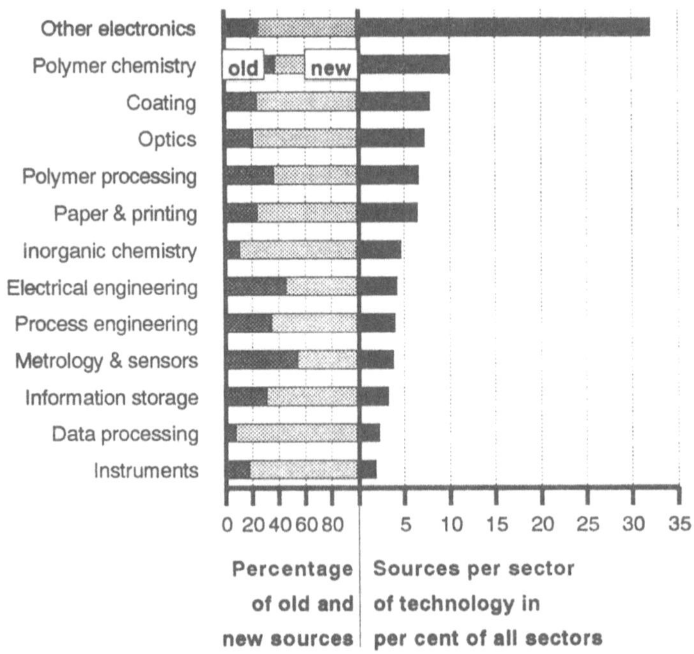

Figure 9-6: Source sectors and age of technological knowledge of the 13 most important sectors in support of electronic appliances based on polyimides as measured by patent-to-patent references in the search reports from the invention period 1985-86 (old sources: 1980 and before; new sources: 1981 and thereafter).

Examining the sector paper & printing, the concerned source technology sometimes deals with surface treatment for semiconductor devices and integrated circuits and hence also with micro-electronics. Thus, in these cases, the patent classification terminology is obviously misleading (Schmoch 1991). Another important source of knowledge from the paper & printing sector contains typewriters and selective printing mechanisms. In particular, these methods comprise thermographic, magnetographic and electrographic procedures. The concerned patent documents describe printing heads wherein semiconductor devices are integrated and which include polyimide films. The major growing knowledge sources for polyimide applications in electronics are indeed the technical fields of paper & printing and inorganic chemistry. These two sectors represent special surface treatment techniques. The speciality studied thus draws on two main sources: polyimide chemistry and electronics in a broad sense.

Another indicator for emerging technical fields is the ratio of new versus old patent citations for a fixed invention period (in this case 1985-86). The respective data are also shown in figure 9-6. In case of the chemical core (polymer chemistry) and polymer processing the share of new sources is about 60 % which can be considered as a baseline for the other fields. Coating, optics, paper & printing (i. e., printing heads and surface treatment), inorganic chemistry and data processing exhibit the highest ratio of new citations (75 to 90 %) pointing to the growing importance of these fields for polyimide applications in electronics. Compared to medical lasers, the core of technology, polymer chemistry, is quantitatively *more important* than in the case of medical lasers, and *older*.

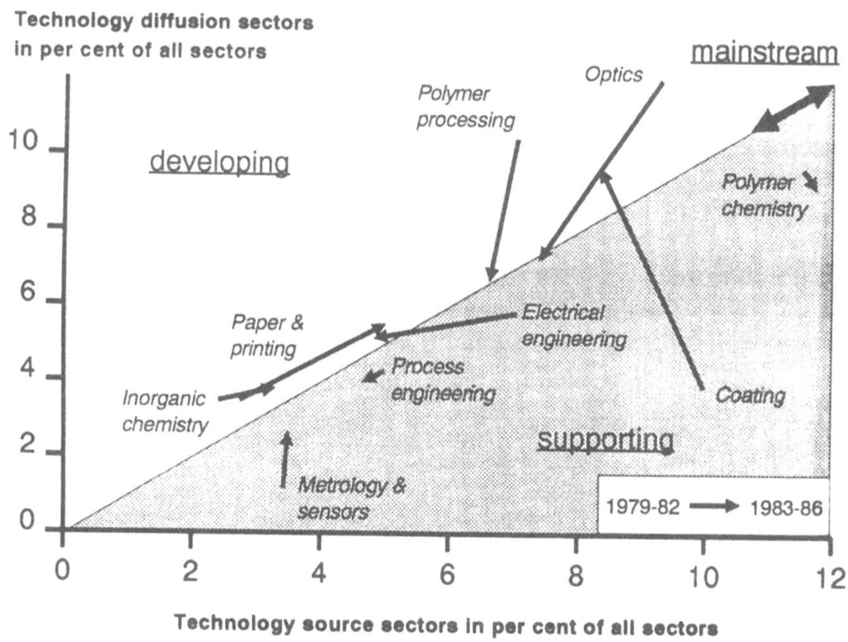

Figure 9-7: Supporting and developing sectors of polyimides in electronics as structural parts of the expansion of technological knowledge. Cross-section comparison of the invention periods 1979-82 and 1983-86 of sample patent applications

In the diffusion analysis of polyimide applications in electronics by patent co-classification, several material and processing related sectors such as polymer chemistry, coating, polymer processing, process engineering, and inorganic chemistry arise. The demarcation of "polyimides in electronics" includes all semiconducting electronic applications (e. g. printed circuits). The non-semiconducting electronic sectors is the most important diffusion area of polyimides in the given demarcation (about 30 % of all diffusion sectors). As for lasers, the developing (diffusive) and the supporting technologies have been grouped into a two-dimensional array (figure 9-7).

The non-semiconducting part of electronic technology is the strongest developing sub-field around polyimide usage in (semiconducting) electronics (not shown in figure 9-7). The main supporting technology is polymer chemistry which leads to the conclusion that the chemical core of polyimides is maintaining its importance as a source of knowledge for this special field of application. An interesting case is coating which is crossing the mainstream line. In older patent documents, coating was the most important source of knowledge besides polymer chemistry whereas in newer patent documents coating additionally has become a major focus of technological development itself. Furthermore, the fields of electrical engineering, of polymer processing, of optics (with patents on the manufacture of integrated circuits by optical means), of inorganic chemistry (with inventions concerning superconductors), and of metrology influence the development of polyimides in electronics more and more as mainstream technologies supporting and benefiting with equal shares.

In conclusion, the combination of these two sets of data seems to be a helpful tool for describing the knowledge expansion in science-based technology. For the particular field of polyimide applications in electronics, like in the medical laser field, the mechanism of technology development does not include a general turning down of basic technology during advanced stages, rather, the development of basic research is going on and continuously incorporated in applied R&D. To advance the particular technical field of polyimides in electronics, several technical side disciplines which can also be used in other technical areas, like for example paper & printing, inorganic chemistry, and coating have to be mastered. In other words, interdisciplinarity is mandatory. Technical break-throughs emerging during development of polyimide applications in electronics did not take place in the chemical core where the relative patent activities remained more or less the same but in other technical fields (inorganic chemistry, paper & printing, coating).Thus, the idea of sequentially pushing core technology (or science) in order to obtain innovations has proven to be inappropriate once again.

9.7 CONCLUSIONS FOR CORPORATE R&D MANAGEMENT AND SCIENCE POLICY

The sequential or linear model of innovation which "undoubtedly goes back to Schumpeter" (Silverberg 1990, p. 177) makes a strong distinction between "innovation" and "diffusion". As science to technology to innovation to diffusion is understood to pass unidirectionally from one to the next, the notion is implicit in the linear viewpoint "that the faster the better, the earlier the time-point of adoption the better, the higher the level of diffusion the better" (op. cit., p. 178).

The purpose of this chapter is to expose any cyclic characteristics in the development of science-based technologies in order to add empirical support to the proposed cyclic or coupled model of science-based innovation. A second tacit purpose was to assess possible new R&D management tools for strategic corporate planning and science policy in the realm of science-based innovation. In particular, time trends, the technological diffusion of sample patents, the pattern of the related patent-to-patent references, and the quota of non-patent literature citations have been discussed theoretically and tested in the cases of lasers, laser medicine, polyimides, and polyimides in electronics. If one passes the various steps in review, their ease of application and their power of explanation seem to be different for the corporate R&D management and for science policy. As in the previous chapters the relation of the empirical findings to theoretical modelling has been discussed in various sections, here some conclusions for corporate R&D management and science policy will be drawn.

A first background picture of what goes on in science-based innovation may be drawn by the establishment of patent time trends combined with a simple model of a distinction between core and applied, i. e. product-related, technologies. In the cases studied the development of core technology is characterized by waves of activity, whereby the initial wave is oriented to technical realization whereas the second take-off is linked to a broader upturn of applied technology. Therefore, the observation of time trends from the early stages can give some indications on the appropriate period to stand-by and watch the international scene, or to actively transfer knowledge between (public) science and (company) technology, or to acquire supporting technologies.

An analysis of non-patent citations in patents did not reveal any trends. This adds plausible support for the hypothesis that a high number of non-patent citations can be considered as an indication for a strong science interface. Further investigations and a sound validation of the results may improve credibility and feasibility of this indicator. In this chapter it was found as in chapter 4 on a more general level that, indeed, scientific fields and technological sub-fields unter investigation differ significantly from the respective average science-intensity.

With respect to the analysis of technology knowledge expansion, the direct analysis of diffusion within the sample by co-classification is as straightforward as the indirect way by references to source technologies, yet the influence of distorting factors in the latter case is not as clear as in the former. Nevertheless, the combination of diffusion data and the respective technology source data leads to an interesting picture of the dynamics between supporting, developing and mainstream sub-fields. The comprehension of a technology field is completed by an analysis of the age structure of the source technologies in order to identify sectors with a high speed of exploition. As a general conclusion, if a new applied field of technology appears as a key technology in its own right, marked interdisciplinarity seems to be required.

With these empirical findings the theoretical distinction between "science-based innovation" and "diffusion" seems to be very misleading since it "obscures the numerous improvements made to a new product or process during diffusion. (...) Innovations require complementary skills, tacit knowledge and all kinds of supporting structures, if they are to diffuse successfully" (Freeman & Soete 1990, p. ix). Science, technology, innovation, and diffusion are never entirely separable at least in cases of science-based innovation and science-based diffusion.

In conclusion, the description of the knowledge transfer between science and technology and within technology seems to be quite complex so that the definition of simple quantitative procedures is not possible. Nevertheless, a network of indicators, which are based partly on sample patents, partly on the respective references, can be established, giving an interesting insight into the interface processes. "The use of patent and other data, (...), modifies case study or historic material and in particular 'conventional wisdom' about innovations in general or the development of a particular innovation." (Walsh 1984, p. 231). All in all, this type of analysis bears good prospects for future use in R&D management and in science policy.

The authors are grateful to their colleagues Thomas Reiss and Edo Albrecht for numerous stimulating discussions. Norbert Kirsch compiled parts of the data on-line. This work was supported by the Stiftung Volkswagenwerk, Hannover, Germany.

REFERENCES

Albert, M. B., Avery, D., Narin, F. and Mc Allister, P. (1991), 'Direct validation of citation counts as indicators of industrially important patents', *Research Policy*, 20, pp. 251-259.

Baré, R. (1981), 'Results of a Statistical Study of the References Cited in the Search Reports Established by the EPO', *World Patent Information*, 3, 56-60.

Beyer, E., Loosen, P., Poprawe, R. and Herziger, G. (1985), 'The Development of Laser Technology and its Importance for Material Processing', *Laser und Optoelektronik*, 3, pp. 274-281.

Callon, M., Courtial, J. P. and Laville, F. (1991), 'Co-Word Analysis as a Tool for Describing the Network of Interactions between Basic and Technological Research: The Case of Polymer Chemistry,' *Scientometrics*, 22, pp. 155-205

Carpenter, M. P. and Narin, F. (1983), 'Validation Study: Patent Citations as Indicators of Science and Foreign Dependence', *World Patent Information*, 5, pp. 180-185.

Claus, P. and Higham, P. A. (1982), 'Study of Citations Given in Search Reports of International Patent Applications Published under the Patent Cooperation Treaty', *World Patent Information*, 4, pp. 105-109.

Coward, H. R. and Franklin, J. J. (1989), 'Identifying the Science-Technology Interface: Matching Patent Data to a Bibliometric Model', *Science, Technology, & Human Values*, 14, pp. 55-77.

Dunlop, L. and Oppenheim, C. (1980), 'The Effect of Recycling on a Patent Citation Network', *World Patent Information*, 2, pp. 100-102.

Ellis, P., Hepburn, G. and Oppenheim, C. (1978), 'Studies on Patent Citation Networks', *Journal of Documentation*, 34, pp. 12-20.

European Patent Office (1989), *Annual Report 1988*, München, The Hague, Berlin.

Freeman, C. (1990), 'Technical innovation in the world chemical industry and changes of techno-economic paradigm', in C. Freeman and L. Soete (eds.), pp. 74-91.

Freeman, C. and Soete, L. (1990), 'Introduction', in C. Freeman and L. Soete (eds.), pp. vii-x.

Freeman, C. and Soete, L. (eds.), (1990), *Explorations in the Economics of Technical Change*, Pinter Publishers, London.

Grupp, H. (1986), 'Aktuelle Entwicklungen in der Lasersicherheit', Report to VDI - Verein Deutscher Ingenieure, FhG-ISI, Karlsruhe.

Grupp, H. (1990), 'The Concept of Entropy in Scientometrics and Innovation Research', *Scientometrics*, 18, pp. 219-239.

Grupp, H. and Hohmeyer, O. (1988), 'Technological Standards for Research-Intensive Product Groups', in A. F. J. van Raan (ed.) *Handbook of Quantitative Studies of Science and Technology*, North-Holland, Amsterdam, pp. 611-673.

Grupp, H., Reiss, Th. and Schmoch, U. (1990), 'Knowledge Interface of Technology and of Science - Developing Tools for Strategic R&D Management', FhG-ISI Report to Volkswagen foundation, Karlsruhe.

Grupp, H., Schmoch, U., Schwitalla, B. and Granberg, A. (1990a), 'Developing industrial robot technology in Sweden, West Germany, Japan and the USA', in J. Sigurdson (ed.), pp. 106-129.

Grupp, H., Schmoch, U. and Kuntze, U. (1991), 'Patents as Potential Indicators of the Utility of EC Research Programmes', *Scientometrics*, 21, pp. 417-445.

Halter, P. (1985), 'The Medical Laser Market Waits for R&D to Deliver', *Lasers & Applications* 10/85, pp. 55-57.

Holland, D. (1988), 'Wechselwirkung von ökonomischer Dynamik und Theoriedynamik, dargestellt am Beispiel der Entwicklung und Anwendung hochpolymerer Materialien', Habilitationsschrift, Humboldt-Universität, Berlin.

Holland, D., Albrecht, E. and Paetzold, G. (1984), 'Innovationsprozesse aus der Sicht des Zyklus Wissenschaft-Technik-Produktion', *Wissenschaftliche Beiträge der Sektion Wissenschaftstheorie und -organisation der Humboldt-Universität Berlin*, 31, pp. 29-63.

Noyons, E.C.M., van Raan, A.J.F., Grupp, H. and Schmoch, U. (1992), 'The Science and Technology Interface: Inventor-Author Relations in Laser Medicine Research', forthcoming.

Pool, R. (1991), 'The social return of academic research', *Nature*, 352, p. 661.

Reuben, B. and Burstall, M. (1973), *The Chemical Economy,* Longmans, London.

Riediger, S. (1984), 'Zum Zeitverhalten neuer technischer Konzeptionen (Basisinnovationen) dargestellt am Beispiel der Entwicklung, Produktion und Applikation neuer, hochveredelter polymerer Materialien', Dissertation, Humboldt-Universität, Berlin.

Schmoch, U. (1990), *Wettbewerbsvorsprung durch Patentinformation, Handbuch für die Recherchenpraxis*, Verlag TÜV Rheinland, Köln.

Schmoch, U. (1991), 'Tracing the Knowledge Transfer from Science to Technology as Reflected in Patent Indicators', in *Scientometrics*, in print.

Schmoch, U., Grupp, H., Mannsbart, W. and Schwitalla, B. (1988), '*Technikprognosen mit Patentindikatoren*, Verlag TÜV Rheinland, Köln.

Schmoch, U., Kirsch, N., Lay, W., Plescher, E. and Jung, K. O. (1991), 'Analysis of Technical Spin-Off Effects of Space-Related R&D by Means of Patent Indicators', *Acta Astronautica*, 24, pp. 353-362.

Sigurdson, J. (ed.), (1990), *Measuring the Dynamics of Technological Change*, Printer Publishers, London.

Silverberg, G. (1990), 'Adoption and diffusion of technology as a collective evolutionary process', in C. Freeman and L. Soete (eds.), pp. 177-192.

Trajtenberg, M. (1990), 'A penny for your quotes: patent citations and the value of innovations', *RAND Journal of Economics*, 21, pp. 172-187.

US Patent and Trademark Office (USPTO), Office of Technology Assessment and Forecast (1976), 'Sixth Report', US Government Printing Office, Washington D.C.

van Vianen, B. G., Moed, H. F. and van Raan, A. F. J. (1990), 'An Exploration of the Science Base of Recent Technology', *Research Policy*, 19, pp. 61-81.

Walsh, V. (1984), 'Invention and innovation in the chemical industry: Demand-pull or discovery-push?', *Research Policy,* 13, pp. 211-234.

10 MAPPING AN EVOLVING TECHNOLOGY CLUSTER: THE COMPOSITION AND STRUCTURE OF FACTORY AUTOMATION

Anders Granberg

10.1 INTRODUCTION

This chapter has been written within the framework of a broadly based, collaborative research effort entitled "Sweden's Technological System and Future Competitiveness." In a first round of investigations, the so-called STS Project focuses on the field of factory automation (FA) in the engineering industry sector, with the aim of elucidating the nature and significance of the Swedish technology base in that particular domain. Special emphasis is placed on the technology acquisition strategies pursued by users and suppliers of FA equipment, on the structural and functional characteristics of the institutional networks in which these actors operate, on the role of the national "R&D infrastructure", and on the broader macroeconomic implications of the conditions and developments explored at the lower levels of analysis.

It stands to reason that the studies undertaken within the STS Project require a shared framework of basic concepts, facts, and conjectures regarding the domain of FA technology itself - a *general* framework, that is, which is relevant to the research topics indicated above, but which does not depend upon, and reflect, the particulars of the Swedish case. The purpose of this paper is to provide such a framework. More specifically, the tasks addressed are the following: to outline a conceptual approach to the description and analysis of factory automation as a complex and diversified technological domain or technology cluster (section 10.2); and to provide an overview - on a general, "transnational" level - of the composition and structure of the FA cluster in terms of the constituent disciplines and specialties and the relationships among them (section 10.3).

The title of the paper merits a comment. The phrase "technology cluster" has been chosen to underscore the point that we are dealing here not with a single, clearly delimited and well-integrated field of technology, but rather with a large and complex agglomeration of diverse technologies brought together by virtue of their ability to contribute, in various ways, to the solving of technical problems in factory automation. Of course, many of these technologies reappear in other clusters centred on other areas of application. The construction of pictures - static and dynamic, verbal and graphic - of the technology cluster is referred to as "mapping." The word is deliberately used to convey the point that (as in cartography, *mutatis mutandis*) there are questions of coverage, selectivity, scale, and projection that cannot be evaded, and, in particular, that there can be no such thing as wholly distortion-free map.

The study is based largely on information drawn from the available literature. An extensive scan of the FA publication field has been carried out and a large number of studies, review articles, state-of-the-art reports, statistical sources, and other material relevant to the mapping task have been compiled, examined, and summarized.[1]

10.2 CONCEPTUAL APPROACH

The approach that will be taken to the conceptualization of technology and technological change in the FA domain closely resembles that which has been developed and applied by the author in the investigation of such fields as fibre optics and biological pest control; for details, see Granberg (1987a, 1987b). For the purposes of this study it may suffice merely to outline some of its basic features.

TECHNICAL KNOWLEDGE AND TECHNICAL PROBLEM SOLVING

Two principal elements can be distinguished. The first and most fundamental is the view of *technology as knowledge* and of technological development as a knowledge-generating process (this may be compared, e.g., with those analytical perspectives which emphasize the artifactual or community aspects of technology).[2] The adoption of a "cognitive viewpoint" (see chapter 2) is generally conducive to the exploration of the links between technology and those societal institutions - most notably science - which

1 The appended list of publications includes only items referred to in the paper and thus covers only a minor portion of the total body of literature surveyed in this study.

2 Technology is, of course, a multifaceted phenomenon. Its various aspects have provided the bases for numerous alternative answers to the question of what is meant by 'technology'. For an overview of definitions and classifications, see Mc Ginn (1978), Mitcham (1978) and chapter 2.

are centrally (and indisputably) concerned with the production, organization, and dissemination of knowledge. Also, tying into the tradition of knowledge-oriented studies of science and technology means that attention is drawn to certain cognitive distinctions and categories which bear on the task at hand; the following should be mentioned:

- know-what and know-why versus know-how
- descriptive versus prescriptive knowledge
- generalized versus particularized (or idiographic) knowledge
- codified versus uncodified knowledge
- communicable versus incommunicable (or personal, tacit) knowledge
- embodied versus disembodied knowledge.

We shall return to these distinctions shortly, when explaining the premises and delimiting the subject matter of the study.

The second major element of the approach - which is closely connected with, and complementary to, the first - is the emphasis placed on the *problem-solving aspect* of technological development. Technology, or technical knowledge, like other forms of knowledge, grows by problem solving - by the identification, specification, and solving of technical problems. It is suggested that the basic problem faced by the technologist is that of determining the means (or tools or devices or technical systems) by which particular functions can be effectively and efficiently performed. Thus, an articulated, ideal-type technical solution may be formally defined as "a set of statements asserting (1) that a technical system, T, materialized and implemented in accordance with a design, D, will, under the conditions of application, C, exhibit a set of operating characteristics, O; and (2) that O is an adequate translation of the assigned function, F" (Granberg 1987a).

It should be noted that technical problem solving - the core of which is the process of design - is understood here as including not only the generation of a blueprint for a device or system, but also the determination of appropriate conditions and rules for implementation and operation. The corroboration of a technical solution requires, of course, the materialization (as a prototype and, subsequently, as a marketable product) and implementation (in laboratory and field tests and in regular operation) of the technical system[3] in question. In a wider perspective, technical problem solving

3 In the following, the term 'technical system' will serve as the general designation of the materialized forms of technical solutions, including tools, devices, machines, etc. as well as larger complexes of such entities.

presents itself as a continuing, iterative process involving numerous design modifications and incremental improvements as well as the occasional radical design innovation (Gardiner & Rothwell 1989). It is important to note that the functional requirements, which basically define the technical problems to be dealt with, change not only in response to changing external, or "extra-technological" needs, but also as a result of technological development itself. It should also be pointed out that the questions which arise in a particular problem-solving context need not all of them be "technical" in the above sense. If a technical problem is analytically decomposed into sub-problems in a relevance tree-like fashion (e.g. Martino 1983), there may well appear problems - concerning, e.g., the basic properties of materials - which can justifiably be termed "scientific." Their classification as scientific or technological research problems is somewhat arbitrary and depends on their context rather than their content (see also chapters 2, 3, 8 and 9). Finally, attention should be drawn to the fact that problem solving, technical or otherwise, is a process which both generates and *feeds on* knowledge, and that it involves a search for relevant information and competences - an examination of the personal, team, or in-house knowledge store acquired through experience and education, as well as a scan of the external knowledge environment (Chakrabarti et al. 1983, Gibbons & Johnston 1974, Kunz & Rittel 1970, and von Wyk 1984). When analyzing the cognitive inputs of a problem-solving process, it may be useful to distinguish between particular pieces of information generated or acquired in the course of the problem-solving effort, and the general background knowledge and personal competence which the problem solver brings to the task.

DELIMITATION OF TECHNOLOGIES

Technical problems and technical solutions provide a basis for the identification and classification of those larger entities referred to as "technologies" (or "technological fields," "disciplines," "specialties" etc.). Thus, technologies can be conceptualized as classes of technical problems or solutions defined with reference to certain shared characteristics. Obviously, there are many possible grounds for the delimitation of such classes. However, we can single out a few key features by observing that a technical solution is generally capable of being analyzed in terms of (i) the *function* it is meant to perform (e.g. the supply of power); (ii) the *natural processes* utilized for this purpose (e.g. nuclear fission) - the satisfaction of any given functional need typically requires the coordinated use of many kinds of processes; and (iii) the *design* which links these processes into a functional whole (e.g. a particular reactor design) (Granberg & Stankiewicz 1981).

On the basis of these features the following major categories of technologies can be distinguished: (i) *Functionally oriented technologies,* or technologies characterized by their concern with the performance of particular functions; the functions are often broadly defined, yielding extensive fields or large clusters - such as energy technology or communications technology - of considerable diversity as regards the design content and the technological and scientific underpinnings of the constituent solutions. (ii) *Basic technologies,* or technologies centred on the control and practical exploitation of particular natural processes; these fields - e.g. semiconductor technology or microbial technology - tend to be closely linked with natural science and are, as a rule, "generic" in the sense of being open-ended on the applications side. The design dimension appears as the defining feature of two groups of technologies: (iii) *engineering fields,* or technologies - including such disciplines as mechanical engineering or electrical engineering - which are oriented towards the design of particular classes of technical systems defined in terms of shared structural features or modes of functioning; and (iv) *generic design technologies,* or technologies - e.g. systems analysis and systems engineering, operations research or cybernetics - which are grounded in formal science and which are concerned with the fundamental properties and optimal configurations and modes of operation of broad classes of technical (and socio-technical; cf. chapter 2) systems. This classification scheme is intended primarily to facilitate and guide the analysis of a complex technological domain. There is no presumption here that every discernible field may be neatly and unambiguously pidgeonholed into an appropriate category. It is also recognized, of course, that technologies may be categorized on grounds other than their content; for instance, the term *core technology* - which will be used in this chapter - is commonly defined as a technology which is central to the technical competence of a particular firm or industry.

INSTITUTIONALIZATION OF TECHNOLOGIES

Before turning to the technologies of factory automation, let us briefly consider an additional facet of technology and technological change which is clearly relevant to the mapping enterprise, viz. the process of institutionalization. It is reasonable to suggest that an area of problem-solving activity can be identified, and meaningfully spoken of, as "a technology" only if it has attained some minimum degree of stability and internal coherence. Progress towards a well-established, fully institutionalized technology is indicated by a series of events, including (in rough order) the appearance of the problem area as a recurring conference topic or theme, as a subject of courses, textbooks, and special journals, and as a basis for professional societies, academic chairs and departments. In this process, the scope - or widening scope - and perceived significance of the focal problems or solutions are obviously important; a narrow or isolated problem area may not progress very far in this respect. Institutionalization is generally associated with certain shifts along the cognitive dimensions indicated above. These shifts result from concerted efforts to articulate and communicate personal, experiential knowledge; to codify hitherto uncodified knowledge; to systematize and extract general principles from idiographic knowledge; and to develop the know-what and know-why underpinning technical know-how or practical, prescriptive knowledge. Also to be noted here is the fact that institutionalization typically involves the evolution of a field of incipient specialization into a Janus-faced entity: a technological discipline one side of which - the R&D field - comprises the processes and products of problem solving proper and the associated industrial-academic networks of collaboration and exchange. The other side - the educational field - presents itself in the activities and results of the compilation, transformation, and organization of corroborated solutions and proven techniques in forms appropriate for teaching and training (Bristol 1986, see also chapter 1).

The question of the institutionalization of technology points up certain difficulties inevitably encountered in a study such as this. First, there are no hard and fast rules for deciding whether or at what stage a given entity might be accepted as a "technology" or an "established technology." Second, institutionalization is a process which takes place partly within the framworks of national R&D and educational systems. This means, e.g., that the ways in which broader fields of technology are subdivided and organized, and particular technologies are delimited and labelled may differ considerably from one country to another. The problem is clearly recognized by this author; however, it is also evident that the issue cannot be satisfactorily treated within the limits of this survey.

10.3 COMPOSITION AND STRUCTURE OF FACTORY AUTOMATION TECHNOLOGY

In this section, which constitutes the major portion of the chapter, an overall picture of the FA technology cluster will be presented. First, as a point of departure, the "basic technical problem" of factory automation will be considered. The identification and structuring of FA-related disciplines and specialties will then be effected in the form of a problem analysis in which subproblems and associated knowledge and competence requirements are noted and displayed. Initially, the problem-solving perspective of a user company (i.e. a user of FA equipment) will be adopted. Subsequently, the scope of the survey will be expanded, covering larger and more "remote" portions of the FA cluster, as the perspectives and needs of other actors - such as equipment suppliers and their subcontractors or suppliers - are taken into account.

THE BASIC FA PROBLEM

Following the conceptual guidelines of the preceding section, we may say that the basic technical problem in factory automation is to determine the means by which particular production functions can be performed with a minimum of human effort.[4] The problem, in other words, is to specify and validate a design - whether a novel design or an adaptive modification of an existing design - of a technical system that effectively and efficiently performs the function, or functions, in question. What are the functions that might be automated, and what are the currently available types of solutions?

Automation in the engineering-industry sector (or, more generally, in the discrete-parts manufacturing industry) pertains to a large variety of functions that fall within, or cut across, the three major spheres of production (cf., e. g., Hunt 1987 and Kaplinsky 1984): (i) *the design sphere,* which comprises such tasks as the basic and detailed conceptualization and drafting of products or production systems, and the analysis and appraisal of design proposals; (ii) *the manufacturing sphere,* comprising, e.g., the setting-up of production equipment, the handling, transportation, and storage of materials, parts, and products, the processing or forming of materials into parts, and the assembling and inspection and testing of parts and products; and (iii) *the management sphere,* comprising a broad spectrum of information handling tasks relating to design and manufacturing as well as to purchasing, marketing, and other business functions of the firm. As suggested by Kaplinsky (1984), three basic types of automation may then be distinguished:

(i) "intra-activity automation," or the automation of particular, separate activities (tasks, functions) within a given sphere;
(ii) "intra-sphere automation," or the automation of a set of interlinked or integrated activities within a given sphere; and
(iii) "inter-sphere automation," or the automation of a set of interlinked activities that cuts across two or three spheres.

---

4 The term 'production function' may, to some readers, seem like an invitation to confusion. Although it is clear from the context that it refers here not to the economist's equation, but to some process, activity, or task of production that is to be automated, the author agrees that the term is best avoided. Note that the phrase 'a minimum of human effort' covers the minimization of mental as well as physical effort. This study is concerned with contemporary, or modern, FA developments the distinguishing feature of which is the progressive automation of measurement, communication, and control functions. While mechanization remains an important element of the change process, it is increasingly dependent upon, and intertwined with andvances in the first-mentioned area.

Overviews of factory automation are often sketched either along horizontal lines - focusing on a sequential arrangement of functions and associated equipment, e.g. from basic design to final inspection - or along vertical lines, from lower to higher levels of functional and systems integration and control (Ranta 1989). The above classification scheme is open or flexible in this respect; it is not committed to either view and thus readily accomodates the great variety of FA problems, philosophies, and development paths that can be observed in different firms and industries. In this chapter, we shall be concerned primarily with intra-activity and intra-sphere automation in design and manufacturing and with the emerging forms of inter-sphere automation.

THE SOLUTIONS OFFERED

The kinds of technical solutions that have been developed in response to these functionally defined types of automation tasks can only be briefly indicated here.[5] In the design sphere, early computer-aided drafting tools and design analysis software can be seen as the embodiments of intra-activity automation technologies, while modern multi-function CAD (computer-aided design) systems are illustrative of intra-sphere automation. In the manufacturing sphere, intra-activity automation is exemplified by single-function, stand-alone industrial robots (IRs), NC/CNC machine tools (numerical control/computer numerical control), and automatic testing equipment; intra-sphere automation here appears in the form of such multi-function or flexible machines and systems as machining centres, FMCs (flexible manufacturing/machining cells), and FMSs (flexible manufacturing systems). It should be pointed out here that AMH (automated materials handling) equipment, which may include AGVs (automated guided vehicles), is an important component of an automated manufacturing system. Inter-sphere automation, finally, is an area which is still at an early stage of development. As examples can be mentioned the so-called CADCAM systems which bridge the spheres of design and manufacturing, and the inter-sphere linkages which may be established by connecting administrative and planning databases in the management sphere - e.g. the databases of such manufacturing management systems as CAPP (computer-aided process planning) and MRP (material requirements planning) - with databases in the spheres of design and manufacturing. Inter-sphere automation which involves all spheres, or at least the management and manufacturing spheres, is often referred to as CIM (computer-integrated manufacturing). It is in this

5 For a comprehensive and detailed presentation of the various forms of automation, including their historical background, see Kaplinsky (1984).

sense that the term will be used in this paper.[6]

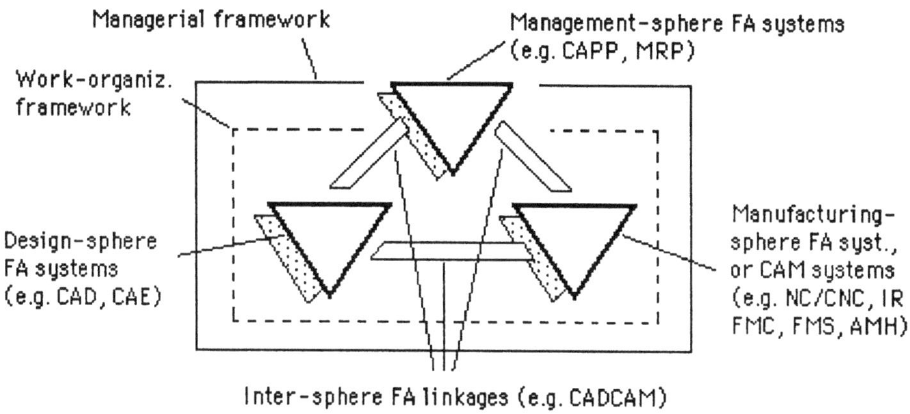

Figure 10-1: Major types of FA systems and organizational frameworks

Figure 10-1 shows, in a highly schematic fashion, the major systems of factory automation and the organizational frameworks in which they function. The inverted triangles indicate the hierarchical structure of FA systems, from the level of the individual machine to the level of comprehensive, multi-machine networks. Twin triangles are used to convey the dual nature of the systems, or the fact that hardware systems are tied to similarly structured software systems of communication and control. Two kinds of organizational settings are distinguished: one is the work organization of the design, manufacturing, and administrative units of the firm; the other is the broader managerial framework of plant and firm-level planning and coordination. It is recognized here that current automation efforts do not stop at the boundary of the plant or firm, as evidenced by the development of computer-based linkups between companies and their subcontractors and customers. The present survey does not, however, attempt to cover the technical issues raised by the construction of such inter-firm automation networks.

---

6 Factory automation is a field replete with acronyms such as CAD, CADD, CAE, CAM, NC, CNC, FMC, FMU, FMS, CIM, MIS, CAP, CAPP, MRP, etc. The reader should be warned that agreement of the semantic scope and precise meanings of these terms is far from complete. This is evidently true of the planning-system designations, the definitions of which may vary considerably. Opinions may also diverge on the exact boundaries between flexible 'cells', 'units', and 'systems'. 'CIM' is perhaps a special case in that it is used not only in the sense indicated here, but also as (i) the designation of a very comprehensive or 'large-scale' FMS, or the technology embodied by such a system; (ii) the name given the long-term goal or vision of the fully automated factory, or 'the factory of the future'; and (iii) a collective term attaching to a comprehensive, fully or highly computer-integrated production system or technology as well as to the various kinds of automated tools and sub-systems or constituent technologies which it comprises.

THE USER PERSPECTIVE

We shall now return to the basic FA problem, approaching it from the viewpoint of a user firm. The firm, it will be assumed, has reached the conclusion that some function or set of functions in its domain of production (whether in design or manufacturing or both) needs to be automated, or that it needs to be more extensively or effectively automated than is currently the case. As suggested earlier, the task to be addressed is that of producing and verifying the design of a technical system that will meet the company's needs. It will be left open here whether the problem is one of intra-activity, inter-activity, or inter-sphere automation, whether the technical system to be designed is a single machine or a set of interlinked machines, and whether the main challenge is posed by the hardware or software parts of system.

The basic problem may be broken down into a sequence of subproblems or tasks as shown in table 10-1. What can be said, with reference to that table, about the situation of the user firm? What is the range or set of problems that the firm will have to deal with, and what is the required technological competence base? To what extent will it have to rely on the problem-solving abilities of other FA actors? Some large and diversified user companies may in fact be capable of solving their basic FA problems by accomplishing all, or nearly all, of the above-listed tasks on the strength of their own R&D and engineering skills and material resources - even when the system in question is a large and complex one. More typically, however, the possibilities open to the user firm will be severely limited. As a rough generalization it may be said that active participation in the problem-solving process by the user firm is indispensible as far as the initial definition of the basic problem, or the specification of functional requirements (1), the justification of investment (7), and the operational evaluation of the installed equipment (10) are concerned. Such participation may be highly desirable, though not perhaps strictly necessary, in basic design (2), the specification of operating rules (6), the installation of the system (9), and the continued adaptive modification of the system design (11). The remaining elements of the problem-solving process will, as a rule, be handled by FA equipment suppliers, possibly with the complementary involvement of commercial purveyors of FA services, academic consultants, industry-association advisors, etc. The breadth and depth of the user firm's involvement will depend not only on the size and the technological and material resources of the firm, but also on the nature (e.g. the complexity and uniqueness) of its production processes. An FA system is always, at least to some degree, application-specific in that it must be adapted to the particular needs of the user. First, it must meet the specific requirements inherent in the production process or processes to be automated.

Secondly, it must mesh - through bilateral adjustments - with the work-organization as well as the higher-level managerial structures of the firm (cf. figure 10-1). Such adaptation evidently calls for user participation in some form. The need for such involvement tends to increase with the comprehensiveness and scale of integration of the FA system.

Table 10-1: Main tasks in FA problem solving

1. Specification of basic problem
2. Basic system design
3. Preliminary appraisal
4. Detailed design
5. Secondary appraisal
6. Specification of data or rules for systematic production and operation
7. Justification of investment[7]
8. Production of system (incl. components or sub-systems)
9. Installation (incl. on-site assembly) of system
10. Operational evaluation
11. Adaptive design modification

Despite the flexibility offered by modern FA systems, investment in manufacturing equipment entails a commitment on the part of the user firm to the future fabrication of certain products. It stands to reason that the appropriate planning perspective will vary with the size and "systemic scope" of the investment and that, e.g., the implementation of large-scale, plant-level FA systems require long-term views and commitments in the product area (Gershwin et al. 1986). By comparison, the time perspective which a user firm will apply to the production equipment itself, and to the question of technical change in that area, is generally more limited (Dallimonti 1986 and Gerwin 1988). Thus, faced with an automation problem, the firm will tend to look for solutions that are well-tried and proven, and that can be adapted to its particular needs without major delays or technical risks.[8]

Let us now consider the technology base that might enable a user firm to accomplish those tasks - indicated above - which it typically has to address; what kinds of

7 In a user perspective, justification of investment of course refers to the user firm's justification of the anticipated costs of FA system acquisition and implementation.

8 Again, our 'typical user' is a company which is unable or unwilling to undertake - by its own in-house efforts and by extensive coverage of subproblems - the development of new FA systems. As has already been pointed out, there are exceptions to the rule; these may well include firms pursuing long-term development goals.

techologies and forms of technical competence are required or clearly desirable? Two points should be stressed at the outset: First, the reader must bear in mind the fact that the boundaries between technical, economic, and social or organizational problems are not sharply drawn and that the tasks at hand are not all of them "technical" in the narrow ("physico-technical") sense of the word; on the contrary, problems of a socio-technical and techno-economic nature are likely to play a prominent role here, the more so the greater the perceived significance of the effective integration of FA systems, organizational structures, and business strategies. Second, it is important to realize that technical problem solving in the industrial enterprise is vitally dependent on the personal, experience-based insights and know-how accumulated by the employees. Such knowledge is, in part, tacit and incommunicable; it is, to a large extent, particularized, firm-specific (or product- or process-specific), and uncodified (Corbett 1989). Here, the significance of these cognitive forms as inputs to technical problem solving and the development of technological disciplines is readily acknowledged. The focus of the present chapter, however, is on cognitive entities or bodies of competence - "technologies," "technological disciplines," etc. - which are more broadly based and which have at least begun to show signs of institutionalization.

Figure 10-2 provides an overview of the fields of technical knowledge and competence that are clearly relevant to the problem-solving efforts of the user firm, particularly in the spheres of design and manufacturing (for a close-up of the right-hand rectangle, see figure 10-1; again, one must bear in mind that the various fields may be differently delimited and designated in different countries or cultural regions). The fields included are for most part well-established, both as educational fields and as general frameworks for R&D. However, certain qualifications should be made concerning some of the entities displayed. Thus, the area labelled "FA justification" has been included here as an example of a small, emergent specialty - set within the broader domains of industrial economics and industrial management - which has developed in response to a particular techno-economic problem, viz. the difficulties encountered when applying traditional return-on-investment calculations to modern FA equipment (Evans & Mantel 1985 and Kaplan 1986). The field named "Man-machine interface" may be characterized as a semi-institutionalized set of research and educational activities centred on the biological, psychological, and socio-organizational aspects of man-machine relations and the attendant issues of adaptive design (Kompass 1989). Software engineering, finally, is a field which appears in variant forms and loci of institutionalization, e.g. as a part or extension or complement of "computer and information sciences," "computer science," "software systems," "programming languages," etc.

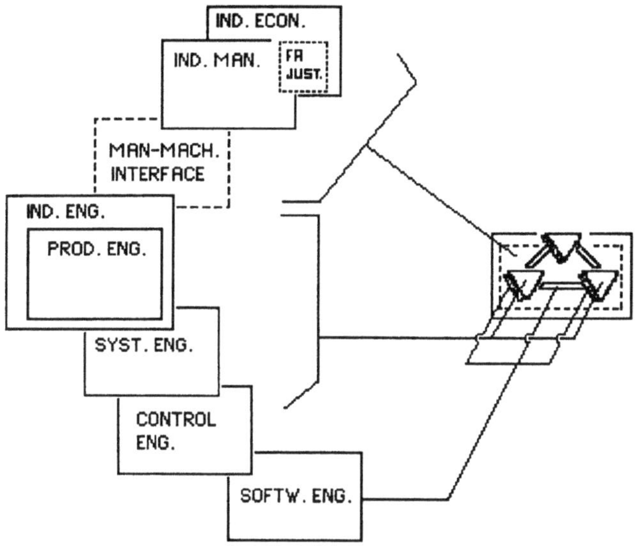

Figure 10-2: User-supportive fields

It is suggested that the fields shown in figure 10-2 are supportive of the user-firm efforts both in their role as generators of new knowledge (i.e. as R&D fields, academic and industrial) of potential relevance to the problems faced, and in their role as compilers, organizers, and disseminators of existing solutions and methods (i.e. as educational fields). The fields of industrial economics and management - on account of their concern with the economic, organizational, and managerial aspects of automation - and the partially crystallized field of man-machine interface pertain mainly to the task of investment justification and the attainment of optimal fit of FA systems and work-organizational and managerial frameworks. The fields of industrial or production engineering (which are not always distinguished), systems engineering, control engineering, and software engineering (which includes here the development or adaptation of integration or communication software as well as application software) are relevant to user participation in basic-problem specification, basic system design, specification of operating rules, installation, operational evaluation, and adaptive design modification. It would, of course, be unrealistic to expect the user firm to possess expert knowledge in precisely those subareas of the above-mentioned fields that match the particular requirements of its FA project; such expertise will often have to be provided by other actors, including FA equipment suppliers. The proposition advanced here is merely that possession of a basic in-house competence in these fields - mainly as part of the educational background of the personnel and as a complement to their firm-specific, process- and product-related expertise[9] - will expand the scope and enhance the effectiveness of the user firm's involvement in the problem-solving process (Warnecke 1988).

THE SUPPLIER PERSPECTIVE

We shall now broaden our view of problem solving in the FA domain by going beyond the role of the user to consider the contributions of the suppliers of automation equipment and associated services. As in the case of the user firm, the field composition of the supportive technology base will be examined. Similarly, one must begin by recognizing the fact that supplier-firm involvement in the problem-solving process *may* be very extensive; thus, turn-key solutions might be provided in response even to complex, high-level FA problems - notably by suppliers of the "systems integrator" species. However, it is again necessary to introduce certain limitations: First, it seems appropriate to focus attention on those elements or phases of the problem-solving process which have not been covered in the user context, more specifically the design (in the strict sense: numbers 2-6 in table 10-1), of FA systems, including machines, machine networks, and the requisite software structures; the process of design validation by user operation will not be considered. Secondly, the question of the production of FA equipment will not be addressed here as it would put us on a path of "regressive inquiry" (the automation of the production of automation equipment, etc.). Third, as regards the design and manufacturing of technically sophisticated components or subsystems - integrated circuits (ICs) or sensing devices, say - it will be assumed that these are tasks which fall largely in the domain of the supplier's subcontractors or sub-suppliers; though, again, it is conceded that some suppliers may well have advanced capabilities in these areas. It is worth pointing out, finally, that the shift from users to suppliers entails a change in the actor's perspective on technical development. The supplier will, as a rule, be more strongly compelled than the user to take a longer-term view of technical change in the FA domain, including the componentry area, simply because the emerging possibilities may have a direct impact - as opportunities or threats - on the company's products.

Figure 10-3 presents a picture of the principal fields that enter into the technology base of the supplier industry. Apart from the "core" fields of the right-hand column, only the major engineering fields and generic design technologies supporting the problem-solving efforts of this actor category are displayed. As in the user chart, intra-

---

9 Note, e.g., the benefits which a user firm might derive - in this context - from product-related strenghts in such fields as mechanical, electrical, and electronics engineering.

management-sphere automation is not considered. Six core fields, central to FA suppliers, are distinguished: CAD (including the adjunct of CADCAM, not shown here), NC/CNC, robotics, AMH, FMS, and CIM - i.e. those parts of the technology base which directly underly the design of CAD equipment and systems, NC/CNC machine tools, industrial robots, automated materials handling equipment and systems, FMSs, and CIM systems, respectively. The first three can be described as fields of technical specialization which have reached a relatively advanced stage of institutionalization; also to be noted here is the fact that major subareas and product categories of these fields are now technologically mature. The fields of AMH, FMS, and CIM, on the other hand, must be seen rather as semi-crystallized or emergent institutional entities (indicated by broken box lines). AMH is a heterogeneous, loosely integrated field, functionally defined by its concern with the design of various devices for the "handling" or transfer of materials, parts, and products. As a provider of mechanical linkage, AMH plays a key role in the construction of manufacturing-sphere machine networks. The design of networks is, of course, the focal task in the FMS and CIM fields. In the case of FMS, signs of institutionalization are readily observable and seem to be increasing. The situation of the CIM field is far more diffuse, reflecting persistent disagreements on the very nature, scope, and purpose of the enterprise. As a brief characterization of the core fields, one can say that NC/CNC, robotics, AMH, and CAD are machine- or tool-centered, hardware-cum-software, fields (with the added qualification that systemic, or networking, aspects are not entirely absent and that they are most salient in AMH and CAD). FMS and CIM, however, are clearly oriented towards network building and systems integration; key problems include the determination of optimal configurations and operating rules, and the development of coordination and integration software.

Any given core field comprises a competence mix which is grounded in a set of technical disciplines - disciplines which, in a supplier perspective, are of particular importance as educational fields, although their R&D functions are not to be slighted.[10] In figure 10-3, seven supplier-supportive disciplines are displayed. In terms of the classification outlined in section 10-2, five of them are identifiable as broad "engineering fields," viz. electronics, electrical, mechanical, software, and production engineering (though the latter also posses some traits of a "functionally oriented technology"). The breadth of these fields is reflected in the fact that they typically comprise a variety of institutionally distinguishable specialties focusing on particular sub-classes of technical systems and applications; this is illustrated most clearly by the old and well-established field of mechanical engineering. Of the remaining fields, one,

10 The term 'discipline' is used here primarily to denote a field which is academically well established and serves a significant educational function.

control engineering, is to be seen primarily as a "functionally oriented technology" (although some features of a "generic design technology" are also discernible). The field of systems engineering, finally, is a prime example of a "generic design technology." The position of the production engineering rectangle calls for a comment. As a field concerned with the design of industrial production systems - large and small, automated and non-automated - and the planning and operational aspects of such systems, it is depicted here as cutting across the other disciplines (Warnecke 1988). This is not to suggest, however, that the latter have equal bearing on the principal tasks of production engineering; witness the comparatively broad relevance of systems and control engineering. Figure 10-3 provides a sketch map of the disciplinary sub-structure of the core fields, focusing on the major "relevance links." The connecting lines indicate, e.g., that a mechanical engineering competence is a vital asset in the design of NC/CNC machines, industrial robots, and AMH equipment, and that a systems engineering background is particularly relevant to the development of FMSs and CIM networks (Sage 1987). Finally, a word of caution on the question of selective coverage is appropriate here. In deciding whether to include or exclude certain fields, some arbitrariness seems unavoidable. Artificial intelligence (AI) is a case in point. The AI field might have been placed among the supplier-supportive technologies of figure 10-3 on the strength of its bearing, most notably, on developments in the core fields of robotics and CAD. Its absence from the chart reflects the judgement that AI knowledge and design skills have not yet become prominent as ingredients in the problem-solving competence of the typical FA supplier.

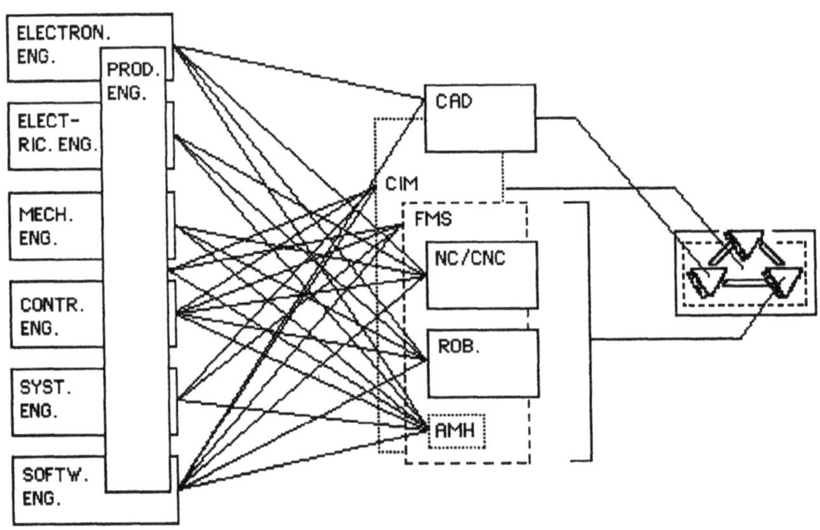

Figure 10-3: Supplier-supportive fields

THE SUBSUPPLIER PERSPECTIVE

The third step in the mapping of the automation cluster consists in a further expansion of the field of vision, now to take in a set of technologies pertaining to certain key components and basic functions of FA systems. Recall here the assumption, introduced earlier, that equipment suppliers will not, as a rule, undertake the design and manufacturing of all important and technically demanding components of their products. Recall also the main functions performed or facilitated by design- and manufacturing-sphere automation equipment, i.e. the drafting, analysis, and appraisal of designs; the handling, forming, processing, assembly, inspection, and testing of materials, parts or products; and the linking and coordination (intra- or inter-sphere) of different production processes. The effective performance of these practical functions depends in turn on the performance of a limited set of basic, information-related functions, viz. (i) the generation of information by sensing and measurement; (ii) the processing of information by arithmetic and logic operations; (iii) the storing of information, including input, holding, and retrieval, by electronic and magnetic media and processes; and (iv) the transfer of information by signal transmission. The crucial link between these basic component functions and the practical, end-use functions mentioned before is the function of control - a function which is fundamental to all modern manufacturing-sphere automation efforts and to the real-time guidance and coordination of inter-sphere production systems.[11]

The performance of these basic component and intermediate-level functions is accomplished by a large variety of devices and systems, including the following: sensors, sensing systems, and metering devices; logic ICs (integrated circuits), controllers, and computers; memory ICs, tapes and disks; transmitters, transmission media, and detectors or receivers; and the basic software or "operating systems" which many of them require. The nature and extent of the FA-system suppliers' - as distinct from the subcontractors or suppliers' - involvement in the development of these components vary considerably. The likelihood that suppliers will participate in such efforts increases, of course, with the need for components and sub-systems that are adapted to the specific requirements of factory automation. Thus, suppliers will have to consider, e.g., the relative technical and economic merits of standard versus special-purpose computers (such as CAD workstations), standard versus application-specific ICs (ASICs), or standard versus application-specific sensing devices, etc. In the following, the focus of attention will be on important FA componentry which is developing

11 Cf. Bell's (1972) suggestion that automation technology can be analyzed into three basic functional components: the *transformation* of material inputs; the *transfer* of materials, parts, and products; and the *control* of the various transformation and transfer processes.

through R&D processes that are largely driven and shaped by other actor categories than that of FA-systems suppliers, although the latter might participate in, and indeed significantly contribute to, the problem-solving efforts.[12]

Figure 10-4 provides an overview of the major component technologies of factory automation. An attempt has been made to cover both the principal fields of component and sub-system engineering and the broader technologies encompassing or underpinning these specialties. The right-hand part of the chart is a graphic illustration of the preceding remarks on the key role in automation of certain fundamental information-handling functions. Note the focal position of control as a "mediator" providing extensive linkage with the core fields of factory automation. The direct connecting lines which have been drawn between some of the primary functions and the core fields are to indicate the special significance of sensing and measurement in robotics and NC/CNC, and of signal transmission (communication) in the systems-integrating fields of FMS and CIM; the CAD connection underscores the particular importance, outside the control context, of information processing in the automation of design activities.

The remaining portion of the chart comprises a large cluster, including subclusters, of disciplines and specialties exhibiting various forms and degrees of institutionalization (not indicated here). The cluster extends from the technologies concerned with the design and manufacturing of those devices and systems which perform the information-handling functions listed on the right - i.e. metering devices, sensors and sensing systems, communication links, computers, and control systems; ICs are, of course, key components in a variety of systems, although the chart puts the emphasis on their role in computers and communication equipment - into the more fundamental, underpinning fields on the left. The picture of the component-technology cluster is largely self-explanatory and will not be fully verbalized here. Again, the mapping is plainly selective: only technologies and linkages judged to be of major importance are displayed. It is worth noting that some examples of so-called "functionally oriented technologies" and "basic technologies" (see section 10-2) are to be found among the fields now covered. Thus, the field of sensor technology on closer inspection presents itself as a large and highly diversified assemblage of technical specialties focusing on devices that operate on radically different principles and perform variant functions - an assemblage which is held together by a shared concern with the general function of

12 This is not to say that the situation is always clear-cut. Consider, e.g., the role of some workstation makers as CAD-system suppliers, or the recent emergence of large computer firms as CIM-system suppliers. On the other side of the line, one may point, e.g., to the prominent role of robot makers in the development of control and sensing systems.

"sensing." Communications engineering is another example of a functionally oriented technology, in this case a technology concerned with the design of various kinds of systems, including wire-based, radio, and optical links, which perform the basic function of communication. Similarly, the field of artificial intelligence (AI), which ties into the cluster partly by way of software engineering, can be seen as a loose agglomeration of research fields based upon the common functional denominator of imparting intelligence-like capabilities to machines. While opinions may legitimately differ on the precise aims and boundaries of the entities referred to as "semiconductor technology," "laser technology," and "fibre optics," the suggestion here is that these fields be understood as basic technologies, or technologies centered on the control and practical exploitation of particular natural processes. In this case, the primary objects of exploration and exploitation are the special conduction properties of semiconductor materials, the phenomenon of the stimulated emission of light, and the process of total internal reflection, respectively. The three fields are all "generic" in the sense of not being *a priori* confined to any particular application or functional area. On the subject of laser applications, it should be pointed out that there are many different ways in which lasers can enter into factory automation; one noteworthy path, which is not marked in figure 10-4, is via laser-based materials processing.

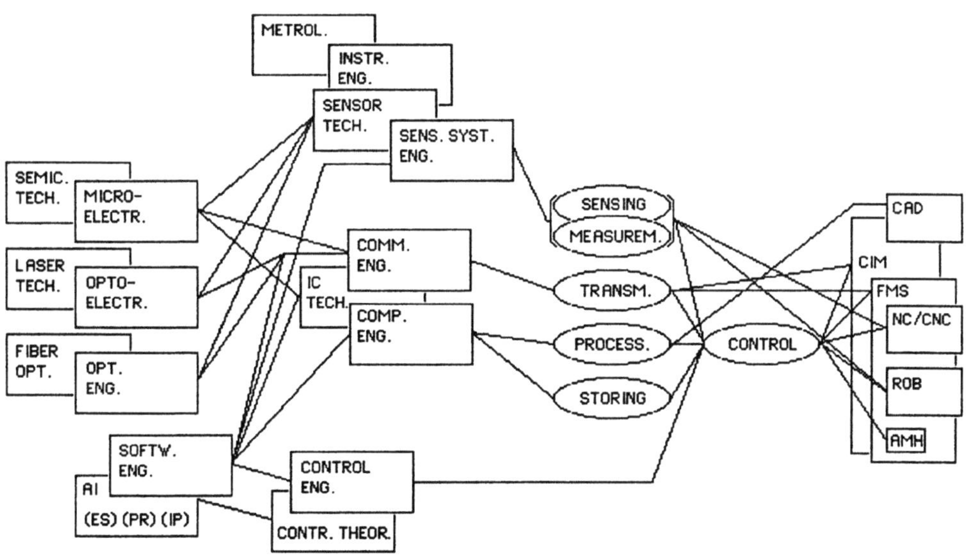

Figure 10-4: Major component technologies and related fields (ES: Expert Systems; PR: Pattern Recognition; IP: Image Processing)

Like the preceding steps, the extension of the survey into the domain of component technology has two major consequences. First, it entails a very considerable expansion of the set of actors contributing to the development of FA equipment. There is a substantial increase not only in the number of relevant actors, but also in the range and diversity of goals and interests, organizational types, and institutional settings. Secondly, it entails a further widening of the technical-change perspective. Thus, researchers investigating fundamental (though FA-related) problems in, e.g., laser technology or AI must accept that the practical, end-use impact of their work, if successful, may lie in a distant future. The appropriateness of longer time scales also applies, of course, to the technology analyst seeking to understand the dynamics of cross-cluster linkages, from the problem-solving efforts undertaken in the scientifico- technological borderlands on the left-hand side of the chart to the solutions which emerge from the FA core fields on the right and which are subsequently modified and perfected in the user context.

THE OVERALL VIEW

As the fourth and final step in the mapping exercise, the partial charts of figures 10-2 to 10-4 will now be assembled into a general picture of the FA technology cluster. The field of vision will be slightly expanded so as to cover also some of the supporting sciences. The result is presented in figure 10-5. By way of summary, the chart provides an overall view of the composition and structure of the cluster. Apart from the set of FA core fields (I), six major subclusters are distinguished:

- a socio-technical cluster (II) pertaining to a wide range of problems in conjunction with the acquisition, user adaptation, and operation of FA equipment;
- a general engineering cluster (III) with broad ties to the technical competence base of both suppliers and users;
- a sensing-metering cluster (IV) of particular relevance to the control (including adaptive control) problems of factory automation;
- an "Information Technology cluster" (V) bearing on the entire spectrum of information-handling problems arising in the FA context;
- a physical-science based technological support cluster (VI) of direct relevance to the hardware-centered problems addressed by clusters IV and V; and
- a formal-science based cluster (VII) providing extensive software and generic-design support to problem solving in all systems, sub-systems, and component areas of factory automation.

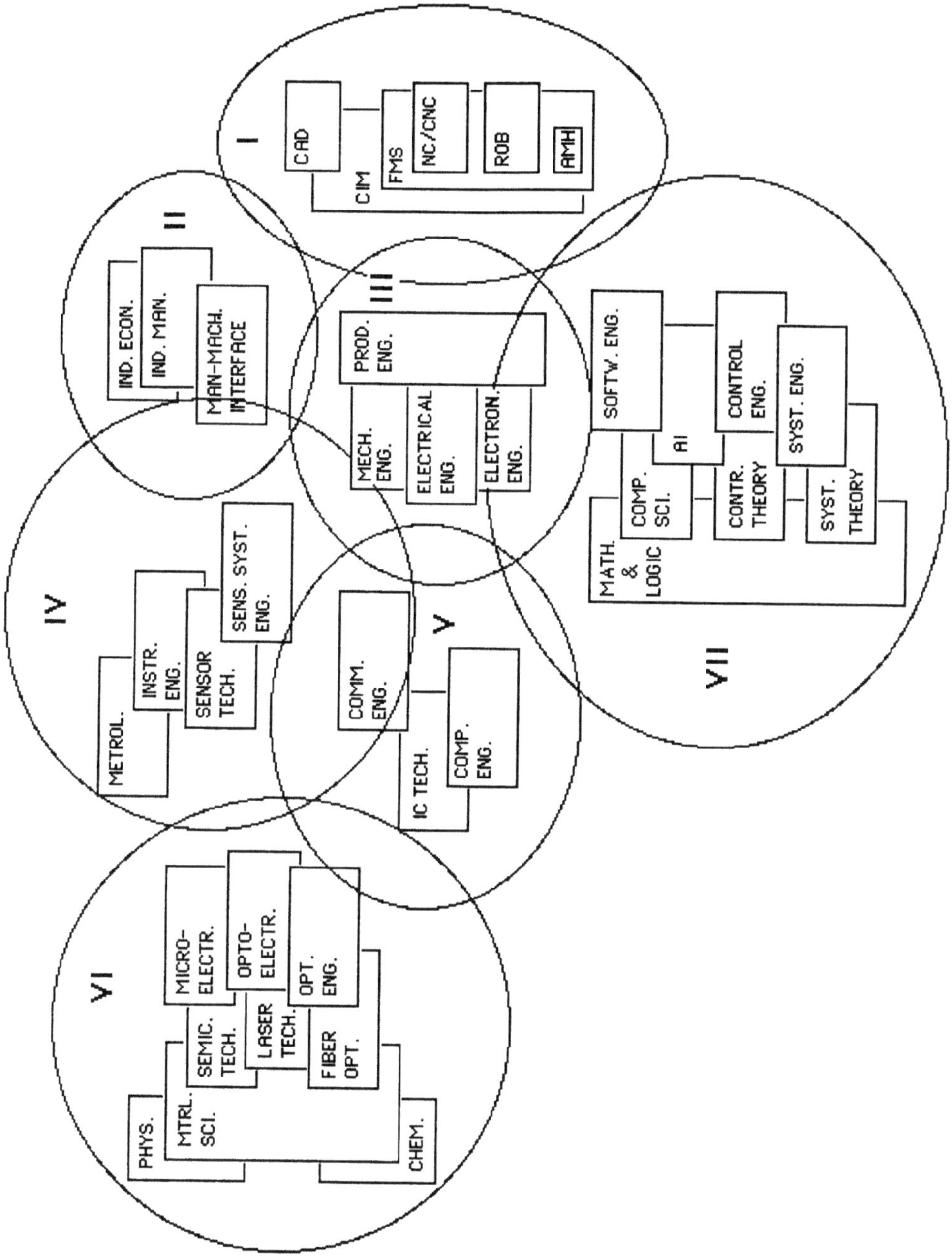

Figure 10-5: Overall view of the FA cluster

The general structure of the "supercluster" reflects the analytical line of approach followed in this paper, beginning with the user- and supplier-supportive clusters (I-III and partly VII) on the right-hand side of the chart, proceeding to the central subsupplier clusters (IV and V), and then moving further left and downwards into the underpinning science and basic-technology clusters (VI and VII). As indicated in the preceding analysis the component clusters of the overall map are variously interlinked through relations of dependence, complementarity, and support in problem solving. The overlaps drawn in Figure 5 serve as graphic, summary illustrations of such linkages. It is clear, however, that the large and complex network of actual cross-cluster connections cannot be adequately portrayed by a two-dimensional representation.

10.4 CONCLUDING REMARKS

It seems appropriate at this stage to return to the starting point, as it were, and to consider once more the general character and purpose of the mapping exercise. The following comments are offered by way of review and conclusion:

- First, as regards the analogy with ordinary cartography, we may note that the technology sketch maps presented in this chapter are plainly selective, the focus being on larger, fully or partly institutionalized fields of major importance and on the essential linkages among them; and that the maps are, inevitably, shaped or distorted by the "projection" employed, the governing perspective being, in this case, that of industry and the practical problems of industrial firms.
- Second, it should be pointed out, as a partial corrective to the above, that the progessive broadening of the problem-solving perspective and the exploration of various supporting disciplines and specialties - many of which have a strong academic base - ensures that a larger and more heterogeneous collectivity of actors is brought into play and that the views and interests of industry do not rule supreme.
- Third, it bears repeating that a major aim of the study has been to elucidate the composition and structure - rather than the dynamics - of the FA technology cluster. Accordingly, the technology maps developed on these pages provide a picture of the cluster that is essentially static. At the same time it is evident that dynamic relationships do in fact enter, at least in the form of implicit assumptions or claims, into the construction of these maps (e.g. the "pull" exerted by assigned or perceived problems and potential applications, and the "push" generated by new solutions and enabling competencies). The question of technological change in factory automation

is a complex one which cannot be explored in this brief chapter, though it is, of course, an important subject of analysis within the purview of the STS project.

- Finally, it must be emphasized that this mapping attempt has not been undertaken for its own sake. Rather, the sketch maps are tools that will be used in our investigation of Sweden's FA technology system or R&D infrastructure. They will serve, e.g., as reference frames or orientation aids in interview studies in which respondents will be asked to indicate the position and role, in the broader FA domain, of their technological specialty or organizational unit, to outline their networks of contacts and collaboration, to identify weak links and missing elements, etc. There is also a possibility that the qualitatively derived cluster charts presented here will be complemented by, and checked against, FA technology maps based on bibliometric surveys.

REFERENCES

Bell, R. M. (1972), 'Changing Technology and Manpower Requirements in the Engineering Industry', Sussex UP, London.

Bristol, E. H. (1986), 'An Industrial Point of View on Control Teaching and Theory', *IEEE Control Systems Magazine*, Feb. 1986.

Chakrabarti, A. K., Feineman, S. and Fuentevilla, W. (1983), 'Characteristics of Sources, Channels, and Contents for Scientific and Technical Information Systems in Industrial R and D', *IEEE Transactions on Engineering Management*, May 1983.

Corbett, J. M. (1989), 'Automate or Innervate? The Role of Knowledge in Advanced Manufacturing Systems', *AI & Society*, July-Sept. 1989.

Dallimonti, R. (1986), 'The Horizon Plant', *Control Engineering*, May 1986.

Evans, J. R. and Mantel, S. J. (1985), 'A new approach to the evaluation of process innovations'. *Technovation*, Oct. 1985.

Gardiner, P. & Rothwell, R. (1989), 'Design management strategies', in Dodgson, M. (ed.), *Technology Strategy and the Firm: management and public policy*, Longman, Harlow.

Gershwin, S. B., Hildebrant, R. R., Suri, R. and Mitter, S. K. (1986), 'A Control Perspective on Recent Trends in Manufacturing Systems', *IEEE Control Systems Magazine*, April 1986.

Gerwin, D. (1988), 'A Theory of Innovation Processes for Computer-Aided Manufacturing Technology', *IEEE Transactions on Engineering Management*, May 1988.

Gibbons, M. & Johnston, R. (1974), 'The roles of science in technological innovation', *Research Policy*, Nov. 1974.

Granberg, A. (1987a), 'A Conceptual Framework for the Study of Generic Fields of Technology', Research Policy Institute, Lund.

Granberg, A. (1987b), 'The Dynamics of Science-Based Technologies', Research Policy Institute, Lund.

Granberg, A. and Stankiewicz, R. (1981), 'The Development of 'Generic Technologies' - the Cognitive Aspects', Research Policy Insitute, Lund.

Hunt, V. D. (1987), *Dictionary of Advanced Manufacturing Technology*, Elsevier, New York.

Kaplan, R. S. (1989), 'Management Accounting for Advanced Technological Environments', *Science*, 25 Aug. 1989.

Kaplinsky, R. (1984), *Automation - The Technology and Society*, Longman, Harlow, Essex.

Kompass, E. J. (1989), 'Man-Machine Interfaces: The Evolution to Computer Control', *Control Engineering*, July 1989.

Kunz, W. and Rittel, H. (1970), 'Systemanalyse eines Forschungsprozesses', in Kunz, W. and Rittel, H. (eds.), *Systemanalyse und Informationsverarbeitung in der Forschung*, R. Oldenbourg Verlag, München-Wien.

Martino, J. P. (1983), *Technological Forecasting for Decision Making*. North Holland, Elsevier, New York.

McGinn, R. E. (1978), 'What is Technology?' in Durbin, P. T. (ed.): *Research in Philosophy and Technology*. Vol. 1, JAI Press, Greenwich, Conn.

Mitcham, C. (1978), 'Types of Technology', in Durbin P. T. (ed.), *Research in Philosophy and Technology*, Vol. 1, JAI Press, Greenwich, Conn.

Ranta, J. (1989), 'The Impact of Electronics and Information Technology of the Future Trends and Applications of CIM Technologies', *Technological Forecasting and Social Change*, 35, pp. 231-260.

Sage, A. P. (1987), 'Knowledge Transfer: An Innovative Role for Information Engineering Education', *IEEE Transactions on Systems, Man, and Cybernetics*, Sept./Oct. 1987.

van Wyk, R. J. (1984), 'Panoramic scanning and the technological environment, *Technovation*, May 1984.

Warnecke, G. (1988), 'Produktionsfaktor Wissen', *VDI-Z*, No. 11.

PART IV

DISCUSSION

11 ANALYSING LINKS BETWEEN SCIENCE, TECHNOLOGY AND INNOVATION - STATE OF THE ART IN SCIENCE RESEARCH AND IN INNOVATION RESEARCH

Knut Koschatzky and Hariolf Grupp

11.1 WHY THIS CHAPTER?

Conferences or seminars are used to combine three major elements: the presentation of the papers, the discussion and more formal or informal talks with colleagues on current research, and other matter. All three elements together determine the quality of a conference and its usefulness in the participants' minds. As this book presents the essence of the conference to a greater audience, one faces the problem that it is indeed very easy to print or summarize the contributions, but it is very difficult to paint a complete picture of what was really going on with respect to at least the three elements mentioned above.

This chapter tries to make a compromise. It will provide information additional to the previous ten chapters by summarizing the discussion of the papers presented within the three sections of the proceedings. It also reviews the contributions with no paper. In many cases, discussions following formal presentations have a clarifying effect, reveal different opinions and give the opportunity to present views to problems or matters which the author only touched and could not discuss in detail. Additionally, by writing down the discussion, the distance between the reader on the one hand and the conference on the other can be bridged. He or she can realize that nothing artificially has happened, but a real conference with its pros and cons, different opinions and theories took place. The discussion might also offer insights into the generation of new research problems and the birth of new theories and methodology. Together with the papers it also gives an impression on the state of the art in a specific field - in this case in combined science and innovation research.

11.2 CONCEPTS AND NEW METHODOLOGY

In chapter 2 "Technology as an Autonomous Socio-Cognitive System" *Rikard Stankiewicz* presents conceptual thoughts about the organization of technological R&D by using the framework of social systems analysis. Since the term "autonomy" plays an important role in Stankiewicz's concept, several subsequent questions related to that issue. It was not totally clear what "autonomy" should really mean as there might be a distinction between cognitive and concrete systems of technology. According to Stankiewicz, the socio-cognitive system of technology (SCST) is autonomous to the extent to which it is self-organizing and capable of constructing and maintaining effective boundaries (chapter 2, section 2.1). It was doubted that the SCST can be totally autonomous. This was also realised by Stankiewicz because he identified factors which weaken or slow down the tendency towards societal differentiation and autonomization of the SCST, e. g. the cultural resistance, the particularism and contextuality of technological praxis, and the tacit character of much of technical knowledge. Stankiewicz pointed out that the term is indeed a tricky one. In his opinion, distinctions have to be made between concrete and cognitive systems within the observable clusters of science and technology, although technical and social elements can not be separated since they are linked closely. Stankiewicz also admitted that it might not be possible to speak of total autonomy between the two systems, but there is always a degree of autonomy.

Another bundle of questions relevant to the topic of the conference dealt with the boundary between science and technology, as both are cognitive systems, and with the interactions of both systems. As a matter of fact, people move between science and technology and also between universities and companies. But, still, up to Stankiewicz there are many differences between science and technology as they are two different systems. But on the other hand, not all what is done in universities is pure science. Universities are the best homes for technologies. They educate and send people out to the companies where the former "scientists" become part of the technological community. Collaborative research in Japan is a form of intensifying the linkages between companies and universities by temporarily sending young industrial researchers as special students to university research groups. This linkages should improve the mobility of people who stick their whole life to one company.

Stankiewicz's last remark opened the floor for questions concerning the role of basic research in the technological system. It is obvious that today's high tec is generated mainly in industry, not in universities primarily. Therefore, sometimes it makes no sense to hire people from other companies with at least the same level of technical knowledge instead of university graduates. Joint research companies may provide training facilities, a service which would be much more difficult, maybe impossible, for universities. But, on the other hand, the transfer of industrial people for doctorate studies to universities may be a more effective way for industry to cooperate with universities. Overall, Stankiewicz pointed out that the universities are falling behind industry in many cases. But this observation raises the question again if the role of basic research is changing and what basic research really is. According to Stankiewicz's concept basic research is cognitive and therefore immanent; it is not negotiable.

Vololona Rabeharisoa presents in chapter 3 a scientometric study on the links between science and technology by analyzing articles published by inventors. She has focussed on French fuel cell patents between 1960 and 1975 and on articles published by their inventors. Furthermore she looked for articles cited by patents between 1976 and 1987 which enabled her to differentiate this "cited research" from the research "produced" by the inventors before 1975.

Because of the methodological basis of the study the first part of the discussion dealt with methodology. It might be important, also on the background of the conference's topic, to draw a line between scientific and technical papers. Technical papers had been defined in Rabeharisoa's study by describing technical devices and by a list of special (technical) reviews. However, it was impossible to differentiate scientific from technological authors, since there was too much differentiation in the oeuvre of one and the same author. A next question asked for the type of journals used in the study. Sometimes companies do not want to apply for a patent, but also do not want that the invention is patented by others. Therefore they publish an article in a special (company) journal which for the company has the same disclosure quality as a patent. These kinds of journals had also been included in the analysis (e. g. a special journal from the Compagnie Française du Pétrole), but were counted as regular articles. It was asked how many of the inventors were authors. In Rabeharisoa's concept all inventors were authors, because this was the basis of the study. On the average an inventor published four to five papers a year.

Following methodological questions, emphasis was laid on the linkages between science and technology. According to Albrecht it was not quite clear what had been science, technology or industrial development in the field of fuel cells. A full insight could only be achieved when scientific papers and patents are put together. The basic structures were visible in Rabeharisoa's analysis, but the picture would appear to be clearer when both patents and papers had been analysed. Based on the results of own studies, this assessment was supported by van Raan. To gain an insight view into the life cycle of a technology, more than one indicator must be used. Patents and publications are indicators for different stages in the technical development process. Rabeharisoa confirmed these statements. Bibliometric analysis alone can not give indications of success or failure.

One participant wanted to know if the study's data proof also for an analysis of science generation by technology. Rabeharisoa found out that sometimes scientific papers cited patents. But the problem she was confronted with was that the field of fuel cells was situated in between science and technology. In the period of study it was not possible to distinguish if and what was still science and what was already technology. Therefore, Stenberg raised the question that it might be interesting to know how the researchers view their own situation. According to Stankiewicz's approach they should have the possibility to position themselves, either to science or technology or to the socio-cognitive system of technology. In many cases they will participate in both concepts, in basic science and in the development of devices. Stankiewicz gave some comments on the development break in 1975 which can be seen very clearly in Rabeharisoa's data. The same happened with single cell proteins (SCP). After some optimistic device oriented developments technical problems occurred which led to a slowdown in activity. But a few years later some basic aspects were picked up, for instance in metabolism of bacteria. As a result, new people and institutions entered that field. Although the short-term commercial prospects diminished, the R&D activities continued. On the basis of these experiences one can conclude that an insight view by only analysing the publications of the initial inventors neglects the activities stimulated from outside. Therefore some model of process, reflecting on optimism and pessimism waves would help in data interpretation. In her answer Rabeharisoa agreed to the last comments. It is not only important to look at one field, but on specific configurations within this field including the connections of the researchers, i. e. who works with whom, who enters, who leaves the field. That would give a clearer picture on the dynamics of technological development.

In chapter 4 on "Perceptions of Scientification of Innovation as Measured by Referencing between Patents and Papers: Dynamics in Science-Based Fields of Technology", *Hariolf Grupp* presents the results of a project on "Performance Indicators for Science-Intensive Technologies" commissioned by the German Ministry for Research and Technology (BMFT). In his contribution Grupp concentrated on the results FhG-ISI derived by using European patent documents, whereas a parallel study was done by Olivastro and Narin on the basis of United States patents. For that reason there were only some short questions of understanding after Grupp's presentation and both papers were jointly discussed after Francis Narin's contribution.

Francis Narin's results of a study prepared by Dominic Olivastro and himself for BMFT on the "Definition of Fields of Close Cooperation between Science and Technology" is published elsewhere (Olivastro & Narin 1989) and not included in this book as an independent chapter in full length. The following *summary* draws on this publication as well as on some selected results contained in a series of figures distributed to the participants of the conference.

This study was part of a project on "Performance Indicators for Science-Intensive Technologies" commissioned by BMFT and jointly worked up by CHI Research Inc. and FhG-ISI (see chapter 4). CHI Research was responsable to collect references in US patent documents to the non-patent literature (NPL) by using the SCI database, whereas FhG-ISI was using European patent documents. Both studies draw on the same methodology, namely time series based on priority dates (and neither on patent nor on grant or application dates) and a 28 fields breakdown of technology based on the International Patent Classification and not the United States Patent Office Classification.

The CHI study consists of two major parts: In the first part, all patents generated in the US patent system which were applied for between the years 1975 and 1984 were analysed. The analysis characterized the science base that is supporting the technology (patents) of West Germany and its four closest competitors, France, the UK, Japan and the US. For this part of the study, the number of "Other References Cited" (out of which approximately 60 per cent are scientific journals) on the front pages of the patents was used as indicator for the science base. In the second part the organizational structure of the science base had to be analysed based on patents issued in one single year, 1987, and assigned to a set of selected companies. FhG-ISI and BMFT picked out 53 West German companies representing a cross-section of chemical, electronic and mechanical industries. CHI Research then developed lists of comparable US and Japanese companies, based on

- presence in the same industry,
- closeness of revenues, using 1987 sales figures
- correlation in patent profile across 42 Standard Industrial Classification (SIC) areas,
- numbers of patents.

As a result a list of 50 Japanese and 53 US companies had been derived.

Olivastro and Narin came out with the following results.

1. The dependence on science within any specified technology field is about the same for all the investigated countries, and is determined mostly by the field itself. The most science-intensive technology is "Genetics & Pharmacy" with two to three NPL-citations per patent in very recent papers. The next most science-intensive fields are chemicals and electronics.
2. Scientific results cited by the electronic patents are the youngest of all the fields, typically only two to three years old. In contrast, scientific roots cited in chemistry are often five to ten years old. However, this data may be somewhat complicated by the fact that the US patent system has been slow in granting patents in the highly science-intensive fields, so that the data are non randomly incomplete, but suffer from a notable lag in approval of patents in the highly science-intensive areas. One important reason for these granting obstacles is that the United States Patent and Trademark Office has had particular difficulty in finding qualified patent examiners in rapidly growing areas such as biotechnology and information processing.
3. The national patterns of science intensity are reflective largely of the national patterns of patenting. West Germany is heavily patenting in the chemical areas with high but old science intensity, and in areas of relatively low science intensity (e. g. mechanical areas). As a result she has an overall pattern of low or slow science linkage. Japan, which is heavily patenting in electronics, shows relatively low science dependence, but is dependent on recent basic results. The US, on the other hand, show a higher science intensity because of her very strong position and very scientific patents in "Genetics & Pharmacy" but one that is not as linked to current science as the Japanese.
4. Very little of the scientific base cited in West German patents is published in German journals, especially outside of chemistry.
5. The largest growth in West German patenting was in the fields of medium science intensity. In contrast, the strongest areas for the UK and US are in the more highly science-intensive biotechnology areas.
6. The patents of the West German companies in all three industries tend to cite science as often as those of Japanese companies, but less than those of US companies.
7. The science base for each country has a strong national component in a way that preferentially the own country's publications are used. In the electronic industry, however, the part of science that is most likely to drive technology primarily originates in the US and Japan. This is especially true for German technology.
8. One intriguing observation is that the technology-science linkages seem to be strongest in the areas where the countries are technologically strongest. For example,

in electronics there is a strong component of citing Japanese research papers by all inventors. In biomedicine there is a significant citing to US papers whereas the same is true to German papers in chemical patents. This component hints that there is an association between the technological and the scientific strengths of the countries.

Thus, the overall picture of West German technology and its science dependence that emerges from this project is, according to Narin, that the science base of Germany with its traditional strength in chemistry, is supporting the traditional German technological strength in chemistry. On the other hand, the relative weakness in both biotechnology and electronics appears to be associated with weak linkages in those areas of West German technology to underlying West German science.

The discussion of both Grupp's (see chapter 4) and Narin's (see above) contributions was opened by Anthony van Raan. He gave some additional remarks on Narin's results. According to his own study on chemical technology in which a science citation intensity matrix was developed (van Vianen, Moed & van Raan 1990), the matrix showed an extremely skewed distribution concerning patents with NPL references. Up to 80-90 per cent of the chemical patents did not have NPL references at all, the other 10-20 per cent did contain a lot of references which led to the uneven statistical distribution. Results parallel to what Narin found out had been derived for the age distribution of NPL references. US and Japanese references were relatively young in contrast to older ones from FRG and also the UK. These differences relate to the type of the cited journals. Youngest references originated from journals like Nature or Science whereas older references came out of real chemistry journals. Since Nature and Science are journals with the purpose of quickly reporting research evidences, articles in chemical (and other) journals face some time-lag because of the fundamentality of the research articles. Therefore older articles do not automatically point to a less alert acquisition of scientific results.

Another remark concerned the average number of references per patent. Studies reveal that non-English speaking countries show statistically lower NPL ratios than the English speaking world. For example one of the lower NPL ratios can be observed for Japan. These general biases between countries have to be considered for data interpretation.

Grupp gave a comment on the basis of the data Narin used in his analysis. One has to distinguish between pure science development (fundamental research) on the one hand and industry related developments (strategic research) on the other. Dealing with patents and with scientific citations in patents one faces the problem that the cited references in most cases have a bias towards technical application. This would mean that the postulated science links are merely links to applied, technical science and not to the one in its puristic form. For that reason and on the background of the assumption which defines the science base by references to scientific journals one has to take into account that also R&D work done by industry and published in scientific journals is science according to the definition used in the study.

Narin agreed with van Raan's remarks. According to own studies about 25 per cent of chemical patents from the years 1987 and 1988 did contain NPL references. On the other hand, a validation study from the early 1980's tested and verified the assumption that citations from a patent to the scientific literature are legitimate indicators of the science used to develop the technology present in the patent (Carpenter & Narin 1983). In the study CHI has determined that, in general, at least half of these references are to scientific publications.

Another methodological problem was raised by Schmoch. He was questioning if the matching of the control group of US and Japanese companies could have been done by just analysing the patent profile of the company as one unit. All patents of each company had been included under one label which might lead to uncertain results in the case of big companies active in various technical fields. Therefore a selection according to patent classes would have produced a better list of comparable companies. According to Narin they were well aware of that problem, but decided for the labelling under company names for time and capacity reasons, although they knew that it would have been better to apply both selection procedures.

A next group of questions dealt with the obvious micro-electronics gap Narin's and also Grupp's study reveal for Germany and for Europe. It is true that not only in the FRG, but also in the UK or France (here except some peaks in telecommunication-related electronics) science and technology in the electronics field are worse compared to Japan or the US. One hypothesis for the German science base weakness in electronics might be that the German industry does not rely so much on the cooperation of universities than in the countries which are technologically stronger in electronics. Narin pointed out that the weakness is a European phenomenon, because not only Germany, but whole Europe, as one of the three major competitive partners, is weak. It was argued that citation analysis can reveal the linkages between science and technology, but has difficulties in the explanation of the underlying structures. Whereas in the US there is a close connection between universities and industry, the weaker links in Europe and in the FRG might be one reason for little scientific supply and industrial demand in the micro-electronics field and also for a weaker technological performance. According to Stankiewicz, specific studies have to be made to trace and differentiate the relations between science and science-based technological developments. New indicators of coupling are needed for a clear analysis of this phenomenon, but they are unfortunately not yet developed.

Grupp finally gave attention to the fact that the data reflect the situation in the mid 80's. In the meantime, German and European policymakers and companies got aware of the problem. As a major European joint initiative, the JESSI programme (Joint European Submicron Silicon Initiative) was formulated as a European answer to the development of a 64 MB chip in the US and Japan and for closing the micro-electronics gap. Within this programme close cooperations between industry and non-industry researchers are planned. One large joint research lab of the Fraunhofer society is in the planning stage at Itzehoe, close to Hamburg. More than 400 researchers from university, Fraunhofer Society and industry will work together. With this initiative the observed situation might change during the coming years. But this joint institute, again, is running into coordination problems between "science" and "technology" actors in Germany and will more likely not be realized.

11.3 CHANGES IN NATIONAL SCIENCE AND INNOVATION SYSTEMS

The first paper of this section was presented by *Edo Albrecht* (see chapter 5). He discussed the structure and dynamics of the former GDR research system with respect to barriers and perspectives for the exploitation of science by the East German industry. For many participants it was the first time to receive competent information about a research system of which many western researchers only knew very little about.

Until today many of Albrecht's anticipations about possible changes in the system became true. The former East German research infrastructure will be adjusted to the (West) German system. Parts of Academy of Science institutes will be integrated into Max Planck Society or Fraunhofer Society, the prospects for the other parts of the academy still remains open. The structure of universities is just being changed according to the (West) German model (e. g. "Fachbereiche" and not "Sektionen" as the upper organizational unit). The "Länder" gain more competence and also receive the responsability for all educational matters, universities included. In industry the "Kombinate" and "VEBs" are being closed, split up, and sold to domestic and international investors. For that reason, Albrecht's paper contains a detailed historical view into the former GDR research system, but it remains still open what new structures will develop in the eastern part of the now unified Germany.

As Albrecht pointed out there were surprisingly close linkages between universities and industry. Roughly speeking, with respect to the conditions of a central planning economy, only about 60 per cent of university funding came from government, the rest from industry. In electronics, industry paid about 70 per cent of the university budget.

A new strategy within technology policy in the new Eastern "Länder" is the development of technological centres for encouraging innovation oriented small sized enterprises. Involved in funding and operation are the Chambers of Commerce, but also companies and universities. Some open questions remained at the time of the conference concerning the affiliation of the centres. A location either near a university or an academy institute was thought to bring the highest possible synergetic effects, but because of the then uncertain future of the academy no detailed location planning of the centres, with the exception of Dresden and Jena, had been possible. The future prospects and success of technology centres is closely linked to the formation and financial capabilities of the new "Länder" in the Eastern part of Germany. Due to many uncertainties at the time of the conference no final answer had been possible.

Most questions dealt with the particularities of the East German research system. Albrecht again emphasized that the universities had to do a lot of applied research because of the small research capacity in industry. Compared to the academy, the universities were more engaged in applied and contract research. For that reason the experience in university - industry cooperation was good, much better than in West Germany. A question was raised concerning the division of labour between the academy and the universities. According to Albrecht the division between academy and universities was not expressed very sharply. Not only academy institutes but also institutes in university did basic research, e. g. in natural science departments. In some cases university researchers were also employed by the academy, professors headed both university and academy institutes. In general, the division between basic and applied research was not so strongly visible as in the FRG. Academy institutes, for instance, were also engaged in industrial development and had small-scale production facilities.

As a result of the political situation East German research was almost isolated from the Western world. This was true not only in high technology fields (COCOM list), but also in "every day science", because publications were hard to receive and only a few selected people was given the opportunity to attend Western conferences. For that reason and also because of the problems to get rare components and other machinery, nearly every institute or structural unit developed an autarchical tendency. They had to care for their own reproduction of their organizational system which served as a quite strong barrier for domestic and international cooperations. Another result of the isolation was the low efficiency in new technologies. Although the policy-makers always claimed that the gap between East and West will diminish (e. g. in the micro-electronics field), it actually increased.

A last question was aiming at the situation of innovation research in the former GDR. To Albrecht's opinion innovation research has to ally with science research. The general problem is that innovation research in Germany is mostly linked to economic disciplines whereas science research traditionally is very close to philosophy or sociology. These differences have to be overcome. In the former section "Science theory and science organization" of Humboldt university, the alliance between science and innovation research already existed. Albrecht hopes that it will be possible to transfer this alliance to the united German landscape of universities and science. The prospects although are shadowy.

In contrast to the well developed linkages between university and industry in the former GDR, *Shinichi Yamamoto's* paper (see chapter 6) on "Linking Science and Innovation in Japan - A Success?" revealed that the university sector in Japan has been losing its share in research performance, although it still provides industry and other sectors with scientists and engineers.

The discussion focused on some details of the Japanese higher education system. According to student enrollment figures, the share of private universities is about 80 per cent of all universities. Their main function is teaching in the field of undergraduate education. Most of the university research is done in national universities (research and education with the function of supply of researchers) and in research institutes attached to universities. Up to Yamamoto, the separation of research and teaching is not as problematic as it may sound, since not all universities can be research universities. Students must also be educated and especially undergraduate education serves to select human resources out of the science system. But it has to be taken into account that beside the large salary difference between industry and university one other important reason for the decreasing research performance of universities is the rigid system of university budgets and the closed circuit of the policy formulation system at the university sector.

Although most of the industrial research is done in-house and not in cooperation with universities, there are close ties between universities and industry. As Yamamoto already pointed out in his paper, there are different forms of cooperation (chapter 6, section 6.3). The most important are donations. They amounted to 39.2 billion yen in 1990, compared to a Monbusho's budget for stimulating joint or contract research of 3.8 billion and 7.1 billion yen respectively. From an outside view it seemed a bit unclear to some participants why on the one hand 63 research institutes attached to universities offer their service for cooperative research and on the other hand companies mostly prefer to give donations to universities. In his answer Yamamoto explained that the 63 institutes serve for cooperative research by "university researchers" and not with industry and additionally emphasized the specific structure of Japanese industry - university relations. In a way Japanese companies use the external research capacity, but one has to take into account that R&D is a very specific field within the activities of a Japanese company. This is why they have strong preferentials to do it in-house. The interests in donations are two-fold. Universities prefer donations not only as an additional source of financing, but also as an essential contribution to the budget of some research universities because of the recent tightness of their regular budget. The companies have interest in close ties with universities because they can transfer parts of their staff recruitment activities to selected institutes and professors. Donations serve as a "payment" for the expertise of professors since graduates can only join the labour force of a company through a professoral recommendation.

A last question by Grupp combined Yamamoto's and Albrecht's paper with the approach presented by Stankiewicz. In the former GDR large parts of universities were paid by industry (which represents the common picture of science-technology links), but compared to the technological performance the result was unsatisfactory. In Japan the linkages are low, but Japan does well in R&D and technology. This phenomenon can only be explained by non-technical aspects, such as by the political and social system. But when, as both cases show, the national aspects are so strong, can one really speak of globalization of technology, as Stankiewicz does in his paper? Stankiewicz pointed out that one should not mix between the analytical concept and reality in a form of real systems. Japan and GDR are indeed two extreme cases; in one technology is included in the social system, in the other it is not. Therefore empirical studies are needed to answer the question to what extent autonomy is managed in different countries. Only after the results are known, Grupp's question can be answered.

A third paper in this section was presented by *Lennart Stenberg* on "Molecular Beam Epitaxy - A Mesoview of Japanese Research Organization" (chapter 7). In his presentation, Stenberg focused on the use of bibliometric data as a means to compare the structure and dynamics of the MBE research system in Japan with corresponding systems in other countries. Bibliometric maps were drawn by analysing some 6500 publications indexed in the INSPEC database during the period 1969-1989.

Again the discussion started on methodology. The demarcation of MBE, as one participant wanted to know, was done by using the term MBE and various derivatives of the term "Molecular Beam Epitaxial Growth". The latter was introduced as a controlled term in INSPEC in 1979. The main reason for using a broader search strategy than that defined by this term was the need to capture references prior to 1979. For the period 1979-1989 the set of publications thus obtained was 40 to 50 percent larger than had only the controlled term been used (chapter 7, section 7.2).

A practical problem was the unification of corporate sources appearing under different names or with different addresses. Stenberg managed this problem by sorting out all the data manually, something which required a certain familiarity with the underlying institutional structure. Another question dealt with co-authorship. Co-authorship reflecting international or inter-institutional cooperation can only be definitely identified when the database provides a list of all authors and their respective addresses. This is the case for SCI but not for INSPEC which only lists the address of one author. According to Stenberg, co-authorship can be found in the MBE field, but accounts for only a small fraction of all Japanese publications. As a matter of fact, in a separate exercise for acquiring detailed data to be discussed in interviews, the names of all authors were identified for each of the Japanese organizations and authors with the same name but with different organizational affiliations were identified. Interviews, covering the largest organizations in terms of publications, revealed that only a part of this multiplicity of addresses was due to co-authorship and that it could safely be concluded that inter-institutional co-authorship was limited to a few percent of all Japanese publications. Narin remarked that the general share of inter-institutional co-authorship varies between 30 and 40 percent. Therefore, as Stenberg pointed out, bibliometrics is not a stand-alone tool, but has to be complemented by other methods and sources of information. It is quite easy to draw a superficial map out of a database, but a correct interpretation will often require additional information from interviews and analysis of such matters as institutional structure and the cognitive structure of the field under study. Such complementary inputs reduce the reliablity problem compared to when the analysis only draws on database information. The optimal strategy will depend on financial and time constraints and on the size of the field. The smaller the field to be analysed the more feasible it will be to do detailed field investigations or to gather complementary information in other ways.

Friborg asked about the possibility to identify research groups by collecting information on installed MBE machines and by juxtaposing this with publication data to draw conclusions about joint use of MBE equipment. So far as Stenberg knew, some MBE machines are being used for production or for other routine tasks, while the great majority of MBE machines are being used for research purposes and consequently can be expected to produce results which show up in publications. To do MBE research requires at least access to an MBE machine. Cases of co-authorship involving researchers without access to an MBE machine had been identified. These authors were usually either physicists characterizing materials grown by someone else using MBE or researchers using MBE-grown materials for the fabrication of devices. The overwhealming majority of authors of MBE publications, including most of the device-oriented researchers, however, work in groups which have one or several MBE machines.

1980-81 marked a turning-point for the diffusion of MBE systems, since at that time they became commercially available. As their ease of operation improved, they became increasingly used also by people outside the core of the MBE research community. Depending on the purpose of the research, the requirements for the quality of the crystals grown by MBE will vary significantly as will consequently the expertise in MBE *per se* required from the researchers. These changes in the composition of the research community having a hand in MBE technology have to be considered when analysing MBE related publications.

Compared to Yamamoto's statements on the generally decreasing relative role of universities in the Japanese research system, Stenberg's study showed a contrary development for the MBE field (chapter 7, section 7.3, sub-section on universities). While during the early years of the 1980s, universities had rather limited contacts with companies, they became a more vital part of the Japanese research community in more recent years. Several distinguished researchers from industry and the Electrotechnical Laboratory (under MITI) have been recruited as professors to build up new research groups. This movement is partly an effect of the career systems in the firms. Upon reaching a certain age, able researchers are faced with the choice between being promoted to higher levels of management, taking them away from daily contact with research, or continuing as researchers in a specialist capacity. For the person set on continuing to do research it appears however that a position as professor at a university is usually more attractive than that of a research specialist in a firm. Again, this example shows that only detailed empirical studies of concrete economic and social processes can reveal the real nature of science - technology links and their changes over time.

In his verbal contribution, *Ben Martin* presented the outline of a research project he is working on together with Diana Hicks and Phoebe Isard attempting to analyse the interface between corporate R&D and academic research. Up to now, no written publication is available. The following outline of the project is based on a series of figures and tables distributed to the participants. This project is using a bibliometric approach to explore the changing linkages in Britain and Japan and has the following major objectives:

- to collect data on trends in publication output of UK and Japanese firms between 1981 and 1990;
- to map bibliometrically the links between corporate and academic research in UK and Japan by analysing the references cited in industrial papers and patents;
- to identify the reasons for the growth in the publication output of firms and to establish whether basic research is playing an increased role in the innovation process;
- to improve our theoretical understanding of the interface between science and technology;
- to assess whether Japanese companies draw more heavily on foreign research; and
- to analyse the relationship of corporate scientific publications to subsequent patenting in order to see if bibliometric data can be used as leading indicators of innovations.

The approach of Martin and his colleagues also involves a field survey which is attempting to obtain a better understanding of the derived bibliometric data. After the selection of 10 - 15 companies in UK and around 30 in Japan and some preliminary interviews to help formulate the hypotheses to be tested, one of the major tasks is to construct a database on research publications from the selected companies. For a sample of papers in the database, the references in each corporate publication will be analysed and interviews carried out with the authors of a selection of the papers on the reasons why they cited particular references. A second major component of the project will be an analysis of the trends in publications and citations, the patterns of institutional collaboration, and the characteristics of the cited references (e. g. cross-field and cross-sector links). To gain an insight into the relationship between publication and patenting, parallel data on patents and on the references cited in patents will be compiled and compared with the publication data in a similar way to that described in chapters 3 and 4, but with an emphasis on selected companies that are not apparently oriented towards specific technologies. To help in analysing and interpreting the empirical results, interviews with industrial researchers and R&D managers on the links between university and industry and between science and technology are planned.

According to Martin, the likely research outputs include:
- obtaining empirical data on corporate publishing in the UK and Japan;
- establishing a rationale for publication and citation;
- empirically comparing basic research activities of UK and Japanese companies and their relative reliance on foreign basic research;
- contributing to the debate on the measurement of the impact of science on technology and whether publishing can be used together with patenting as indicators of diversification;
- identifying indicators of leading-edge corporate research which are useful to government policy-makers;
- creating a tool for companies to use in obtaining intelligence on research by their competitors.

After giving an overview on the objectives and methodology of the project, Martin presented some preliminary results. A review of the literature concerning the reasons why scientific authors cite other researchers has revealed three types of studies. The first deals with empirical evidence on citation practices, the second focuses on criticisms of citations as indicators, and the third group comprises attemps to construct a theory of citations. According to the literature, it seems quite clear that there is a wide variation in referencing practices with the result that establishing a clear citer motivation is problematic. All the various "theories" of citation have weaknesses in dealing with specific questions of citation practice which means that each empirical study must establish the function of citations in the research area it is concerned with. For a discussion of the problems in understanding the role of citations in patents, see chapters 3, 4, and 9.

A preliminary analysis of the institutional origins of scientific work cited in the research publications of Philips (Redhill) and Shell in 1988 showed that the majority of references are of university origin (41 per cent and 54 per cent respectively). 20 per cent each are government sources whereas 39 per cent of the references in Philips papers are from industry compared to only 12 per cent for Shell papers. Concerning the nationality of the references, the data for 1988 show a wider variety for Philips. In the company's research publications, references from the USA dominate with 39 per cent (compared with 43 per cent for Shell), UK comes second with 19 per cent (28 per cent in the case of Shell), but in third place are papers from the USSR with 11 per cent of the citations compared to 0 per cent in Shell publications. Other countries whose scentifie work is drawn upon are Japan, France and the FRG.

A problem in the analysis of the data files relates to the absence of information (in the original source papers) concerning the institutional and national origins of the cited references. Therefore all of the references have had to be checked manually for their address and organizational status.

Using SEPSU (Royal Society) patent data (compiled by Collins & Wyatt 1988), Martin and his colleagues have started to analyse the relationship between publication and patenting. From a keyword search, Collins and Wyatt identified 366 genetics patents granted in the US over the period 1980-1985. They also looked up the examiner and applicant citations and the institutional addresses for references in the basic research literature. In this way, data on inter-institutional and inter-country citation links can be derived. Analysis of the country matrix on actual versus expected numbers of (examiner and applicant) citations by genetics patents to scientific papers reveals that UK papers are cited at an above average rate by French, Japanese, US and other patents (even after allowing for the home advantage which can be observed for all countries), whereas for US papers the actual number of citations exceeds the expected figure only in the case of French patents. This suggests that UK publications are drawn upon as scientific references at an above average rate in international genetics patenting.

The discussion focused mainly on questions of methodology and database analysis. Concerning the technological fields to be analysed in the study, no firm decision had been made so far except for the use of the genetic patent data from SEPSU. In general a population of about 1500-2000 industrial research papers for the UK and over 3000 papers for Japan could be expected per year. Given that English language articles (e. g. from the UK) are highly cited in US patents, the question arises of whether there is a significant bias in the database towards English language articles (see also chapter 4). It was pointed out that Japanese companies, for instance, do not always publish in English, and that on the other hand Japanese patents are cited less frequently by examiners at the USPTO because of the language bias. Although Francis Narin pointed out that only 10 % of the Japanese-origin papers covered by SCI are written in Japanese whereas 90 % are in English, there is still certainly a problem of language bias when making international comparisons. This problem has to be openly recognised, but the only way of dealing with it is by adopting the pragmatic approach of 'learning by doing'.

Another important point is the different nature of publications and patents. Publications generally have a world-wide distribution which means that one publication is equally "valid" in all countries. The validity of a patent, however, is legally restricted to one country only. Patent protection of the same invention in more than one country is only possible by applying for that specific patent in each of the countries (Grupp, 1990). For that reason, many Japanese patents, for instance, have an English language duplicate (e. g. for an application at the USPTO or EPO). In the disclosure process, the patent examiner tends to look for patent references written in his or her native language which again produces a bias towards the English-speaking world (chapter 9). As Grupp pointed out, the methodology for analysing paper citations in patents must therefore be different from that of paper citations in papers.

One last remark on Martin's presentation related to the different structure of the scientific literature in the US and Japan. According to research work carried out by Stankiewicz in the field of genetics (Stankiewicz 1990), there are fewer scientific articles in Japan but more applied articles in trade journals which are often not covered in publication databases. This could cause a problem in Martin's study since the Japanese coverage of the database which he and his colleagues are using - namely, the SCI - is not be best. These unresolved questions of validity underline again the exploratory character of this most interesting investigation. It seeks to clarify many open-ended questions spelt out elsewhere in this book.

11.4 CROSSROADS IN SELECTED TECHNOLOGIES: CASE STUDIES

This section contains three case studies on selected technologies. *Anthony van Raan* and *Ulrich Schmoch* presented results of a collaborative project between the Centre of Science and Technology Studies, Leiden, and FhG-ISI, Karlsruhe, on the diffusion of knowledge in laser medicine and polyimide chemistry. The Leiden group (Ben van Vianen, Anthony van Raan; later joined by Ed Noyons) focused on the science side by using bibliometric data whereas FhG-ISI (Ulrich Schmoch, Hariolf Grupp, Thomas Reiss) analysed the technology side by means of patent data. The results are presented in chapters 8 and 9 and only a few questions concerning methodological issues were raised. These mainly concentrated on the bibliometric study part.

The application of the databases used in the Leiden study for the fields to be investigated and the amount of time needed for data clearing and unification was subject to the first questions. To find out which database fits best for laser medicine, van Vianen and van Raan carried out searches in PASCAL, BIOSIS, SCISEARCH, and INSPEC. PASCAL is a multi-disciplinary database which is supposed to cover all scientific fields, although the rather large number of French publications indicate a clear bias towards French literature. One major problem of searches in PASCAL is the annual adjustment of the classification scheme. An important modification became effective in 1984. In order to create search strategies over time, a concordance of classification codes is needed. Unfortunately such a list is not officially available. A concordance for general codes from an internal database at CNRS, the host of the database, helped van Vianen and van Raan to overcome this problem. Within the project they developed a recategorization of PASCAL classification codes which now is available to everybody. In BIOSIS, no major changes in the classification system occured. However, many concept codes are attributed to one article. Their average amount varies between 10 to 14. Additionally it is not possible to differentiate among countries. Searches in BIOSIS within the subfields "Therapy" and "Radiation and Isotope Techniques" which present applied and basic research should have indicated a time lag between these two subfields (like results from PASCAL and Chemical Abstracts show). However, their growth curves seemed quite parallel which led van Vianen and van Raan to the conclusion that the high amount of concept codes attributed to each article caused the lack of clear separate trends in the data. In SCISEARCH the number of retrieved publications for the years 1973 to 1982 was very small. It was concluded that the laser-medicine field could not be properly defined by means of a title-word search only (chapter 8, section 8.6). In INSPEC the country of origin of the publication is not available. Additionally, the numbers of publications were too low for further analysis. For that reason, the findings of the study were based upon PASCAL data.

As already pointed out, unification of databases and creation of concordances is a very important step in bibliometric research. In Leiden about half a man-year of permanent staff is needed for this purpose. But, according to van Raan, it pays off at the end. Since database problems play an important role in the kind of research presented at the conference, Schlie-Roosen encouraged the database users to organize themselves and to address their wishes and problems to the producers and hosts of databases as one group. That might be more successful than individual and cumbersome approaches.

The third case study was presented by *Anders Granberg* on "Mapping an Evolving Technology Cluster: The Composition and Structure of Factory Automation" (chapter 10). In contrast to other approaches presented at the conference, Granberg yields his results by a more qualitative technique. The knowledge for defining clusters in the field of factory automation was not based on automated bibliometric database searches, but on the interpretation of selected literature and on expert's interviews. Compared to Engelsman's study (see below) it was interesting to learn that also a qualitative approach can produce valid maps of a specific technology field. However, having in mind that all the other papers presented at the conference do contain automated bibliometric or patent analysis, it was asked if Granberg plans a bibliometric complement to his study. Granberg was not sure because to his opinion the results were quite stable and it would only be worthwhile to carry out a bibliometric study if additional results could be expected. More important to him is the question of how to develop an organic picture of the social needs and requirements incorporated in factory automation. The analysis and mapping of social - technical links and interdependences create methodological problems which are still an unsolved question and need further research.

Eric C. Engelsman presented as a forth case study the results of a patent-based study of knowledge diffusion he worked on together with Anthony van Raan titled "Relational Structures between Fields of Technology". The results are published elsewhere (Engelsman & van Raan 1991) meanwhile and are not reprinted in this volume. This *summary* draws on short lecture notes by Engelsman & van Raan distributed to the participants.

According to Engelsman, during the last decade patent data became available in electronic databases. As a consequence, the interest in linking these patent data with economic or R&D-related data increased strongly. But as it is the case with all technology indicators, one has to take the utmost care in using patent statistics for assessing strengths and weaknesses in technology (Pavitt 1988). Nevertheless, patents do contain a wealth of information on R&D, the companies or countries involved, and time and market-related information.

Diffusion of technological knowledge is the flow of knowledge from one particular technology field ("core field") to the enhancement of knowledge and the support of innovations in another technology field ("absorbing field"). In terms of patenting activities, technology diffusion is studied by uncovering the multiple relations between patents, in particular the relations between patents of the absorbing and of the core field respectively.

In principle there are three ways in which technology diffusion can manifest itself when using patents as the data source:

1. *Citation Analysis Approach:* Citations in the search report of a patent document in an absorbing field may refer to a patent (or to scientific or technical literature) in the core field.

2. *Co-Classification Approach:* The classification of a patent (as defined by the International Patent Classification, IPC) may refer to two or more different technology fields thus revealing the relationship between these fields.
3. *Co-Word Approach:* Field-specific terminology (keywords) originating from the core field may be used in the title or abstract of a patent application in an absorbing field.

These tokens of knowledge diffusion show different levels of uncertainty with respect to the *direction* of the diffusion process. In the first case, focussing on patent-to-patent citations, the relation between the citing and the cited patent is -in principle- quite clear: the cited patent is the knowledge supplier, and the citing patent the knowledge absorber (see the pioneering work done by Narin 1987). When patent classifications are being used (co-classification approach) only a statement can be made whether an undirected relation between two technology fields does exist or not. Within the co-word approach the use of field-specific terminology gives rise to problems related with *standardization* and *semantics*. Open questions concern the specificity of the terminology for a particular field or the different meanings of specific terms in different fields.

The three cases also represent different levels of *intensity*. Patent-to-patent citations may indicate a strong, direct relation between specific patents. Classification codes also indicate a strong relation between the core and the absorbing fields since the codes are based on specific knowledge inherent to the patent. In the third case a strong relation is not necessary. Terminology can refer to knowledge or processes which are common to many different technology fields. If in such cases obvious relations between fields exist, one can hardly speak of diffusion of technological knowledge.

The patent data used for this study are extracted from two major databases:
- European (EPO) patent documents from EPAT, containing all (published) patent applications at the European Patent Office
- international data from WPI/L, containing about 8 million patent documents from 30 industrial countries.

The data are aggregated into 28 technology fields as they are defined in chapter 4 in this book; see table 4-1. Regarding the *co-classification approach* the number of patent applications in the database EPAT were classified in each of these 28 technology fields, and the number classified in two different fields at the same time were counted for the years 1980 and 1987. The same was done for the database WPI/L. The results can be described in a set of four co-occurences matrices. Regarding the *co-word approach* all indexed title words for European patents in WPIL in all 28 technology fields were collected for 1987. From these data a correlation matrix for the 28 fields was constructed, based on about 10.000 words.

To construct an abstract visualisation of links in modern technology (for a general introduction see van Raan & Tijssen 1990), multi-dimensional scaling technique was used to map the results of the co-classification and co-word analysis. Their map is also used in chapter 4 and essentially reproduced in figure 4-2.

The first striking detail in the map for 1987 based on the classification of all European patent documents is the grouping of the technology fields in three main clusters (see figure 4-2): a chemistry, a mechanics, and an electronics cluster. The chemistry cluster is connected through the fields "Inorganic Chemistry" and "Coating" with the electronics cluster, while the mechanics cluster remains slightly more isolated. Comparing this result with the map for 1980 (not shown here) some slight changes in time can be observed. Within the chemical cluster the relations between "Organic Chemistry" and "Genetics & Pharmacy" on the one hand and "Nutrition" and "Genetics" on the other have increased. This is also true for "Polymer Chemistry" and "Synthetic Resins". Laser technology is now stronger related to "Optics" and closely located to "Information Storage". There is also an increased relation between "Printing" and "Handling" and between "Electrical Machinery" and "Coating". Strangely enough, the relations with "Telecommunications" have decreased, especially towards "Data Processing", although both are still on the right side of the map.

In the corresponding map constructed from international (WPI/L) data, the same three clusters can be noticed for 1980, but this time the chemistry cluster looks more compact, what could be caused by a weaker relationship with the electronics cluster. The map for 1987 reveals how the mechanics cluster moves closer to the chemistry cluster. In contrast to the EPO map for 1987 (figure 4-2) there are also strong relations between "Image Transmission", "Information Storage", "Data Processing", "Optics" and "Telecommunication". The co-word map for 1987 (not shown here) also shows a structure which has clear parallels with the co-classification maps.

The results derived so far show that the strongest individual relationships can be indicated with the co-classification approach, although it is impossible to indicate a direction of diffusion. In future comparisons, therefore, maps based on citation analysis are also planned, since in this case the direction of the diffusion process is known.

Engelman's contribution pointed out that a momentary view of diffusion patterns is highly useful in *policy making* since it gives an indication of so-called "crossroad" technologies. These crossroad technologies (e. g., coatings) indeed should draw major attention, since they are the nodes in the modern technology network through which knowledge is being transferred.

The discussion focused on the methodology of the study and on the application of the maps for policy decision-making. Comparing Engelsman's study with the approach Granberg used (chapter 10), structural differences, but also similarities can be observed. Granbergs approach was of qualitative nature. The latter analysed the field of factory automation by reading selected literature, interviewed specialists and by this identified technology clusters. Engelsman used statistical data drawing on patent examiners' view of technical relations and closeness. Although the type of work is different, the aim of both studies is the same, namely to produce an image of technological relations.

Grupp recalled the generation of the 28 technology fields Engelsman used in his analysis. The list was derived by interviewing technology experts of Fraunhofer Society and by defining 28 technological areas as homogeneously as possible. Engelsman's maps reveal that the experts had been right, expecially for the laser field which had been defined separately because of its unclear position between electronics and optics.

The maps paint a "landscape" of technological fields in a multi-dimensional system. Some fields are closer together than others. Stankiewiez was undecided whether the maps really represent the intrinsic structure of technology or merely mirror the structure of the patent classification system, which might also be multi-dimensional. Stankiewicz pointed out that this may be checked by help of a list of technical clusters which is independent of classification systems. According to Engelsman, the argument of "imprinted" multi-dimensionality might be true for the co-classification approach, but the co-word approach is completely independent from any classification system and again produces fairly similar multi-dimensional relations.

The question of centrality is a very important one since it restricts the application of the maps. The problem is what can be defined as "central" since the system is multi-dimensional (only the illustration is two-dimensional). The same is true for "peripheric" technological fields. For that reason it will be problematic if policy-makers use these maps as tools for decision-making, e. g. for resetting research priorities, without having an interpretation of the data. Another question in this context relates to the selection of maps in a way to provide a "best" map fitting well for a specific policy goal. Again, there cannot be a "best" map, since the term "best" is undefinable in a multi-dimensional system. What the maps can be used for is to show which of the technological subfields are closely linked and which are distinct and to feed this evidence into a decision-making process for supporting technology links and the technology network.

11.5 CONCLUSIONS: STATE OF THE ART IN SCIENCE AND INNOVATION RESEARCH

As can be seen from the summary of the discussion in this chapter, a major part of the questions was devoted to methodological issues. Although a lot of items had been discussed, questions mainly circled around four major fields of interest:
- Database problems,
- Validity issues,
- Institutional versus cognitive aspects,
- Policy relevance.

They do not reflect all aspects relevant to science and innovation research, but represent a selection of scientific focal points worthwhile to illustrate in short as concluding remarks of this chapter.

DATABASE PROBLEMS

For all research work done in the field of patent or publication analysis, peculiarities in database structures make it necessary to deal in detail with contents and structures of databases. Important elements to look at before selecting a database are - among others - origin, completeness and kind of journals offered by the database producer (see also Grupp 1990, p. 449). A lot of misinterpretation can be avoided when a matching process is carried out at the beginning of a study. Van Vianen's and van Raan's experimental searches in several databases for the field of laser medicine proofed the value of such reliability testing.

The most striking challenges in database use are adjustments and unifications to be necessary when data records do not provide sufficient information, e. g. on institutional addresses or co-authorship, and to cope with classification changes or breaks without having access to concordances. This, for example, was true for the PASCAL database for which the Leiden group had to build up an own concordance based on internal information from the host. Other problems are associated with the amount of concept codes, restricted search possiblilities like title-word search only, or language biases as a result of selective journal coverage. Database related questions should not be neglected. In Leiden it takes about half the time of a permanent staff per year to match research needs and database restrictions.

VALIDITY ISSUES

Bibliometric and patent analysis are commonly being used to gain an insight view in structures and dynamics of science and technology. When publication and patent data are combined it has to be realized that papers and patents have a different quality with respect to their legal status. Each patent is only valid within one country whereas a paper can be distributed world-wide (see also Grupp 1990, p. 450). For a world-wide protection of an innovation the inventor or applicant has to apply for a patent at each national patent office separately (with some exceptions in the cases of the European Patent Office and the World Intellectual Property Organization). This, for example, gives place for strategic patenting (patent applications in selected countries only) which is not possible for papers.

Combining patent and publication methodology by analyzing non-patent references (NPL) in patents, the restricted validity of patents can create a language bias since the patent examiner tends to look for references which are written in his native language. Additionally, according to validation studies in this field, the average number of NPL references in a patent depends on the language used in the country of application. English language patents show a higher amount of NPL references than non-English language patents.

In research about the scientification of technology some other methodological issues are also important to look at. The age of NPL references sometimes plays a role in evaluating the maturity stage of science and technology. But one has to pay attention not to draw hasty conclusions out of deviations in publication years of cited references with respect to different countries or fields of technology. More recent references might be more strongly related to short research notes published in weekly or monthly journals like *Science* or *Nature* whereas older articles might be more fundamental and published in reviewed journals. Nevertheless, papers dealing with this issue show that there is indeed a positive correlation between new science and new technology.

National peculiarities in the patent system or publication strategies are also important to know about and to take into account. The high amount of domestic patent applications in Japan depends on the lower level of inventiveness as compared to world standards. In Japanese genetics, for example, the number of applied articles published in so-called trade journals is higher than the number of scientific articles in reviewed journals. Again, this leads to biases in the results of database searches since not every "grey" journal can be covered.

When using NPL references as an indicator of the science base of a technology, one has to be cautious in the definition of what will be science and what will be technology. Selecting certain scientific journals as indicator for scientification of technology might be tricky because it seems possible that R&D work done by industry and published in such a journal will then be classified as "science", although it is pure "technology".

The discussion revealed that bibliometrics should not be used as a stand-alone tool to analyse science-technology links. To obtain insights in science-technology structures and dynamics, indicators representing different maturity stages in technological development are needed. Therefore bibliometric studies should be accompanied by patent analysis. Furthermore, bibliometrics as a stand-alone tool only provides a very rough picture of the research landscape. When bibliometrics is complemented by other validity methods or by a detailed field study, the reliability problem of database supported analysis will be reduced and retrieved data can be matched according to expertise derived from interviews or other information.

According to case studies presented at the conference, the mistake should not be made to define the field under investigation and the search strategy to identify relevant actors too narrow. Like it was shown in the MBE study, the diffusion of technology gave scientists from outside MBE research the opportunity to use this technology for own purposes. These people started publishing research articles on MBE, although they were not originally engaged in the development of the MBE technique. If a study concentrates on the initial inventors only, technological diffusion, changes in technological application and the dynamics of science-innovation links will be neglected.

Another interesting methodological issue refers to the different approaches within the mapping techniques. Landscapes of technology can be generated by using automated database search procedures, but also by expert interviews and analysis of relevant literature. A scan over the whole area of technology might only be possible by electronic data processing, but a detailed description of one very specific field (e. g. factory automation) can at least as well be done by field studies.

INSTITUTIONAL VERSUS COGNITIVE ASPECTS

A differentiation between specific technology and the socio-cognitive subsystem of technology, "SCST" (a complex of activities centered on the production, assembly and organization of knowledge enhancing man's instrumental control of his physical, biological and social environment), was postulated. During the discussion it remained open to what extent the SCST is autonomous to other systems (e. g. to the system of concrete technology). It was suggested that empirical studies should contribute to this question and should focus on the relationship between cognitive dynamics and the institutionalization pattern of technology. Up to date only few attemps have been made.

Another complex of discussion focused on the role of universities in the research system. It was agreed upon that research work in universities must not automatically be science or science-based, because applied research work, process and product developments are also carried out by professors or in certain institutes. Using universities as the only (institutional) indicator for the science-base of technologies in bibliometric studies would draw a wrong picture of real science-technology links. Experiences from Japan (and also from other countries) reveal that in general universities are falling behind industry in many research issues. Nowadays technology is generated in industry, not in university. But there are also exceptions to this common picture. It was shown for molecular beam epitaxy research that some Japanese universities increased their research share by recruiting industrial researchers as professors. The recruitment strategy can be viewed as a result of the dual carreer system in companies. This example disclosed non-technical rationality and makes it obvious that detailed additional studies in boardering social and economic fields are needed for the interpretation of changing research performances.

A comparison between the Japanese and former GDR system of university-industry collaboration generated a very interesting result. Although very close ties between industry and university existed in the former GDR, her technological performance was quite weak. In Japan, a weak coupling, namely donations to universities, play an important role in industry-university cooperation, but Japan shows a strong technological performance. No general conclusions should be drawn from this example, but it gives some indication that also non-technical aspects are relevant in the explanation of structures and dynamics between science and innovation.

POLICY RELEVANCE

Not only scientists, but also officials from science and technology ministries of different countries attended the conference. Thus the policy relevance of the presentations was an important issue which had been discussed at the end of the conference. Unfortunately there was no time to draw general conclusions for policy makers, but on the example of the technology maps some policy related questions were discussed. It was commonly agreed upon that using a sophisticated methodology which can produce hundreds of different maps within a very short time, scientists have to be very cautious in giving maps out of hand to policy makers without any interpretation of the data and any assessment of the various traps and possibilities of manipulation. A "raw" map should only serve scientific purposes but not as an isolated management tool for decision-making alone.

Although no precise tools had been supplied to policy-makers, the conference was a good example of interactive science-policy links. By attending the conference and getting deep insights into methodological and other research questions and by contributing to the discussion, the transparency of the neeeds of decision-makers and the possibilities and contrains in innovation research increased. Both sides have a much clearer picture now on what is needed and what are the limitations in innovation and science research. This result of the conference is at least promising.

REFERENCES

Carpenter, M. and Narin, F. (1983), 'Validation Study: Patent Citations as Indicators of Science and Foreign Dependence', *World Patent Information*, 5, pp. 180-185.

Collins, P. and Wyatt, S. (1988), 'Citations in Patents to the Basic Research Literature', *Research Policy*, 17, pp. 65-74.

Engelsman, E. C. and van Raan, A. F. J. (1991), *Mapping of Technology*, Report CWTS-91-01, Centre for Science and Technology Studies (CWTS), Leiden.

Grupp, H. (1990), 'On the Supplementary Functions of Science and Technology Indicators', *Scientometrics*, 19, pp. 447-472.

Narin, F., Carpenter, M. P. and Woolf, P. (1987), 'Technological Performance Assessments Based on Patents and Patent Citations', in: H. Grupp (ed.), *Problems of Measuring Technological Change*, TÜV Rheinland, Köln, pp. 107-119.

Olivastro, D. and Narin, F. (1989), *Definition of Fields of Close Cooperation between Science and Technology*, Report submitted to BMFT, CHI Research / Computer Horizons, Inc., Haddon Heights.

Pavitt, K. (1988), 'Uses and Abuses of Patent Statistics', in: A. F. J. van Raan (ed.), *Handbook of Quantitative Studies of Science and Technology*, Elsevier, Amsterdam, pp. 509-536.

Stankiewicz, R. (1990), 'Basic Technologies and the Innovation Process', in: J. Sigurdson (ed.), *Measuring the Dynamics of Technological Change*, Pinter, London, pp. 13-38.

Van Raan, A. F. J. and Tijssen, R. J. W. (1990), 'Numerical Methods for Information on Aspects of Science: Scientometric Analysis and Mapping', in: Ch. Oppenheim, J.-M. Griffiths, and Ch. L. Citroen (eds.), *Perspectives in Information Management, Vol. 2*, Butterworths, London.

Van Vianen, B. G., Moed, H. F. and van Raan, A. F. J. (1990), 'An Exploration of the Science Base of Recent Technology', *Research Policy*, 19, pp. 61-81.

LIST OF FIGURES

Page

Page

Page

LIST OF TABLES

Page

Page

BIBLIOGRAPHIC SKETCH OF CONTRIBUTORS

Edo Albrecht, Professor of Theory of Science, Ph D in economics, is Dean of the faculty of Theory and Organization of Science (WTO) at the Humboldt University, Berlin, Germany. His research comprises various studies concerning the development of the sciences in the XX. century and especially the interaction between the new scientific and techno-economic paradigms.

Anders Granberg, Ph D, is Senior Researcher at the Research Policy Institute (RPI) at the University of Lund, Sweden. His main field of interest comprises the cognitive and institutional structure and developmental dynamics of science-based technologies. Current activities include studies, in a comparative international perspective, of Swedish efforts in factory automation, powder technology, and superconductivity.

Hariolf Grupp, Ph D in physics, is Head of the Technological Change Department at the Fraunhofer Institute for Systems and Innovation Research (FhG-ISI), Karlsruhe, Germany. In the beginning of the 1980's he served as senior researcher at the Office for Science and Technology of the FRG Parliament. His main present focus is on the economics of technical change, quantitative studies of science, technology assessment, and science policy. He is a visiting lecturer at Humboldt university.

Knut Koschatzky, Ph D in regional economics, is Head of the Research Policy section at the Fraunhofer Institute for Systems and Innovation Research (FhG-ISI), Karlsruhe, Germany. He earlier served as assistant of the President of Hanover university and in the Bavarian Ministry of Economic Affairs. Current activities include studies on food technology, technometric comparisons in the German and Israeli sensor technology and on the regional impacts of technological change.

Anthony van Raan, Ph D in physics, is Professor of Science Studies and Director of the Centre for Science and Technology Studies (CWTS) at the University of Leiden, The Netherlands. He was a lecturer and researcher (atomic and molecular physics) in Utrecht, Bielefeld, and Leiden, and is a visiting scientist in several universities and research institutes in the US, UK, and France. His main present focus is on quantitative methods in research performance analysis, in the study of the science and technology interface, and in mapping of strategically important R&D fields.

Vololona Rabeharisoa, Ph D, is a researcher at the Centre de Sociologie de l'Innovation of the Ecole Nationale Supérieure des Mines in Paris. Her research work is focused on the description of links between science, technology and the market in large techno-economic networks. Current studies are concerned with the analysis of interactions between private and public R&D policies and market strategies in energy saving activities in France.

Ulrich Schmoch, graduated as mechanical engineer with a Ph D in social sciences, is head of the research section Development of Science and Technology Indicators at the Fraunhofer Institute for Systems and Innovation Research (FhG-ISI), Karlsruhe, Germany. Before joining ISI Schmoch was patent engineer at a large patent attorney's office. His work covers issues related to international economy and mechanical engineering. He is now interested in patent analysis and technology genesis.

Rikard Stankiewicz, Ph D, is Associate Professor of R&D Policy and Management at the Research Policy Institute (RPI) at the University of Lund, Sweden. His research activities have ranged across several fields including the organization and leadership of academic research units, interactions between universities and industry and the R&D dynamics of new basic technologies. At present he is engaged in a major study of Sweden's Technological System.

Lennart Stenberg, M. S. degree in engineering physics, worked as a researcher in the Resource Policy Group, Oslo, Norway, before he returned to the Swedish Board for Technical Development (STU) as a Principal Administrative Officer in the agency's Planning Department. During the period 1988-90 he was on partial leave from STU, doing a comparative study of the research system dynamics in Japan and other countries as a Visiting Researcher at the National Institute for Science and Technology Policy (NISTEP), Tokyo, and at the University of Tokyo.

Ben van Vianen is completing his Ph D degree in psychology as a junior researcher in the Department of Experimental Psychology at the University of Leiden, The Netherlands. Until May 1990 he was a junior researcher at the Centre for Science and Technology studies (CWTS) at Leiden. His research activities at CWTS focused on the science base of modern technology and on knowledge diffusion in applied research.

Shinichi Yamamoto is Associate Professor at the Graduate School of Policy Science at Saitama University in Japan. He had been working for the Ministry of Education before joining the faculty of the School. His main present focus is on comparative studies of higher education systems including their research function and administrative organization.

INDEX

(Abbreviations annotated are referred to in passages where they are explained. They may be referred to elsewhere in their full form. Pairs of words used as standard terms nowadays are listed alphabetically as single items, e.g., "Bell Labs" instead of "Labs, Bell", or "Eastern Germany" instead of "Germany, Eastern".)

MIX
Papier aus verantwortungsvollen Quellen
Paper from responsible sources
FSC® C105338

If you have any concerns about our products,
you can contact us on
ProductSafety@springernature.com

In case Publisher is established outside the EU,
the EU authorized representative is:
Springer Nature Customer Service Center GmbH
Europaplatz 3, 69115 Heidelberg, Germany

Printed by Libri Plureos GmbH
in Hamburg, Germany